PHYSIOTHERAPY in Neurological Conditions

With Assessment and Treatment Protocols

Gowrishankar Potturi PhD PT

Lecturer and Head
Department of Physiotherapy
Faculty of Paramedical Sciences
Uttar Pradesh University of Medical Sciences
Saifai, Etawah, UP

CBSPD

CBS Publishers & Distributors Pvt Ltd

New Delhi • Bengaluru • Chennai • Kochi • Kolkata • Lucknow • Mumbai
Hyderabad • Jharkhand • Nagpur • Patna • Pune • Uttarakhand

Disclaimer

Science and technology are constantly changing fields. New research and experience broaden the scope of information and knowledge. The author has tried his best in giving information available to him while preparing the material for this book. Although, all efforts have been made to ensure optimum accuracy of the material, yet it is quite possible some errors might have been left uncorrected. The publisher, the printer and the author will not be held responsible for any inadvertent errors, omissions or inaccuracies.

PHYSIOTHERAPY in Neurological Conditions
With Assessment and Treatment Protocols

ISBN: 978-93-86478-19-1

Copyright © Author and Publisher

First Edition: 2018

Reprint: 2020, 2021, 2022, 2023, 2024, 2025

All rights reserved. No part of this book may be reproduced or transmitted in any form or by any means, electronic or mechanical, including photocopying, recording, or any information storage and retrieval system without permission, in writing, from the author and the publisher.

Published by Satish Kumar Jain and produced by Varun Jain for

CBS Publishers & Distributors Pvt Ltd
4819/XI Prahlad Street, 24 Ansari Road, Daryaganj, New Delhi 110 002, India
Ph: 011-23289259, 23266838 Website: www.cbspd.com
e-mail: delhi@cbspd.com

Corporate Office: 204 FIE, Industrial Area, Patparganj, Delhi 110 092
Ph: 011-4934 4934 Fax: 011-4934 4935 e-mail: publishing@cbspd.com; publicity@cbspd.com

Branches

- **Bengaluru:** Seema House 2975, 17th Cross, K.R. Road, Banasankari 2nd Stage, Bengaluru 560 070, Karnataka, India
  Ph: +91-80-26771678/79   Fax: +91-80-26771680   e-mail: bangalore@cbspd.com
- **Chennai:** 7, Subbaraya Street, Shenoy Nagar, Chennai 600 030, Tamil Nadu, India
  Ph: +91-44-26680620, 26681266   Fax: +91-44-42032115   e-mail: chennai@cbspd.com
- **Kochi:** 42/1325, 1326, Power House Road, Opp KSEB, Power House, Ernakulam 682 018, Kerala, India
  Ph: +91-484-4059061-65   Fax: +91-484-4059065   e-mail: kochi@cbspd.com
- **Kolkata:** 147, Hind Ceramics Compound, 1st Floor, Nilgunj Road, Belghoria, Kolkata-700056, West Bengal, India
  Ph: 033-25633055, 033-25633056   e-mail: kolkata@cbspd.com
- **Lucknow:** Basement, Khushnuma Complex, 7-Meerabai Marg (Behind Jawahar Bhawan), Lucknow 226001, India
  Ph: 0522-4000032   e-mail: tiwari.lucknow@cbspd.com
- **Mumbai:** PWD Shed. Gala no. 25/26, Ramchandra Bhatt Marg, Next to JJ Hospital Gate no. 2, Opp. Union Bank of India, Noorbaug, Mumbai-400009, Maharashtra, India
  Ph: 022-66661880/89   e-mail: mumbai@cbspd.com

Representatives

- **Hyderabad** 0-9885175004
- **Jharkhand** 0-9811541605
- **Nagpur** 0-8692091830
- **Patna** 0-9334159340
- **Pune** 0-9664372571
- **Uttarakhand** 0-9716462459

Printed at Goyal Offset Works Pvt. Ltd., Kundli, Haryana, India

to _____

Dr. Kiran Bhat
my teacher who is my role model

Other CBS books by the same authors

- Textbook of Electrotherapy: Theory and Practice
- Principles of Exercise Therapy
- Physiotherapy in Medical and Surgical Conditions
- Physiotherapy in Orthopedics and Trauma
- A Guide to Physiotherapy Assessment

Preface

I believe neurology is the king of medical sciences. For years, I was trying to accomplish my skills and knowledge to share with the physiotherapy fraternity. For the past 10 years, I was teaching neurological physiotherapy at various levels and was keenly observing the students' views on referring textbooks and the hindrances they face through. There are a few books available in the market which emphasize the physiotherapy management of various neurological disorders and which are useful for both the students and practitioners. I have addressed most of the problems that a physiotherapist would encounter in his/her clinical as well as academic career. This book would be a primer for treating various patients with neurological disorders. The language of the book is kept as simple as possible. Care is taken not to overburden the students with various concepts of pathophysiology and investigations. This book is designed with the tailor-made protocols of management of various neurological disorders. The model question papers given at the end of this book would encourage the reader to check whether the learning objectives are achieved or not. I have tried my level best to avoid mistakes and errors, but as "to err is human", if you found any mistake, kindly bring it to my notice and shall be corrected in the next printing/edition. Last but not the least, I would like to thank all my students of past and present who have motivated me to dream about this project and my family and friends who supported me to make this dream come true. I thank you for choosing this book and wish you all the best for your bright academic and clinical career.

With best wishes and regards.

Gowrishankar Potturi

Acknowledgements

I would like to render my sincere thanks to the following:

1. **Dr (Brig) T Prabhakar**
   Vice-Chancellor
   UP University of Medical Sciences
   Saifai, Etawah, UP

2. **Dr Arun Nagrath**
   Head, Department of Obstetrics and Gynaecology
   UP University of Medical Sciences
   Saifai, Etawah, UP

3. **Dr Suraj Kumar**
   Head, Department of Physiotherapy
   UP University of Medical Sciences
   Saifai, Etawah, UP

4. **My teachers at**
   SDM College of Physiotherapy
   Sattur, Dharwad, Karnataka

5. **My family members**

6. **My students:**
   Present and past

7. **Pictures courtesy:**
   Mr Shivam Katiyar and
   Mr KB Ranjeet Singh Chaudhary

8. **Proofreader:**
   Mr Arun Kumar Shukla

Contents

Preface	v
1. Review of Anatomy and Physiology of Nervous System	1
2. Principles of Assessment and Treatment in Neurological Conditions	17
3. Cerebrovascular Accidents	50
4. Head Injuries and Comatose Patients	77
5. Brain and Spinal Cord Tumors	101
6. Infections of Nervous System	109
7. Ataxia	126
8. Parkinsonism	137
9. Spinal Cord Injuries	152
10. Motor Neuron Disease	165
11. Multiple Sclerosis and Demyelinating Diseases	176
12. Transverse Myelitis	183
13. Syringomyelia	195
14. Myasthenia Gravis	201
15. Peripheral Nerve Injuries	205
16. Polyneuropathies	237
17. Pediatric Neurology	251
18. Vertigo, Dizziness and Vestibular Rehabilitation	282
19. Physiotherapy in Neurosurgery	288
Model Papers	295
Index	301

1. Review of Anatomy and Physiology of Nervous System

LEARNING OBJECTIVES

At the end of this module, the students will be able to:
1. Describe the physiological anatomy of neural tissue, brain and spinal cord.
2. Illustrate the various functions of brain and spinal cord emphasizing on the areas of cerebrum.
3. Explain briefly the properties of neural tissue.
4. Demonstrate reflex arc with a neat labeled diagram.
5. Explain various types of reflexes and their elicitation.
6. Give the differences between UMN (upper motor neuron) and LMN (lower motor neuron) lesions.

INTRODUCTION

- Nervous system is a multi-response region which involves receiving the sensations, interpretation and to conduct nerve impulses for action.
- All the higher functions like memory, intelligence, consciousness, behavior and orientation are seated in nervous system.
- It is the chief regulatory and coordination center for bodily activities.

DIVISIONS OF NERVOUS SYSTEM

Anatomical Division

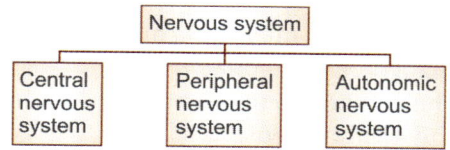

Physiological Division

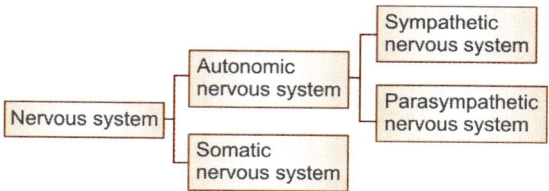

NERVOUS TISSUE[1,2]

Nervous tissue is composed of mainly two kinds of cell:
- Neurons
- Neuroglia

NEURON[3]

Neuron is the structural and functional unit of nervous system (Fig. 1.1).

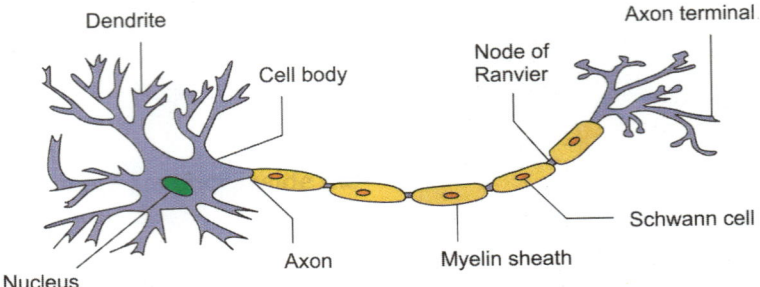

Fig. 1.1: Structure of neuron

Each neuron consists of the following parts:
- Cell body
- Axon
- Dendrites

Cell Body

1. Cell body varies from 5 to 120 μ in diameter.
2. The shape may be rounded, flask-shaped or polygonal.
3. The cell body consists of single, central spherical nucleus.
4. The cytoplasm is surrounded by a cell membrane.
5. Within the cytoplasm, it contains cell organelles like mitochondria, Golgi apparatus, and lysosomes.
6. It also contains granular masses present in the body and dendrites called *Nissl's granules*.
7. Nissl's granules are absent at the site of origin of axon from the cell body, this region is called *axon hillock*.
8. Nissl's granules are involved in the protein synthesis.
9. The cytoplasm also contains fine threads of variable length called *neurofibrils*.

Dendrites

1. The free short branching processes containing Nissl's granules are called *dendrites*.
2. They also contain mitochondria and neurofibrils.
3. Dendrites do not contain myelin sheath and neurilemma and are meant mainly for receiving incoming impulses and carry them towards cell body.

Axon

1. Each neuron has only one axon.
2. It is long, slender, thread-like process.
3. It arises from cell body at axon hillock.
4. It is devoid of Nissl's granules but contains mitochondria and neurofibrils.
5. An axon may give rise to branches at right angles called *collaterals*.
6. The terminal branches are called telodendria.
7. The cytoplasm of axon is called *axoplasm* and surface membrane is called axolemma.
8. The axon conducts nerve impulses away from cell body.

Myelin sheath: Most of the axons are surrounded by an insulating sheath of fatty myelin.

The whitish appearance of peripheral nervous system is due to myelin sheath.

Nerves having myelin are called *myelinated nerves* and those do not are called *non-myelinated nerves*.

Outside the sheath layer, Schwann cell is present. It is covered by an additional layer called *neurilemma*.

The myelin sheath is not continuous throughout the axon, it is interrupted at intervals called *nodes of Ranvier*.

Classification of Neurons

The neurons are classified on the basis of number of nerve processes and on the basis of functions and on the basis of axon length.

On the Basis of Nervous Process

1. *Unipolar neurons*: Neurons which have only one protoplasmic processes, e.g. V cranial nerve nucleus
2. *Bipolar neurons*: Neurons which have two processes an axon and a dendrite attached to the opposite end of the cell-body, e.g. nucleus of VIII cranial nerve
3. *Pseudo-unipolar neurons*: These are bipolar cells with two processes arising in T manner, e.g. dorsal root ganglia of spinal nerves
4. *Multipolar neurons*: Neurons which have many processes, an axon and many dendrites, e.g. motor neurons.

On the Basis of Axon Length

1. *Golgi type 1 neurons*: These neurons have long axons which form fiber tracts within the CNS or passout of it as peripheral nerves.
2. *Golgi type 2 neurons*: These neurons have short axons which lie within the CNS and mostly form the association fibers.

On the Basis of Functions

1. *Sensory neurons (afferent)*: These neurons carry impulses from sense organs to CNS.
2. *Connector neurons*: They connect and correlate the activity of sensory and motor neurons.
3. *Motor neurons (efferent)*: They carry impulses away from CNS to the effector organs, muscles and glands.

NEUROGLIA

There are mainly three types of neuroglia:
- Astrocytes
- Oligodendroglia
- Microglia

Astrocytes

These are stellate cells with round or oval nuclei and numerous processes. They are mainly concerned with nutrition to the nervous tissue.

Oligodendrocytes

These resemble astrocytes, the only difference is they possess smaller and a few processes. They are present in both grey and white matter of CNS. They form the myelin sheath in CNS.

Microglial Cells

These are small, amoeboid cells with scanty cytoplasm, they are meant for phagocytosis and play very important role in defence mechanism.

CENTRAL NERVOUS SYSTEM

Physiological Anatomy

The brain and spinal cord together constitute central nervous system. They are completely covered by 3 membranes called *meninges*, enveloping the brain and spinal cord. The meninges from outside to inwards are:
- Dura mater
- Arachnoid mater
- Pia mater

BRAIN

The brain is a massive collection of neural tissue and lies in the cranial cavity.

The different parts of the brain are:
- **Forebrain:** Cerebrum, thalamus and hypothalamus
- **Midbrain**
- **Hindbrain:** Pons, medulla, cerebellum

Cerebrum

1. It is the largest part of the brain and occupies anterior and middle cranial fossa of cranial cavity.

2. The cerebrum consists of two cerebral hemispheres—right and left.
3. Both hemispheres are separated by a fold of dura mater called *falx cerebri* and are connected by corpus callosum.
4. Each cerebral hemisphere consists of outer gray matter and a central core white matter.
5. The surfaces of cerebral hemispheres are irregular due to the presence of sulci and gyri.[4]
6. The elevations are called *gyri* and the depressions are called *sulci*.
7. Each cerebral hemisphere is divided into four lobes:
   - Frontal
   - Parietal
   - Temporal
   - Occipital
8. The bounderies are marked by the sulci, they are central, lateral and parieto-occipital sulci.
9. Each lobe physiologically divided into areas depending on their functions (Fig. 1.2).

Table 1.1: Areas of cerebrum[5]

Area	Function
Precentral area	Contraction of voluntary muscles
Premotor area (Broca's area)	Motor speech area
Postcentral area	Sensory
Parietal area	Accurate knowledge of objects
Sensory speech area (Wernicke's area)	Perception of spoken word
Auditory area	Hearing
Olfactory area	Smell area
Taste area	Taste
Visual area	Vision

Functions of Cerebrum[5]

1. Helps in sensory perception for pain, temperature, touch, sight, hearing, taste, and smell.
2. Responsible for initiation and control of voluntary muscle contractions.
3. Concerned with higher complex activities like memory, emotions, thinking, reasoning, will judgment, personality, intelligence.

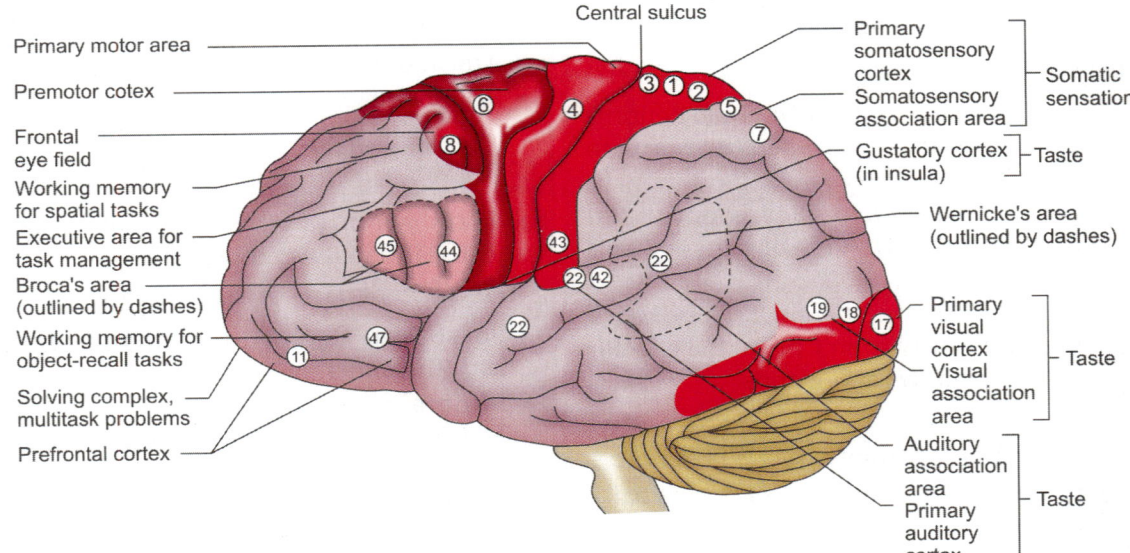

Fig. 1.2: Functional areas of cerebrum

4. Together with hypothalamus, it governs emotional aspects of behavior like pain, pleasure, anger, rage, fear, sorrow, sexual feelings, solicity and affection. Hence, the cerebrum is called "emotional brain".

Thalamus[6]

1. Thalamus is two large masses, mainly made of gray matter, situated one on each side of III ventricle, between the cerebral hemispheres.
2. The thalamus is a relay station for sensory system.
3. There are 7 main groups of thalamic nuclei:
   - Anterior nuclear group
   - Medial nuclear group
   - Lateral nuclear group
   - Ventral nuclear group
   - Intralaminar nuclear group
   - Midline nuclei
   - Reticular nucleus of thalamus

Functions

1. Relay station in the pathway of somesthetic senses from the opposite side of the body.
2. Relay station for impulses coming from opposite side of cerebellum.
3. Relay station for the impulses coming from reticular formation.
4. Relay station for the auditory and visual pathways.
5. Responsible for maintaining consciousness and altering rest.[7]
6. Responsible for recent memory, emotion and language.

Hypothalamus

1. It is the ventral division of diencephalon.
2. It is the highest integrating center for autonomic nervous system.
3. It is present below thalamus and above pituitary gland.
4. It also controls pituitary gland functions.

Functions

1. Regulation of body temperature.
2. Regulates anterior lobe of pituitary gland.[8]
3. Formation of posterior pituitary hormones.[8]
4. Control of circadian rhythm.
5. Control of hunger and thirst, temperature, heart rate, etc.[9]
6. Regulates feelings of rage, aggression, pain, pleasure, behavioral patterns of sexual arousal.[10]
7. Regulates activities of eating and drinking.
8. Regulates diurnal rhythm.
9. The suprachiasmatic nucleus of hypothalamus is called **biological clock**.

Midbrain

1. Also called *mesencephalon*.[11]
2. It connects the forebrain to the hindbrain.
3. It contains cerebral aqueduct, that communicates third and fourth ventricles of the brain for the passage of cerebrospinal fluid (CSF).
4. The cerebral aqueduct divides the midbrain into two parts:
   - Ventral part consisting of a pair of cerebral peduncles
   - Dorsal part called *tectum*.
5. The cerebral peduncle consists of:
   - Crus cerebri
   - Substantia nigra
   - Tegmentum
6. Tectum consists of two pairs of rounded elevations termed colliculi or corpora quadrigemina. These are the centers for reflexes of movements of the head, trunk, in response to auditory stimuli.
7. The midbrain also consists of left and right red nuclei which assists the basal ganglia and cerebellum to coordinate muscular movements.

Functions[11]

1. It is concerned in exerting and coordinating influences over the muscular activities through the medulla oblongata and spinal cord.
2. It is concerned with the exerting regulatory mechanisms to maintain the body temperature and equilibrium.
3. Regulates muscle tone.
4. Through III cranial nerve, controls the movements of the eyeball.
5. Through the mesencephalic nucleus of trigeminal nerve, it helps in transmission of the proprioceptive sensation from the head, face, and neck areas.

Basal Ganglia[12]

1. Basal ganglia is the term applied to the following multiple subcortical nuclei on each side of the brain:
   - Caudate nucleus
   - Putamen
   - Globus pallidus
   - Subthalamic nucleus
   - Substantia nigra
   - Red nucleus
2. Caudate nuleus and putamen together called *corpus striatum*.
3. Putamen and globus pallidus together form the lentiform nucleus.
4. Basal ganglia is another accessory motor system like cerebellum, that functions not by itself, but always in close association with cerebral cortex and corticospinal motor system.
5. Basal ganglia, thereby receive all their inputs from cerebral cortex and in turn returns almost all their output signals back to cortex.

Functions

1. It plays a very important role in controlling reflex activities, concerned with maintenance of muscle tone.
2. It is also responsible for production of certain associated movements like swinging of arm, when the subject is working.
3. Checks abnormal involuntary movements.
4. Red nucleus is a relay station and coordination center for many motor and sensory impulses and acts as a center for righting.
5. Substantia nigra is regarded as a center for the coordination of these impulses which are essential for skilled movements.

Pons

1. It is situated in front of the cerebellum, below the midbrain and above the medulla oblongata.
2. It consists of descending tracts passing between the higher levels of brain and the spinal chord.
3. The pons consists of nuclei of four pairs of cranial nerves (V, VI, VII and VIII) originate from here.
4. Apart from the cranial nuclei, it contains two respiratory centers—pneumotoxic and apneustic centers which control the respiration.

Functions

1. It forms a major sensory relay system which provides information to the different parts of the brain.
2. Gives origin and control four pairs of cranial nerves.
3. Through the respiratory centers, it controls and coordinates respiratory cycles.
4. It is also a relay center for motor impulses, many descending tracts synapse at pons.
5. Also plays an important role in maintaining consciousness and arousal response from sleep.

Medulla Oblongata

1. The medulla oblongata is situated in the posterior cranial fossa behind the basilar

part of occipital bone and is directly continuous with the lower part of the pons and with the spinal cord inferiorly.
2. It is pyriform in shape and is divided into two halves and each lateral half is fur-ther divided into three zones—anterior, lateral and posterior.
3. The anterior zone is pyramidal in shape and is called *pyramid*.
4. Medulla oblongata contains vital centers that lie in the deepest layer:
   - Cardiac center
   - Respiratory center
   - Vasomotor center
   - Reflex center for coughing, sneezing, swallowing and hiccuping.[13]

Functions

1. It is the reflex center for coughing, sneezing, and swallowing.
2. It controls the vital functions of the body like heart beat, breathing, blood pressure, etc. through the vital centers—cardiac, respiratory and vasomotor centers.
3. Many motor and sensory tracts cross from one side to another over here.

Decussation of the Pyramids

1. The descending tracts, while coming out of pons, the scattered fibers are re-united and enter the medulla as a thick bundle.
2. It occupies the most anterior part of medulla producing a distant bulge pyramid.
3. In the lower part of medulla, the majority of fibers cross, while the rest passed down on the same side.
4. Thus, the motor activity of the right side of the body is controlled by left hemisphere of the cerebrum and vice versa.

Cerebellum[14] (Fig. 1.3)

- Cerebellum is the largest part of the hindbrain and lies behind the pons and medulla oblongata.
- The cerebellum is situated in the posterior cranial fossa and is covered superiorly by tentorium.
- It is the largest part of hindbrain.
- The cerebellum is ovoid in shape and consists of two cerebellar hemispheres joined by vermis.

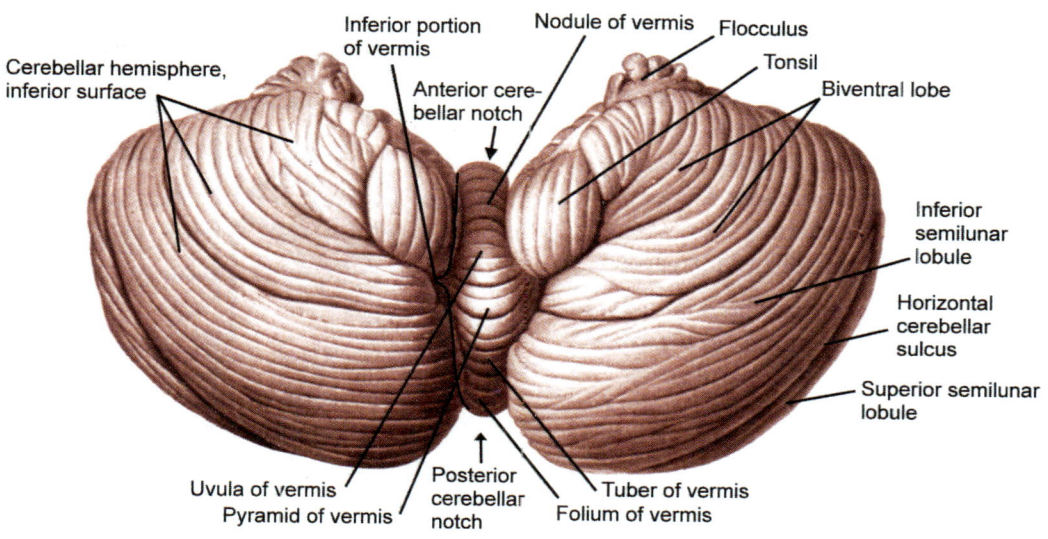

Fig. 1.3: Cerebellum

- The cerebellum is attached to brainstem by three bundle of nerve fibers—superior, middle and inferior cerebellar peduncles.
- The cerebellum is divided into three main lobes—anterior lobe, middle lobe and flocculonodular lobe.
- The anterior lobe is seen on the superior surface and is stereotype from posterior lobe by a wide V-shaped fissure called primary fissure.
- The middle lobe is the largest and is present between primary and uvulonodular fissures.
- The flocculonodular lobe is situated posterior to uvulonodular fissure.
- The cerebellum contains folds which are very thin, called *folia*.
- It is composed of outer gray matter called cortex and inner white matter.
- There is anterior cerebellar notch where we find brainstem.
- There is posterior cerebellar notch where we find the falx cerebelli.
- On the inferior side of cerebellum, in the midline there is vallecula.
- On either side of vallecula, there are rounded elevations called cerebellar tonsils.

Functions

1. It is mainly concerned with the equilibrium and posture of the body.
2. It maintains the muscle tone and center for stretch reflex.
3. It helps in coordination of voluntary movements.

PROPERTIES OF NERVOUS TISSUE[15]

The main properties of nervous tissue are:
1. Excitability
2. Conductivity
3. Refractory period
4. Summation
5. Accommodation
6. All-or-none law

Excitability

It is the ability of the nerve fiber to respond to a stimulus.

The stimuls may be electrical, mechanical or chemical.

Excitability depends on the following factors:
- Strength of stimulus
- Frequency of stimulation
- Duration of current
- Presence of injury

Conductivity

Impulse traveling along a nerve fiber may be propagated in both directions, this property of propagation is called *conductivity*.

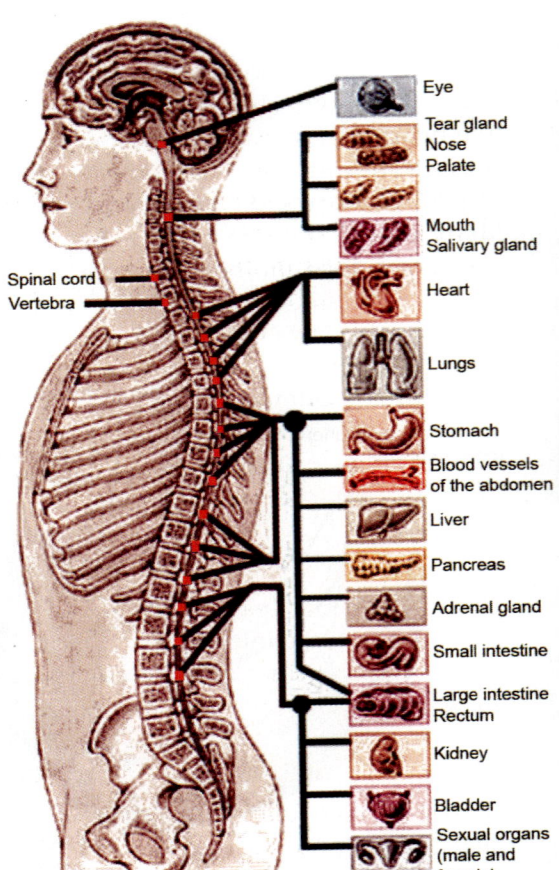

Fig. 1.4: Various organs and innervations by ANS

The mechanism of impulse conduction in myelinated nerve fibers is called "saltatory conduction", i.e. jumping of impulse from one node of Ranvier to other. The advantage of saltatory conduction is that it conserves energy and the velocity of conduction is increased.

Refractory Period

It is the altered state of excitability during response to a stimulus. It is of two types:
1. **Absolute refractory period (ARP):** The nerve fiber does not respond to the second stimulus, however, the strong stimulus is given. This period of inexcitability is known as ARP.
2. **Relative refractory period (RRP):** During the later phase, only a strong stimulus can elicit a response. This is called RRP.

Summation

A subthreshold stimulus will not give a response, but two subthreshold stimuli can give a response, it is called summation.
- **Temporal summation:** Two subthreshold stimuli are given at same point on the nerve one after another, then there is a response.
- **Spatial summation:** Two subthreshold stimuli are given at two different points of the nerve at a time, then a response is seen.

Adaptation or Accommodation

Nerves become less responsive when a stimulus of same intensity or a slow rising current given for a prolonged period of time.

All-or-None Law

Nervous tissue follows all-or-none law, that means the neural tissue shall respond when a complete threshold stimulus is given, if not, no response is seen.

SPINAL CORD[16]

Physiological Anatomy

1. Spinal cord is a downward continuation of the medulla oblongata.
2. It begins at the foramen magnum and descends through the vertebral canal.
3. It ends at the lower border of L1 vertebra (first lumbar).
4. Below the L1 vertebra though the spinal cord is absent, the nerve fibers which have merged from it continue.
5. A cross-section of the cord shows outer white matter and inner gray matter.
6. The outer white matter consists of ascending and descending tracts. The inner gray matter looks like letter "H". There are two posterior horns, two anterior horns and two lateral horns.
7. Lateral horns are seen only in the segments from 1st thoracic to L2 segments from which sympathetic fibers come; lateral horns are also seen in S2, S3, and S4 segments from which the parasympathetic fibers arise.
8. In the central part of the H-shaped gray matter is a canal called *central canal*. It contains CSF (cerebrospinal fluid).
9. From the anterior horn, nerve cells, axons emerge to form the anterior root which is motor.
10. The sensory fibers bringing the impulses from the periphery enter the posterior horn. The nerve cell bodies of these sensory fibers are situated outside the spinal cord and are called *dorsal root ganglion*.
11. The two roots, motor and sensory join outside the spinal cord to form the mixed spinal nerve.
12. In between the two vertebrae, a pair of mixed spinal nerve emerges out of the vertebral canal one on each side.
13. That part of the spinal cord, which has a pair of sensory nerve roots coming into

it and a pair of motor roots going out of it is called *spinal segment*.
14. There are 31 such segments: Cervical—8, Thoracic—12, lumbar—5, sacral—5, coccygeal—1.

Functions of Spinal Cord

1. It forms the informative highway for the bodily neural messages.
2. Carries the sensations to the brain.
3. Distributes the motor impulses to various glands, organs and muscles of the body.
4. It acts as reflex center for many spinal level reflexes.
5. It hosts the sympathetic and parasympathetic nerves.

CEREBROSPINAL FLUID (CSF)[17]

1. The fluid contained in the central canal of spinal cord and the subarachnoid space and cerebral ventricles is known as cerebrospinal fluid.
2. Approximately 150 mL of CSF is present (range is 100–200 mL).

Formation of CSF

1. It is formed by the choroid plexus situated within the ventricles.
2. It is chiefly formed in the lateral ventricles and also in the third and fourth ventricles.
3. Choroid plexuses are tuft-like capillary projections in the ventricles covered by pia mater and ependymal layers, which are fused and contain connective tissue, vessels and nerves.
4. The CSF is formed from the blood flowing through the choroid plexus.

Composition

- 99.13% water
- 0.57% solids
- Cells: 0–5/cu mm (lymphocytes)
- Solids constitute proteins, glucose, chlorides, sodium, and potassium.

Circulation of CSF (Fig. 1.5a)

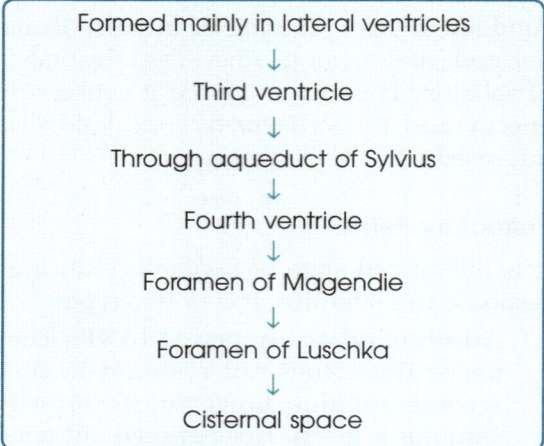

A portion circulates through spinal subarachnoid space.

Greater portion pass upwards to subarachnoid space surrounding cerebral hemispheres.

PERIPHERAL NERVOUS SYSTEM

1. It consists of 31 pairs of spinal nerves and 12 pairs of cranial nerves.
2. Most of the peripheral nerves are mixed nerves.

Types of Nerve Fibers[18,19]

Classification	Types
Structural	Myelinated Nonmyelinated
Functional	A fibers: $\alpha, \beta, \gamma, \delta$ B fibers C fibers
Depending on type of impulse they carry	Motor Sensory
Developmentally	Somatic Visceral
Chemically	Adrenergic Cholinergic
Source of origin	Cranial Spinal

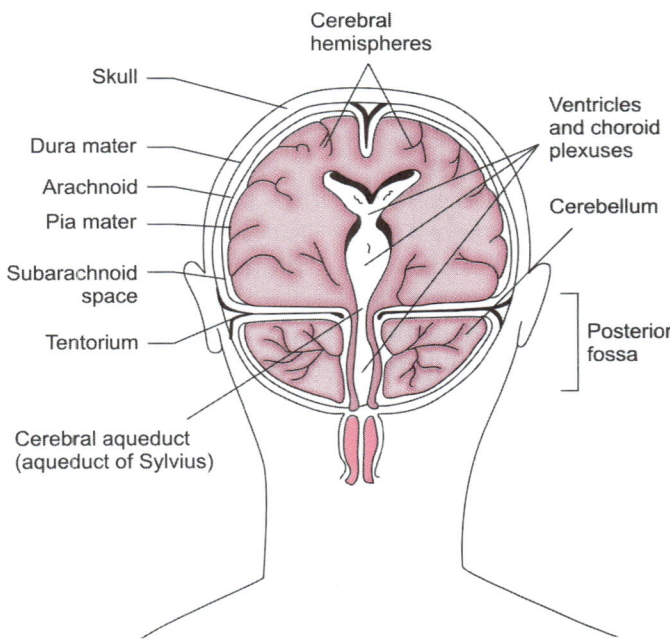

Fig. 1.5a: Ventricles of brain and CSF circulation

Motor unit: One motor nerve fiber with all its muscle fibers it supplies constitutes "motor unit".

Cranial Nerves

There are 12 pairs of cranial nerves originating from nuclei of brainstem. Some of them are sensory, some are motor and others are mixed.

The list of cranial nerves in craniological order and their functions are given in Table 1.2.

Table 1.2: List of cranial nerves in craniological order and their function[22, 23]			
No.	Name	Type	Function
I	Olfactory	Sensory	Olfactory (smell) information from nose
II	Optic	Sensory	Visual information from eyes
III	Oculomotor	Motor	Eye movement, pupil constriction, lens shape
IV	Trochlear	Motor	Eye movement
V	Trigeminal	Mixed	Sensory information from face, mouth, motor signals for chewing
VI	Abducens	Motor	Eye movement
VII	Facial	Mixed	Sensory for taste, efferent signals for tear and salivary glands, facial expression
VIII	Vestibulocochlear	Sensory	Hearing and equilibrium
IX	Glossopharyngeal	Mixed	Sensory from oral cavity, baro- and chemoreceptors in blood vessels, efferent for swallowing, parotid salivary gland secretion
X	Vagus	Mixed	Sensory and efferents to many internal organs, muscles and glands
XI	Spinal accessory	Motor	Muscles of oral cavity, some muscles in neck and shoulder
XII	Hypoglossal	Motor	Tongue muscles

AUTONOMIC NERVOUS SYSTEM (ANS)

The autonomic nervous system control the functions of body automatically. It is divided into two parts:
- Sympathetic
- Parasympathetic

Sympathetic Nervous System

It consists of three neurons, conveying impulses from their origin in hypothalamus, reticular formation and medulla oblongata to effector organs and tissues.

Parasympathetic Nervous System

Two neurons are involved in the transmission of impulses from their source of origin to effector organ. The first neuron is situated in the brain or spinal cord.

Role of ANS on Various Organs

1. **Effects of sympathetic stimulation:**[20,21]

Organ	Effects
Iris muscle	Pupils are dilated
Blood vessels in head region	Constricted
Salivary gland	Secretion is inhibited
Oral and nasal mucosa	Mucus secretion is inhibited
Skeletal blood vessels	Dilated
Heart	Heart rate and force of contraction increased
Coronary arteries	Dilated
Trachea and bronchi	Slight constriction
Bronchial muscle	Relaxed
Stomach	Peristalsis reduced, sphincters closed
Intestine	Peristalsis and tone decreased
Liver	Glycogen-glucose conversion increased
Spleen	Contracted
Adrenal medulla	Adrenaline and noradrenaline secretions increased
Large and small intestines	Peristalsis reduced, sphincters closed
Kidney	Urine secretion decreased
Bladder	Wall relaxed, sphincter closed
Sex organs	Generally blood vessels are constricted

2. **Parasympathetic stimulation:**[20,21]

Organ	Effects
Iris muscle	Pupils constricted
Lacrimal glands	Tear secretion is increased
Salivary glands	Increased secretion
Heart	Heart rate and force of contraction are decreased
Coronary arteries	Constricted
Trachea and bronchi	Constricted
Stomach	Increased gastric secretion and motility
Small intestine	Digestion and absorption are increased
Liver and gallbladder	Blood vessels dilated, secretion of bile increased
Pancreas	Increased secretion of pancreatic juice
Small intestine	Increased secretion of intestinal juice and motility increased
Large intestine	Secretion and motility increased, sphincter relaxed
Sex organs	In males, vasodilatation and erection of penis and in females vasodilatation and erection of clitoris

REFLEX ARC[24]

Definition: A reflex is a stereotypic, sudden, involuntary response to stimulus, e.g. touching the cornea of the eye gives to reflex closure of the eyelids.

Reflex arc consists of:
1. Receptor
2. Afferent pathway
3. Reflex center
4. Efferent pathway
5. Effector organ

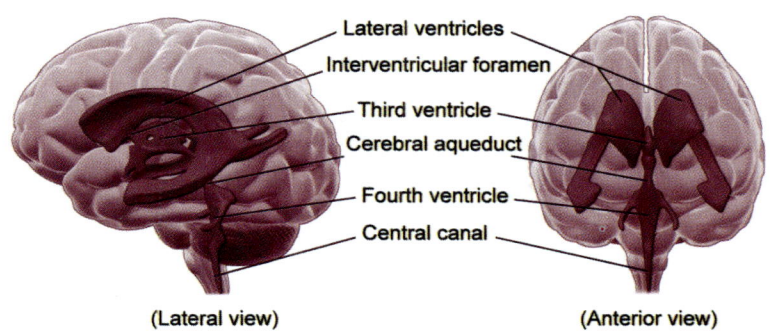

Fig. 1.5b: Ventricles of the brain

UMN and LMN Differences

Upper motor neuron (UMN)	Lower motor neuron (LMN)
Consists of a motor neuron whose cell body is located in the motor area of the cerebral cortex and whose processes connect with motor nuclei in the brain stem or the anterior horn of the spinal cord	Lower motor neurons (LMNs) are the motor neurons connecting the brainstem and spinal cord to muscle fibers, bringing the nerve impulse from the upper motor neurons out to the muscles. A lower motor neuron's axon terminates on an effector (muscle)
Lesions in UMN lead to hypertonicity in muscles	Lesions in LMN leads to hypotonicity in muscles
Reflexes are exaggerated	Reflexes are lost or diminished
Clasp knife mechanism is seen	Not seen
No muscle wasting is seen	Muscle wasting is seen
Babinski is positive showing extensor response	Babinski is absent with no response

Neurogenic Bladder[25,26]

It is the dysfunction of neurologic origin where a person lacks control over the bladder due to brain, spinal cord or nerve conduction.

Innervation of Human Urinary Bladder

The urinary bladder is innervated by the parasympathetic and sympathetic nerve fibers.

The parasympathetic supply is through the 2nd, 3rd, and 4th sacral spinal segments and they help to contract the bladder wall and relax the sphincter.

The sympathetic nerve supply is through 10–12th thoracic and 1st, 2nd lumbar spinal segments. In human beings, the sympathetic supply has no active role in the act of micturition.

Physiology of Micturition[27,28]

- Voiding is otherwise called micturition.
- When the bladder is full, the person feels the sensation of voiding
- The sensation is carried from the receptors present in the bladder wall and are carried by S2–S4 sensory parasympathetic fibers to the concerned spinal segments.
- Though the sensation of fullness of the bladder is felt, the person can withhold or control this sensation for some more time till he reaches a suitable place to void.
- Efferent impulses carried by S2–S4 motor parasympathetic fibers leading to contraction of bladder wall and relaxation of sphincter.

Tracts and Functions

Pathway	Direction	Origin	Crossover	Destination	Function
Corticospinal	Descending	Motor cortex	90% as lateral CST; 10% as anterior CST	Alpha motor neurons	Skilled motor to extremities
Reticulospinal	Descending	Reticular formation	Variable levels	Alpha and gamma motor neurons	Cortical control of voluntary motor function
Rubrospinal	Descending	Red nucleus	Immediate	Alpha and gamma motor neurons	Facilitate extensors and inhibit flexors (for balance control)
Vestibulospinal	Descending	Vestibular nucleus	None	Alpha and gamma motor neurons	Facilitate extensors and inhibit flexors (for balance control)
Lateral spinothalamic	Ascending	Free nerve endings	Immediate	Posterior central gyrus	Pain, temperature
Anterior spinothalamic tract	Ascending	Free nerve endings	Variable levels	Posterior central gyrus	Light touch and pressure
Posterior column: Fasciculus gracilis and cuneatus	Ascending	Meissner's corpuscles, Pacinian corpuscles, muscle spindle, tendon organs	None	Posterior central gyrus	Discriminative touch, vibration sense, proprioception

- When the acting of voiding is ready, the sphincter muscles relax and allow the free flow of urine to the exterior.
- The sphincter muscles both external and internal relax and contract of detrusor muscle results in voiding.

Lesions Affecting Bladder Function

1. **Lesions at the level of the reflex arc (LMNL):**
   a. Lesions in the afferent fibers, sensory atonic bladder, characterized by:
      - Absence of the sense of fullness of the bladder.
      - Retention of urine associated with a huge size of the bladder.
      - Dribbling of urine every now and then because of overflow.
   b. Lesion in the efferent fibers, motor atonic bladder, characterized by:
      - Preservation of the sense of fullness of the bladder.
      - Retention of urine associated with a moderate size of the bladder.
      - Inability to evacuate the bladder voluntarily.
      - Catheterization is usually quickly done.
   c. Lesion in both afferent and efferent fibers or in the spinal center, autonomic or autonomous bladder, characterized by incomplete, irregular, involuntary evacuation of the bladder as the evacuation of the bladder depends on its myogenic contraction.

2. **Lesions above the level of the reflex arc (UMNL):**
   a. Acute: Retention with overflow.
   b. Gradual:
      - Partial lesion—precipitancy of micturition.
      - Complete lesion, automatic bladder—This is characterized by complete and regular evacuation of the bladder which works by the spinal reflex arc.

NB: For any disturbance in bladder function (due to a nervous lesion) to occur, the lesion should be bilateral.

REFERENCES

1. Waymire Jack, "Organization of Cell Types". Neuroscience Online. The University of Texas Medical School, Retrieved 27 January, 2015.
2. Verkhratsky Alexi, Butt, Arthur. Glial Physiology and Pathaphysiology (First ed.), Chinchester, UK; John Wiley & Sons. 2013;p. 76.
3. Al Martini, Frederic Et. Anatomy and Physiology, Rex Bookstore Inc. ISBN 978-971-23-480705. 2007; p. 288.
4. Kandel ER, Schwartz JH, Jessel TM. Principles of Neural Science, McGraw-Hill Professional. ISBN 978-0-8385-7701-1, 2000;p. 324.
5. Gerard J Tortora, "Principles of Anatomy and Physiology", 12th Edition, p. 519.
6. Herrero, Maria-Trinidad, Barcia, Carlos, Navarro, Juana. "Functional anatomy of thalamus and basal ganglia". Child's Nervous System 2002;18(8):386–404.
7. Steriade, Mircea, Llinas, Rodolfo R. "The Functional States of the Thalamus and the Associated Neuronal Interplay". Physiological Reviews. 1988;68(3):649–742.
8. Melmed S, Polonsky KS, Larsen PR, Kronenberg HM. Williams Textbook of Endocrinology (12th ed.), ISBN 978-1437703245. Saunders. 2011; p. 107.
9. Theologides A. "Anorexia-producing intermediary metabolites". PMID 178168. Am J Clin Nutr 1976;29(5):562–68.
10. Swaab DF. "Sexual orientation and its basis in brain structure and function". PNAS 2008;105(30): 10273–74.
11. Breediove, Watson and Rosenzweig. Biological Physiology 6th ed., 2010, pp. 45–46.
12. Wayhenmeyer, James A, Gallman, Eve A. Rapid Review of Neuroscience. ISBN 0-323-02261-8. Mosby Elsevier 2007; p. 102.
13. Robert K Clark. Anatomy and Physiology: Understanding the Human Body. Jones & Bartlett Learning, 2005.
14. AMR. Agur, Arthur F, Dalley. Grant's Atlas of Anatomy. Lippincott Williams & Wilkins, 2009.
15. Lauralee Sherwood, Fundamentals of Human Physiology (4th ed.), Cengage Learning, 2011.
16. Walter Hendelman, MD. Atlas of Functional Neuroanatomy, CRC Press, 2000.
17. Vendelin Slavik, Tereza Dolezal, Cerebrospinal Fluid: Functions, Composition and Disorders, Nova Biomedical, 2012.
18. Peter L, Williams et al. Gray's Anatomy (27th ed.), ISBN 0-443-04177-6. Edinburgh: Churchill Livingstone. 1989.
19. Guyton, Arthur C. Textbook of Medical Physiology (7th ed.). Philadelphia: Saunders. 1986; ISBN 0-7216-1260-1.
20. Pocock, Gillian. Human Physiology (3 ed.). Oxford University Press, 2006; ISBN 978-0-19-856878-0.
21. Moore KL, Agur AM. Essential Clinical Anatomy 2 ed. Lippincott, 2002.

22. Vilensky Joel, Robertson, Wendy, Suarez-Quian, Carlos. The Clinical Anatomy of the Cranial Nerves: The Nerves of "On Olympus Towering Top". Ames, Iowa: Wiley-Blackwell. 2015; ISBN 978-1-118-49201-7.
23. Standring Susan, Borley Neil R. "Overview of cranial nerves and cranial nerve nuclei". Gray's Anatomy: The Anatomical Basis of Clinical Practice (40th ed.) (Edinburgh): Churchill Livingstone/Elsevier. 2008; ISBN 978-0-443-06684-9.
24. Ganong WF. Review of Medical Physiology, McGraw-Hill Publishing, New York, 2001; p. 123, ii.
25. Dorsher PT, McIntosh PM, Neurogenic bladder. Adv Urol 2012; 2012:816274.
26. Taweel WA, Seyam R. Neurogenic bladder in spinal cord injury patients. doi: 10.2147/RRU.S29644, eCollection 2015. Res Rep Urol 2015 Jun 10;7:85–99.
27. Dasgupta R, Kavia RB, Fowler CJ. "Cerebral mechanisms and voiding function". BJI Int 2007;99(4): 731–35.
28. Kinder MV, Bastiaanssen EH, Janknegt RA, Marani E. "Neuronal circuitry of the lower urinary tract, central and peripheral neuronal control of the micturition cycle". Ana Embryol 1995;192(3): 195–209.

2. Principles of Assessment and Treatment in Neurological Conditions

LEARNING OBJECTIVES

At the end of this chapter, you will be able to:
1. Assess the patient with neurological dysfunction.
2. Identify the various problems and provide a diagnosis.
3. Differentiate the diagnostic interventions with the help of proper assessment techniques.
4. Plan out the treatment protocols based on the assessment.

INTRODUCTION

Assessment is a key factor for identifying the problem, designing the treatment protocols and to judge the prognosis. Most of the patients with neurological problems have complex problems and is not always possible to make a full assessment in any one session.

The clinical assessment usually involves questioning, observing and examining the client about the nature, duration and severity of the client's problems.

A good assessment should cover the clients all possible physical assessment.

The assessment usually contains two basic areas:

1. Subjective assessment
2. Objective assessment

A subjective assessment is a method of documentation about the patient by directly asking the patient or relatives or obtaining from the clinical data, e.g. admission notes.

An objective assessment is one in which the clinician shall perform various skills to identify the pathology that leads to the chief complaint/s of the patient.

SUBJECTIVE ASSESSMENT[1-3]

A brief demographic history must be taken either directly interviewing the patient or guardians or taken from medical records. It includes name, age, sex, occupation, socio-economic status, dominance, and address of the patient.

History taking: History taking should be patiently, and tactfully done. A good physiotherapist always spends much time in taking the detailed history as it often holds key to diagnosis.

Chief complaints: These are the main problems faced by the patient when admitted to the hospital. These are written by the medical person in patient's own words in chronological order; that is in the sequence of appearance of symptoms that appeared first. Medical terminologies should not be used. Chief complaints can be gathered directly from the patient or can be from patient's attendant.

History

This means the events took place from starting of the condition till present. This can be gathered directly through the patient or through the attendant.

- **Present history**: It describes patient's condition in chronological order.
  - *Mode of onset*: Sudden or gradual condition/disease is suddenly appeared or appeared slowly.
  - *Mechanism of injury*: Through what way patient was injured. While gathering the information, the examiner should determine the magnitude and direction of the injuring force.
  - *Symptoms* that make patient uneasy, restless in day or night or throughout 24 hours.
  - *Relieving/aggravating factors*: Activities which give relief and activities which increase the condition.
  - *Duration/period*: Time from when he/she suffering from the condition. If the condition started initially, it is acute or if it is of longer duration, it is chronic in nature.
  - *Treatment/medication* if received recorded mention the dosage, procedure, timing.
  - Patient whether conscious or not when came to the treatment center/hospital.

Example: Patient was apparently well till this afternoon, when he met with a road traffic accident while driving the motorcycle. He was hit by a truck from back. He was thrown onto the side of the road with severe injury to pelvic region. He became unconscious and was brought to a nearby clinic where first aid was given and referred to AIIMS. X-ray and other investigations were done. He was diagnosed with fracture hip bone and was operated for the same. Medication and physiotherapy are given as per the orders.

Past history: This details about any history of the same or other condition which the patient met earlier.

Example: Patient suffered from an RTA in which he had fractured his right radius ulna around 10 years back.

Medical history: Means any history of medical problems and any previous surgeries the patient undergone of any region of the body and its medication if presently continuing, should be in knowledge of the examiner. This helps the examiner to take care while prescribing any exercises or investigation.

Personal history: This indicates patient's *personal habits* like chewing tobacco, smoking, alcohol consumption, or any habits which are hindering his/her health.

Example: If a person smokes, then this is harmful for his lungs, may result in lung disease.

Marital status: If a patient is married then their partner can look after the patient by giving physical and mental support.

Family history: This means whether patient has a nuclear or joint family. It includes the spouse, children in a nuclear family, or children, mother, father or any other members in a joint family.

In a joint family, patient receives a better care and treatment than the nuclear family due to more number of family members.

Economic history: This indicates the economic status of the patient. How much is the income, its source and how much is the expenditure. It plays an important role in his/her rehabilitation.

Example: If patient is a poor man, he will not be able to take adequate treatment due to lack of funds.

Therapist has to design treatment plan according to the fund available with the patient or relatives.

Social history: This deal with patient's social and education status.

Social status means patient's self-presentation in society.
- Does patient like to meet people in the surroundings?
- Does he respect and understand the values of other individuals?
- Whether he/she adapts among many people?

Education status: If patient is educated, he/she will be able to communicate politely with others and judge them.

OBJECTIVE ASSESSMENT[4]

Observation skills:
- Observe the general built of the patient.
- Check for the abnormal attitude of the patient (Claw hand, foot drop, drooped shoulders, etc.).
- Observe for wasting of the muscles.
- Observe for any skin changes.
- Check whether any swelling in the involved area or any gross swelling which may be relevant.
- Observe for any scars or unhealed wounds or skin infections.
- Identify for any involuntary movements like tremors, athetosis, etc.
- Observe for any contractures and deformities.

On palpation:
- Palpate the several individual muscles to see for muscle wasting, hypertrophy, or fasciculations and in cases of suspected myositis just palpate the affected site of the muscles to see if there is tenderness.

  Tenderness is graded as:
  - Severe: If patient shows grimace on face with touch.
  - Moderate: If patient shows grimace on face to hold.
  - Mild: If patient shows grimace on face to press (Fig. 2.1).

Note: The examiner should never ask the patient about the experience of pain during checking of tenderness, but should identify by facial expressions of the patient.

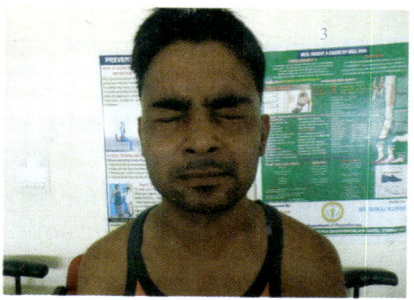

Fig. 2.1: Grimace

- Palpate the edema and grade it under the following headings:
  - Location and site
  - Indurated or non-indurated
  - Pitting or non-pitting type.

On examination:

Examination of higher functions: Examine the higher functions—memory, intelligence, level of consciousness, behavior, orientation and speech.

The level of consciousness is assessed by GCS[5] as shown in Table 2.1.

Memory:
a. *Immediate recall and attention:* By name and address test (note number of errors in repeating the name and address and after how many attempts it is repeated clearly).
b. *Short-term memory:* After 5 minutes, again ask the patient to repeat the name and address used in immediate recall.
c. *Long-term memory:* In this test factual knowledge, e.g. first Prime Minister of India.

Physiotherapy in Neurological Conditions

Table 2.1: Glasgow Coma Scale

	1	2	3	4	5	6
Eye	Does not open eyes	Opens eyes in response to painful stimuli	Opens eyes in response to voice	Opens eyes spontaneously	–	–
Verbal	Makes no sounds	Incomprehensible sounds	Utters inappropriate words	Confused, disoriented speech	Oriented, converses normally	–
Motor	Makes no movements	Extension to painful stimuli	Abnormal flexion to painful stimuli	Flexion/withdrawal to painful stimuli	Localizes painful stimuli	Performs voluntary action on command

Intelligence: To be tested by simple mathematical calculations.

Speech: The patient is tested for his ability to speak, understand spoken sounds or words. The clarity of speech, tone, smoothness, articulation needs to be assessed.

Behavior: Check the behavior of the patient when you assess the patient. The patient may be irritable, hostile, over anxious, resistive, shy, tense, agitated, fearful, inhibited, negative, frank, social or friendly.

Orientation: The topographical orientation of the patient is checked by place, by person and by time, and report oriented or disoriented.

CRANIAL NERVE ASSESSMENT[6]

There are 12 pairs of cranial nerves present in human beings. Cranial nerves constitute a part of peripheral nervous system. One or more nerves can be affected depending on the cause, e.g. space occupying lesions (SOL), myasthenia gravis, multiple sclerosis, etc.

The cranial nerves are tested as follows.

Olfactory Nerve

The main function is smell.

Procedure

1. Wash your hands properly. Introduce yourself to the patient.
2. Explain the procedure clearly and in a simple way to obtain consent of the patient.
3. Make the patient sit in front of you with eyes closed.
4. Ask the patient to close one nostril.
5. Now offer something familiar to the patient to smell and ask to identify by keeping the object 10 cm away from the nostril.
6. Usually, a lemon peel, garlic, ginger, coffee can be offered.
7. Now ask the patient to close the other nostril and identify the smell.

Interpretation

1. If the patient is able to identify, the olfactory is functioning normal.
2. If the patient is unable to identify, then the following can be the reasons:
   a. Prior upper respiratory infections
   b. Head injuries
   c. Nasal and paranasal sinus disease causing damage to olfactory neuroepithelium
   d. Neurodegenerative diseases
   e. Iatrogenic interventions, e.g. septoplasty, rhinoplasty, radiation therapy, etc.
   f. Intracranial tumors or intranasal neoplasms
   g. Epilepsy, hypothyroidism, renal diseases
   h. Some studies have shown that loss of smell is seen in disorders associated

with cerebellar degeneration, e.g. Friedreich's ataxia.
i. It is a hallmark sign in schizophrenia.
j. Some cases of migraine.
k. Viral infections.

Optic Nerve

It is the second cranial nerve. It is a sensory nerve and is mainly involved for vision. The optic nerve is tested for:

a. Acuity of vision
b. Color vision
c. Field of vision
d. Pupillary reflex
e. Fundoscopy examination

Acuity of Vision

- The acuity of vision is tested with the help of Snellen's chart.
- The normal acuity of vision is 6/6 or 20/20 in humans.
- If the patient wears glasses or contact lens normally, then the test must be checked with and without vision aids and report separately.
- If you are checking a handheld Snellen's chart, then hold the chart at 14 inches distance from the patient. Hold the chart at the level of eye of the patient.
- Ask the patient to close one eye each time and read aloud the smallest letter they are able to see.
- If you are using a wall mounted Snellen's chart, then the distance between the patient and the chart should be 20 feet.

Interpretation:
- Visual acuity is expressed as a fraction.
- The numerator or the top number refers to the distance you stand from the chart.
- The denominator or bottom number indicates the distance at which a person with normal eyesight could read the same line you correctly read.
- For example, 20/20 is considered to be normal in humans. A 20/40 reading reports that the patient could correctly read at 20 feet away which can be read by a person with normal vision from 40 feet away.
- If the patient cannot identify all items correctly, number missed is listed after a "−" sign, e.g. 20/80 − 2 for 2 missed on 20/80 line.

Color Vision

- Color vision is tested using Ishihara plates which help to identify the patients who are color blind.
- This test is the most widely accepted test and used for testing red–green color vision deficiency and contains 38 plates of circles created by irregular colored dots in two or more colors.
- The patient is asked to close one eye for testing and vice versa.
- The plates are kept in front of the patient and will be asked to identify the number on the plate.

Field of Vision

Visual fields are tested by asking the patient to look directly at you while you wiggle one of your fingers in each of the four quadrants of visual field of each eye.

Procedure:
- Face the patient approximately 1–2 feet apart.
- Close your right eye and ask the patient to close his/her left eye.
- Move your left arm out and away keeping it at equal distance from both of you.
- Rise your index finger, such that it should be just outside your field of vision.
- Wiggle the index finger and bring it in towards the noses. You and your patient should be able to detect it at the same time.

- Repeat the moving finger in each direction. Use other hand to check the medial field, starting in front of the closed eye.
- Repeat the entire procedure for the other eye as well.

Pupillary Reflex

The pupillary reflex involves adjustments in pupil sizes with changes in the intensity of light.

The optic nerve is involved in the pupillary reflex.

Procedure:
- Make sure that the room is dark and pupils are dilated a little.
- The room should not be completely dark that you cannot observe the pupil reaction.
- Make the patient sit calmly on a chair and ask him to look straight.
- Now shine the light in right eye and observe the pupil being constricted.
- Now repeat the same procedure on left eye.

Interpretation:
- Usually, the normal response of pupillary reaction is described as PERRLA (pupils equal, round, reactive to light accommodation).
- Abnormal responses are observed secondary to direct or indirect damage to optic nerve, parasympathetic injury or to the sympathetic neurons.
- Some sympathomimetics, like cocaine, will dilate the pupil and some narcotics, like heroin, will constrict the pupil.

Fundoscopy Examination

Fundoscopy, also known as ophthalmoscopy, is a test that allows a health professional to see inside the fundus of the eye and other structures using an ophthalmoscope or fundoscope.

It is a part of routine physical or complete eye examination.

It is used to detect and evaluate symptoms of various retinal vascular diseases or eye diseases such as glaucoma.

It also indicates increased intracranial pressure which could be due to hydrocephalus or space occupying lesions in brain, etc.

Oculomotor/Trochlear and Abducens

a. The oculomotor, trochlear and abducens are checked all together, as they are involved in eyeball movements supplying the extraocular muscles.
b. These nerves allow smooth and coordinated movement in all directions of both eyes, simultaneously.
c. Abducens supplies lateral rectus muscle and moves the eyeball down.
d. Trochlear supplies the superior oblique muscle and moves the eyeball down and rotates internally.
e. Oculomotor supplies all other muscles of eye movement and also helps in raising the eyelid and mediates pupillary constriction.

Procedure to Test

- Ask the patient to keep the head immobile.
- Ask him/her to follow your finger with his/her eyes as you trace the letter H.
- Alternatively, direct them to follow finger with their eyes as trace a large rectangle.
- Eyes should move in all directions, in coordinated, smooth and symmetric fashion.
- Hold the eyes in lateral gaze for a few seconds to look out for nystagmus.

Trigeminal Nerve

The trigeminal nerve is involved in sensory supply to the face and motor supply to the muscles of mastication (temporalis and masseter)

There are three sensory branches of trigeminal nerve:
- Ophthalmic
- Maxillary
- Mandibular

It also plays a role in the corneal reflex.

Hence, the trigeminal nerve is tested for:
1. Sensations over the face
2. Motor and jaw jerk
3. Corneal reflex

Sensory

- Ask the patient to close the eyes.
- Touch each of the three areas (ophthalmic, maxillary and mandibular).
- Around the jawline, on the cheek and on the forehead lightly with the help of cotton.
- Later check with a pin for pain sensations.
- You can also check the temperature sensation, by using two test tubes of hot and cold water placed over these areas and ask the patient to identify the sensation.

Motor

- Ask the patient to clench the teeth together.
- Observe and palpate the bulk of the masseter and temporalis muscles.
- Now ask the patient to open the mouth and you resist it.
- Jaw jerk: Place your non-dominant index finger on the patient's chin.
- Strike your finger over the patient chin by a tendon hammer.
- The response is slight protrusion of the jaw.

Corneal Reflex

- Ask the patient to sit with open eyes and look straight.
- Lightly touch the cornea with a sterile cotton wool.
- The response will be shutting of the patient's eyelids.

Facial Nerve

The facial nerve is the seventh cranial nerve and supplies motor branches to the muscles of facial expression.

It also supplies anterior two-thirds of the tongue by its cauda tympanic branch which is sensory.

This nerve is tested by asking the patient to perform certain expressions on the face as follows:
- Make crease on the forehead
- Close the eyes tightly
- Puff out the cheeks
- Smile by showing the teeth

Vestibulocochlear Nerve

It is the eighth cranial nerve and provides innervations to the hearing apparatus of the ear and also transmits information about balance and equilibrium to brain.

The vestibular and cochlear parts of this nerve are tested separately.

Cochlear part is involed in hearing, several tests are present to test this nerve.

Crude Tests of Hearing

Make the patient relax and rub your fingers next to either ear or whisper some words and ask the patient to repeat.

Weber test: It is an easy screening test for hearing. This test is done to detect unilateral (one side) conductive hearing and unilateral sensorineural hearing loss.

How to do: For this test, a tuning fork of 512 Hz is used as it is within the range of normal hearing.
- Make the patient sit comfortably and explain him the entire procedure and take the consent.
- Get the tuning vibrate by striking ends against a rubber cork.

- Place vibrating fork on the midline of skull.
- The sound should be symmetrically heard on right and left ears equally.
- If there is conductive hearing loss (which may be because of obstructing wax in the canal), it is heard louder on the defective side.
- If there is sensorineural, then it is heard louder on the normal side.

Rinne test: The Rinne test is a hearing test and primarily used for testing the loss of hearing in one ear. It compares perception of sound transmitted by air conduction and by bone conduction.

How to do:
- For this test, a 512 Hz tuning fork is used.
- Ask the patient to sit in relaxing position. Explain the procedure to the patient and take consent.
- Get the tuning fork vibrate by striking ends against a rubber cork.
- Place the vibrating fork handle on the mastoid process behind the ear to be tested.
- Ask the patient to tell or raise hand when he stops hearing the sound.
- Once the hearing is stopped, place the tines of the fork next to the ear.
- In normal, the patient should hear the sound again as the air conduction is better than bone conduction.
- If the bone conduction is better than air conduction, this suggests a conductive hearing loss.
- If the air conduction is better than bone conduction, then it suggests a sensorineural hearing loss.

Vestibular Function

The vestibular apparatus of the inner ear is helpful to detect the movement of head and its position in space. If there is a damage to the vestibular system, then the chief symptom would be vertigo.

If the patient complains of vertigo, then **Dix-Hallpike test** is done. It is a diagnostic maneuver used to identify benign paroxysmal positional vertigo (BPPV).

How to do: Make the patient sit upright on the examination table with legs extended.

The patient's head is then rotated to one side by 45°.

Then you assist the patient to lie down backwards quickly with the head held in approximately 20° of extension.

Now observe the patient's eye for about 45 seconds for onset of nystagmus, if there is a nystagmus, then the the test is positive.

Glossopharyngeal Nerve

It is the IX cranial nerve and is sensory for taste sensation supplying the posterior one-third of the tongue and palate. It is also involved in the gag reflex.

The glossopharyngeal nerve is best tested by checking the gag reflex.

How to do: Touch the uvula with a tongue blade or a piece of cotton plug. The response will be reflex contraction of the back of the throat.

Vagus Nerve

Vagus is the X cranial nerve. It provides motor supply to the pharynx. Vagal lesions produce palatal and pharyngeal paralysis, laryngeal paralysis and abnormalities of esophageal motility and other autonomic dysfunction.

How to check:
- Listen to the patient talk as you are taking the history. A hoarseness in the voice, whispering, nasal speech or the complaint of aspiration or regurgitation of foods, liquids through the nose is a striking feature of vagal lesions.
- Ask the patient to open the mouth and say "Ahh" as long as possible. You can use a tongue blade to depress the base of the tongue gently to observe.

- Observe the palatal arches as they contract and soft palate, in normal individuals, there is rise if palate and uvula and uvuala is in central position and does not deviate.
- In vagal lesions, there is no elevation or constriction on the affected side.

Spinal Accessory Nerve

The spinal accessory nerve is the XI cranial nerve and gives motor supply to the sternocleidomastoid and trapezius muscles.

The main functions are:
- Rotation of the head to opposite side of the contracting muscle.
- Tilting of the head to the same side of the contracting muscle.
- Flexion of the neck by both sternocleidomastoid muscles.
- Elevation of the shoulder.
- Drawing the head back such that the occiput tilts towards the acromion.

To check this nerve, the muscle power of these two muscles is checked.

How to check:
1. Palpate the trapezius and sternocleidomastoid muscles.
2. Ask the patient to perform the above mentioned movement first without resistance and later with resistance.
3. Compare both the sides.

Any weakness or paralysis of these muscles indicates the lesion of the nerve.

Hypoglossal Nerve

It is the XII cranial nerve and provides motor supply to the muscles of the tongue. The lesions of the hypoglossal nerve results in paralysis and atrophy of the tongue muscles. There will be fasciculations of the tongue on the involved side.

How to check: Ask the patient to protrude the tongue out. Usually, in normal individuals, the tip of the tongue will be in center, if there is a lesion, the tip of the tongue will point to the normal side due to unopposed normal tone of unaffected side.

Also observe the tongue for atrophy and fasciculations.

SENSORY EXAMINATION[7-10]

Check the sensations over all the dermatomes and report which is affected. Usually, in case sheets, the affected dermatomes are colored or shaded (Fig. 2.2).

Superficial sensations: Check for pain, fine touch, crude touch and temperature in all the dermatomes and compare with normal.

Fine touch is tested by using cotton or wool. Ask the patient to close the eyes and count numbers whenever he feels the sensation.

Crude touch is checked by the tip of your finger.

Temperature is checked by using two test tubes filled with hot and cold water.

Pain is checked by a sharp object like sterile injection needle.

Always remember, the patient can use guess work, so try to trick him, do not ask him "Do you feel? What you feel?" questions.

Deep sensations: The deep senses that are to be assessed are vibrations, pressure, joint position sense and joint kinesthetic sense.

Test: Use your thumb or fingertip to apply a firm pressure on the skin surface.

Response: Ask the patient to indicate the recognition of stimulus by saying yes or no.

To avoid bias, trick the patient, by not giving the stimulus now and then.

To find out the symmetry of sensation, compare with the normal and ask the patient to conform the symmetry of sensation.

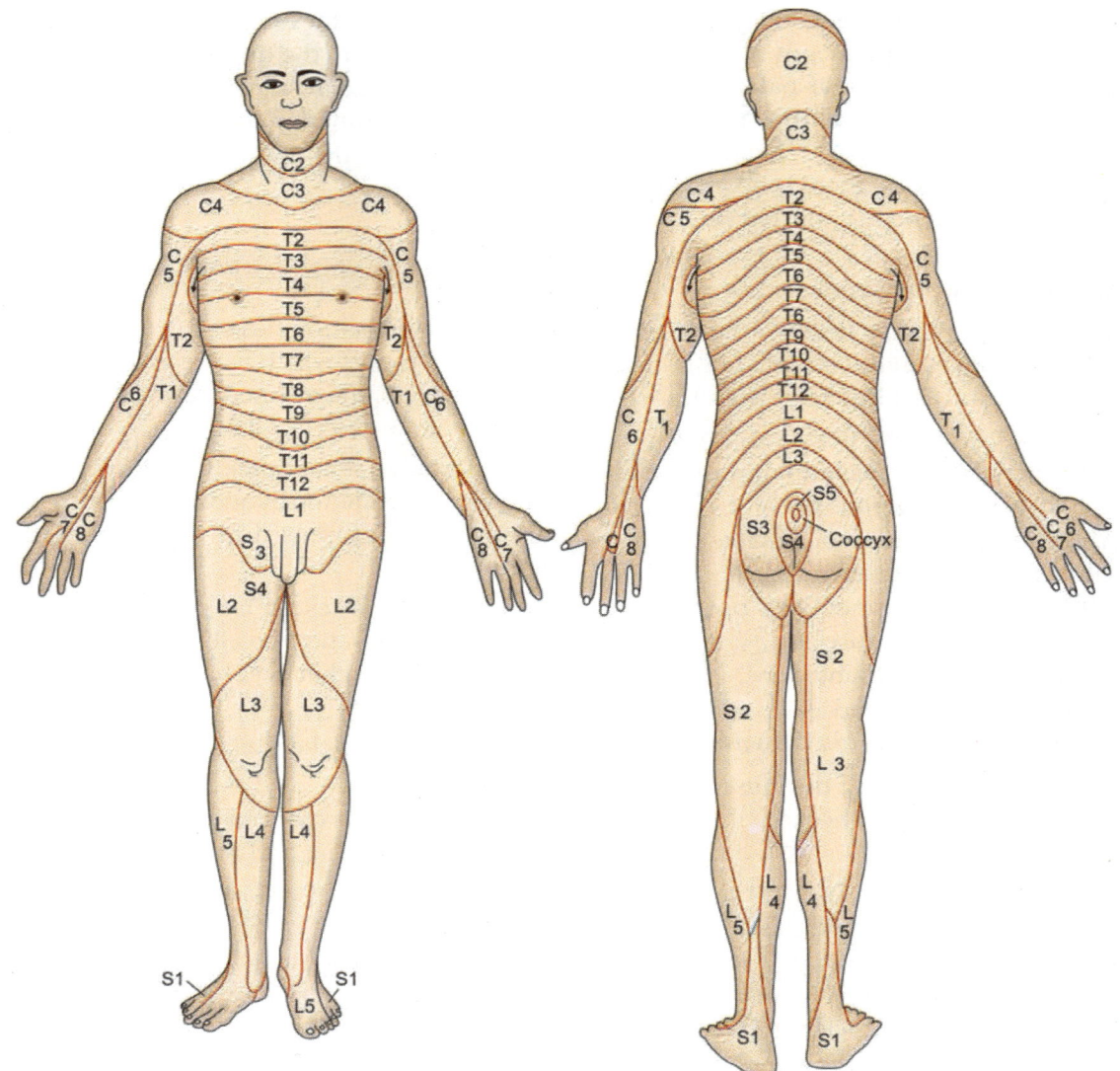

Fig. 2.2: Body chart for sensory examination

Vibration

Test: A 128 Hz tuning fork is used. Randomly place the vibrating and non-vibrating tuning fork on a bony prominence.

Use ear phones to avoid auditory clues.

Response: Ask the patient to respond by saying vibrating or non-vibrating.

To avoid bias, trick the patient, by not giving the stimulus now and then to find out the symmetry of sensation, compare with the normal and ask the patient to conform the symmetry of sensation.

Joint Position Sense (Proprioception)

Test: Move the joint or extremity to be tested through ROM and hold in a static position.

Caution should be used with hand placements to avoid tactile stimulation.

Response: Ask the patient to describe the position of extremity verbally or to duplicate the position of extremity with contralateral extremity.

To avoid bias, trick the patient, by not giving the stimulus now and then.

To find out the symmetry of sensation, compare with the normal and ask the patient to conform the symmetry of sensation.

Joint Kinesthetic Sense (Awareness of Movement)

Test: Passively move the joint or extremity to be tested through a small range of motion.

Response: Ask the patient to indicate verbally the direction of movement while the limb is in motion or to duplicate simultaneously the movement with opposite limb.

To avoid bias, trick the patient, by not giving the stimulus now and then.

To find out the symmetry of sensation, compare with the normal and ask the patient to conform the symmetry of sensation.

Cortical Sensations

The cortical sensations that are to be checked are tactile localization, two-point discrimination and stereognosis.

Tactile localization: It is to check the patient ability to exactly point the area that is either pricked or touched by the examiner.

Ask the patient to close the eyes and instruct him with the help of index finger point out the place where the examiner touches or pricks.

Two-point discrimination: It is a measure of the smallest distance between two stimuli (applied simultaneously and with equal pressure) that can still be perceived as two distinct stimuli.

Use reshaped paper-clip with blunt ends or two-point discriminator.

Apply the two tips of instrument to the patient's skin simultaneously.

Gradually, reduce the distance between two tips until the stimuli are perceived as one.

Measure the smallest distance between the stimuli that is still perceived as two distinct points and record it.

Response: Ask the patient to identify the perception of 'one' or 'two' stimuli.

Steriognosis: It is the ability to identify the object by tactile sensations of the hand or by manipulating the object in the hand.

The patient is asked to close the eyes and a coin or pen or any familiar object to him is placed in his hand and asked to manipulate and identify.

REFLEX EXAMINATION

A reflex is a stereotypic response to a stimulus.

Reflex testing helps to assess the relation and functions of both sensory and motor pathways and also connection between the peripheral nervous system and central nervous system.

Reflex testing helps us to identify the level of lesion in the nervous system.

It also helps to differentiate the UMN type of lesion to LMN type of lesion.

In neurological diseases, normal physiological reflexes may be increased, decreased or lost and abnormal pathological reflexes may appear, especially with UMN lesion.

Superficial Reflexes[9,11,12]

Most of the superficial reflexes are cutaneous reflexes where a tactile stimulus to a localized area of skin or mucous membrane results in a motor response except pupillary reflex where the pupil is stimulated by light. Some reflexes are cranial nerve mediated and some are spinal nerve mediated.

Which reflexes do you need to check?
1. Pupillary reflex
2. Corneal reflex
3. Palpebral reflex

4. Gag reflex
5. Abdominal reflex
6. Cremasteric reflex
7. Plantar reflex (Babinski sign) (it can also be noted as a pathological reflex)

Pupillary Reflex

Stimulus: Light focused on the pupil.

Root valve: Cranial nerve II is afferent and III is efferent.

Response: Constriction of pupil to light

How to check?
- Make sure that the room is dark and pupils are dilated a little.
- The room should not be completely dark that you cannot observe the pupil reaction.
- Make the patient sit calmly on a chair and ask him to look straight.
- Now shine the light in right eye and observe the pupil being constricted.
- Now repeat the same procedure on left eye.

What do you interpret?
- Usually the normal response of pupillary reaction is described as PERRLA (pupils equal, round, reactive to light accommodation).
- Abnormal responses are observed secondary to direct or indirect damage to optic nerve, parasympathetic injury or to the sympathetic neurons.
- Some sympathomimetics, like cocaine, will dilate the pupil and some narcotics, like heroin, will constrict the pupil.

Corneal Reflex

Stimulus: Touching the cornea with a cotton plug.

Root valve: Cranial nerve V is afferent and VII is efferent.

Response: Blinking of eyes.

How to check?
- Ask the patient to sit with open eyes and look straight.
- Lightly touch the cornea with a sterile cotton wool.
- The response will be shutting of the patient's eyelids.

What do you interpret?
- An absent corneal reflex can be due to damage to trigeminal or facial nerve.
- It is also absent in facial muscle myopathy, Bell's palsy or brainstem disease.
- Chronic progressive external ophthalmoplegia.
- Contact lenses may diminish the reflex.
- In unconscious patients, this reflex will indicate the functioning of the lower brainstem.
- You cannot diagnose "brain dead" until the corneal reflex is absent.

Palpebral Reflex

Stimulus: Touching the eyelids or eyelashes.

Root valve: Cranial nerve V is afferent and VII is efferent.

Response: Blinking of the eyes

How to check?
- Ask the patient to sit with open eyes and look straight.
- Touch the eyelid or eyelash.
- The response will be shutting of the patient's eyelids.

Gag Reflex

Stimulus: Touching the soft palate around the tonsils area.

Root valve: Cranial nerve IX is afferent and X is efferent.

Response: Elevation of the palate and contraction of the muscle in the back of the throat (laryngeal spasm)

How to check?
- Ask the patient to sit with mouth opened.
- Ask the patient to say AAH.
- Touch the roof of the mouth with a cotton bud.

What do you interpret?
- In glassopharyngeal nerve palsy, the reflex is absent.
- Absence of gag reflex in stroke patients is an indication for dysphagia.

Abdominal Reflex

Stimulus: Stroking over the skin on all quadrants of the abdomen around the umbilicus.
   Root valve: T7 to T12 spinal segments.
   Response: Contraction of the abdominal muscles and retraction and deviation of the umbilicus towards the stimulus.

How to check?
- Explain the entire procedure to your patient and take consent.
- Make the patient in supine lying.
- Expose the abdominal region of the patient in privacy.
- With the help of a blunt object (the tail-end of the reflex hammer will be beneficial), stroke firmly on the different quadrants of the abdomen.

What do you interpret?
- Abdominal reflex is reported as present/absent.
- An absent abdominal reflex can also be a physiological response in obese patients, less tolerable patients, children, multiparous lax abdominal wall.
- It is pathological and can be due to neurological disturbances like spinal cord lesions involving T7 to T12 segments, multiple sclerosis, motor neuron disease, etc.

Cremasteric Reflex

Stimulus: Stroking on the inner aspect of the thigh.
   Root valve: L1 and L2 spinal segments.
   Response: Elevation of the ipsilateral testicle.

How to check?
- Explain entire procedure to the patient and take consent.
- Have privacy settings, and make the patient lie supine with inner and upper part of the thigh exposed.
- Stroke in the inner and uppermost part of the thigh with a blunt object.

What do you interpret?
- The reflex may be absent in the testicular torsion, LMN lesions and L1, L2 spinal cord injuries.
- In ilioinguinal nerve injuries, the reflex is absent.
- It is a very helpful reflex for testicular emergencies.
- In genitofemoral nerve injuries, the reflex can be absent.

Plantar Reflex

Also known as Babinski reflex. It is a very reliable reflex to differentiate UMN and LMN lesions.
   Stimulus: Stroke the sole of the foot with blunt surface from lateral surface of the foot directed towards great toe.
   Root value: L5–S1.
   Response: The response for the plantar reflex is described in the table below.

Babinski	Response	Interpretation/result
Positive (+)	Extension of great toe with fanning of other toes	UMN lesion
Negative (−)	Flexion of the toes	Normal
Absent	No response is seen	LMN lesion

Deep Reflexes

Deep tendon reflexes are also called *myotactic reflexes* or *muscle stretch reflexes*.

They are elicited by the neurological hammer tap on a tendon which results in the brief or single contraction of the muscle.

What is the physiology behind? When you tap the tendon of the muscle, it causes passive stretching of the muscle and neuromuscular spindles.

This activates Ia sensory fibers which are afferent and results in the sufficient depolarization of the alpha motor neurons which are anterior horn cells at that level of the spinal segment.

The depolarization of motor nerve leads to the contraction of the muscle fibers which through the efferent nerve causes the muscle to contract.

What is a neurological hammer? A neurological hammer, which is also called *percussion hammer* or *knee hammer* in some instances, is an instrument used to assess the deep tendon reflexes.

They come in different models.

Note: When you tap the tendon, immediately you should leave the touch of the hammer over the tendon to get a better response.

How you report the response? Usually, in clinical scenario, the reflex response of the deep tendon reflexes is reported as follows based on the response.

Grade	Response
0	No response, reflex absent
1	Decreased or diminished response but response is present
2	A normal response
3	Exaggerated response or hyper-response
4	Reflex elicitation results in clonus (repetitive shortening of the muscle after a single stimulation)

Evaluation of Reflexes at Different Levels of Dysfunction of Nervous System

Biceps jerk:
1. Explain the entire procedure to the patient.
2. Make the patient sit on a chair or attain a supine lying, if patient cannot sit.
3. Support the patient forearm in flexed elbow at 90°.
4. Palpate the tendon of biceps in the cubital fossa and place your thumb in it.
5. Now tap on your thumb.

Response: Flexion of the elbow.

Brachioradialis jerk:
1. Explain the entire procedure to the patient.
2. Make the patient sit on a chair.
3. Support the patient forearm in flexed elbow at 90° in mid-pronated forearm.
4. As the tendon is not so easily palpated, so tap on the forearm just approximately 10 cm proximal to wrist.

Response: A normal response is flexion of the arm at elbow with supination.

Triceps jerk:
1. Explain the entire procedure to the patient.
2. Make the patient to sit or attain a supine lying, if the patient cannot sit.
3. In sitting position, pull the patient arm off the body such that the arm forms a rough right angle at the shoulder. Flex the elbow.
4. Now feel the triceps tendon as a broad structure above the olecranon process.
5. Now tap on the tendon of the triceps and observe the response.
6. If the patient is lying, just make the patient arm adducted and flex the

Principles of Assessment and Treatment in Neurological Conditions

elbow to 90° such that the forearm rests on the abdomen of the patient.

7. Now feel the tendon of the triceps and tap.

Response: There will be extension of the forearm.

Knee jerk/quadriceps jerk:

1. Explain the entire procedure to the patient.
2. Make the patient in sitting on a high chair, feet off the ground or make the patient lie supine on the examination couch.
3. In sitting position, palpate the tendon of the quadriceps just below the patella.
4. Now tap directly over the tendon and observe the response.
5. In lying position, flex the knee, by placing your fist or a roll of towel under the knee so that the knee is flexed.
6. Now feel the tendon and strike.

Response: In normal response, there will be extension of the knee.

Ankle jerk/TA jerk:

1. Explain the entire procedure to the patient.
2. Make the patient in sitting on a high chair, feet off the ground or make the patient lie supine on the examination couch.
3. In sitting position, dorsiflex the foot so that it makes a right angle with the lower leg.

Now tap on the tendon above the calcaneum on the posterior part of the leg.

Response: Plantar flexion of the foot.

Important points to note:

1. In sensory neuron disorders, like diabetic polyneuropathies, there will be delay in the transmission of impulse to the reflex center (spinal cord), hence there will be diminished or lost responses observed.
2. In peripheral motor neuropathies, which are LMN lesions, there will be lost or diminished reflexes because of disorder in the motor neuron.
3. A disease of the neuromuscular junction or myopathies (muscle disorders), there will be loss of reflexes.
4. Some diseases which are systemic like thyroid disorders, may have their impact on the reflexes, e.g. in hyperthyroidism, the reflexes will be exaggerated.

MOTOR EXAMINATION

The motor examination in neurological patients involves assessment of tone, measurement of muscle girth, power of the muscles, voluntary control of the muscles, presence of abnormal associated movements.

Tone of the Muscles

The tone of the muscles is assessed by repeated passive movement of the limbs.

The tone of the muscles can be graded by modified Ashworth scale of spasticity as shown in Table 2.2.

Table 2.2: Modified Ashworth scale of spasticity

Grade	Description
0	No increase in muscle tone
1	Slight increase in muscle tone with catch and release at the end
1+	Slight more increased range as catch and followed by release for half of the range
2	Resistance is increased throughout the range during passive movements but part moved freely
3	Passive movements difficult with considerable increased muscle tone
4	Part remain rigid in flexion or extension

Table 2.3: Voluntary control testing

Grade	Description
0	No contraction
1	Initiation of contraction or flicker contraction
2	Half range of motion in pattern
3	Full range of motion in pattern
4	Initial half range in isolation and the later half in pattern.
5	Full range of motion in isolation but goes into pattern when resistance offered
6	Full range of motion in isolation against resistance.

In spasticity the voluntary control of movements should be assessed. The brain appreciates movement performed and not the muscles. The voluntary control should be assessed at all the joints and for all movements (Table 2.3).

Muscle Girth

Muscle girth is the circumference or thickness or diameter of the limb. Girth or circumference of all the major muscles is measured. Usually in UMN lesions, there will be no wasting.

Muscle Power

Muscle power is checked in LMN type of lesions and MMT grading is done (Table 2.4).

Table 2.4: Manual muscle testing (MMT) grades

Grade	Description	Qualitative grade
0	No flicker of contraction	No trace
1	Flicker of contraction	Trace
2	Full ROM in gravity eliminated position	Poor
3	Full ROM against gravity	Fair
4	Full ROM against gravity against half the maximal resistance	Good
5	Full ROM against gravity against the maximal resistance	Normal

Tests for Coordination

The coordination is checked by equilibrium and non-equilibrium tests.

Non-equilibrium tests: These tests assess both static and mobile components of movement of body when it is not in upright position. These tests assess both fine and gross movements. The various tests are:
1. Finger to nose (patient)
2. Finger to therapist finger
3. Finger to finger (patient)
4. Alternate nostril to finger (patient)
5. Finger opposition (patient)
6. Tapping of foot and hand (patient)
7. Heel to shin test (patient)
8. Drawing a circle.

Equilibrium tests: These tests assess both gross and fine movements. It also assesses both static and dynamic components of posture and balance when the body is in standing position. The tests include:
1. Rhomberg's test: Ask the patient to stand with eyes closed. If he cannot maintain standing balance with eyes closed, then the test is positive.
2. Standing with feet together.
3. Standing with one foot directly in front of other.
4. Standing on one foot.
5. Walking by placing the heel of one foot directly in front of toe of other foot [tandom walking].
6. Walking along a straight line sideways, backwards.
7. Walk on a circle on heels and later on toes.
8. Walk on a figure of eight.

Other tests and scales to assess coordination are explained in the Chapter "Ataxia".

Tests for Balance

Various tests of balance are as follows:
1. **Romberg test:** Here therapist instructs patient to stand straight with both feet

together with eyes open and close. Therapist should stand adjacent to the patient so that the patient should not fall. This test is positive, if patient sways side-to-side or front and back with closed eyes.
2. **Single leg standing:** Here therapist instructs the individual to stand on one leg with eyes open and close. On eye closing, if patient sways in this position side-to-side or front and back, then the test is positive.
3. **Alternate single leg standing:** Here therapist instructs the patient to stand on single leg alternatively with eyes open and close. If patient falls down while close eyes, then the test is positive.
4. **Wobble board:** Here therapist instructs the patient to stand on a wobble board and start moving the board on right and left side. If patient is able to maintain himself from falling down, it improves his/her balance.
5. **Sharpened Romberg test:** In this test, the patient is made to stand with the feet in a tandem stance attitude, heel of one foot directly in front of the toes of other, arm folded across the chest and stand for one minute without sway, if the patient sways, then the test is positive.
6. **Timed stance:** Timed stance test consists of making the patient stand in various foot positions in eight different positions with eyes open and then eyes closed and the patient is asked to make balance for about 30 sec.
7. **Postural sway test:** It is a computerized test where a patient is asked to stand on a computer driven force plate for about 20–30 sec, sway with both eyes open and with eyes closed if often measured. The computer provides the graphical and numerical quantification of balance.
8. **Nudge test:** The patient is given perturbations in various positions and checked for balance.
9. **Functional reach test:** In this test, the patient is asked to reach an object from standing position as far as possible without losing the balance.
10. **Get up and go test:** This test detects the problems of balance in ADL activities. In this test, the patient is asked to sit in a chair with both feet touching the floor and then asked to stand, walk for about 3 meters, turn around and return and sit. Performance is graded as (1) normal, (2) very slight abnormality, (3) mildly abnormal, (4) moderately abnormal, and (5) severely abnormal.

FUNCTIONAL ASSESSMENT/ASSESSMENT OF ADL

The physiotherapist's ultimate aim is to rehabilitate the patient back to his near normal or normal lifestyle. ADL assessment is very important tool of assessment to identify the functional disability.

An ADL assessment is simply an assessment to analyze a person's ability to perform his personal care and general activities, he/she needs to do on a routine basis in and around home.

A physiotherapist by assessing the ADL assessment determines the level and type of assistance a person requires in order to live an independent life as possible.

This assessment will also able the therapist to plan for the modifications that the patient may be deemed necessary for making him independent.

What are the areas to assess for ADL?
Basic ADL:
- Self-care activities
- Mobility
- Communication

Self-care:
- Feeding
- Grooming
- Dressing

- Bathing
- Toilet activities

Mobility:
- Bed mobility
- Wheelchair mobility
- Transfers from bed to wheelchair and vice versa
- Ambulation

Communication:
- Writing skills
- Speaking
- Using various communication devices like telephone, mobile, etc.

Feeding: Assess the patient how he/she is feeding, the type of set up, utensils and spoons used, ability to chew and swallow.

Grooming and hygiene: Assess how the patient is able to do or how far he needs assistance in oral care, hair dressing, bathing, shaving, etc.

Dressing: Assess the patient, her/his ability to wear the clothes, buttoning, lacing the shoes, buckling the belt, etc.

There are many scales available to assess the ADL performance of the patient. To brief, FIM (functional independent measure) can be used. It has the following rubric measurements:
- 7—independent
- 6—modified independent
- 5—supervision/set up modification
- 4—minimal assistance
- 3—moderate assistance
- 2—maximal assistance
- 1—total assistance

Apart from this, an evidence-based assessment can also be done by Barthel Index,[13,14] Katz index of independence[15] in ADL, etc.

BLADDER AND BOWEL ASSESSMENT

The bladder and bowel assessment is very important aspect in neurological assessment to identify the neurogenic bladder.

A neurogenic bladder is a dysfunction of urinary bladder, where a patient loses his control over the bladder due to disease or damage/injury to the brain, spinal cord or nerves.

UMN Neurogenic Bladder

Also known as spastic bladder or reflex bladder. Spinal cord injuries that occur above conus medullaris leave the S2–S4 levels of the spinal cord segments leaving them intact. This means that all the spinal and autonomic nerves located in these levels can still be functional in a reflex way (unconscious way).

The detrusor muscle is likely to signal your brain with only smaller amount of urine because of this.

There is a detrusor muscle dyssynergia meaning the timing between bladder muscle and the external sphincter contract at the same time leading to development of outpouchings of the bladder wall and reflux of urine into the ureters and kidneys.

LMN Neurogenic Bladder

Also known as flaccid bladder. Spinal cord injuries that occur at or below the conus medullaris damage the nerves that are involved with the bladder function. In these injuries, there is no reflex nerve activity and the bladder no longer functions and is in flaccid state.

There will be incontinence due to loss of tone in the external sphincter.

How to Assess the Patient?

- The assessment of the patient with bladder and bowel dysfunction starts with the taking consent from the patient.
- Take a clear history that includes medical and surgical history, history of medications used or being used, history of neurological dysfunction, OBG history.

- You may need to ask the following questions:
  - How many days are you suffering from the symptoms?
  - What are the aggravating factors?
  - What is the frequency of the passing urine in 24 hr?
  - Do you need to rush to the toilet?
  - When you pass the urine, is it coming continuously or interrupted?
  - Do you get the feeling that you have completely emptied the urine?
  - Do you ever leak when you cough, sneeze, laugh or jump?
  - Do you have a history of urinary tract infection in the past 6 months or a regular history of UTI?

After that check the patient's bowel habits.

A fluid input/output chart can be reviewed from the nurses notes.

A bladder and bowel scan can be helpful in identifying the problem.

GAIT ASSESSMENT

This is very important aspect of neurological examination. The gait parameters are checked and compared with normal.

Observe the symmetry of the gait, ability to walk with a narrow base, the length of the stride when walking at a normal pace, and ability to turn with a minimum of steps and without loss of equilibrium. Check for tandem walking (ability to walk on a straight line). Further for the sake of evidence-based practice, Functional Gait Assessment Scale[16] can be used to assess the gait in stroke patients

Psychological assessment: The patients with neurological disorder often experience anxiety and depression. It is the major secondary complication and more attention is required. To assess, HADS[17] (Hospital Anxiety and Depression Scale) was developed by Zigmond and Snaith in 1981.

PRINCIPLES OF MANAGEMENT OF NEUROLOGICAL DYSFUNCTION

Various approaches in neurological physiotherapy are given below.

Bobath Approach

The Bobath approach is known as neurodevelopmental treatment (NDT) and was developed by Mrs Bertha Bobath from 1942 onwards. Dr Karel Bobath, her husband, who is a neurologist, contributed to this theory by giving theoretical explanations.

Rationale

NDT is a holistic approach dealing with the quality of patterns of coordinated movements. It deals with not only sensorimotor problems but also problems of development, perceptual-cognitive impairment, emotional, social and functional problems of daily living in patients with neurological dysfunction.

The concept of NDT is based upon the Bobath approach which mainly concentrate to prevent the synergic patterns and facilitate normal movement. It also promotes to learn how to control postures and movements.

The main strategies of this concept are *therapeutic handling* whereby the patient's movements are influenced by facilitation and inhibition techniques. *Facilitation* techniques to promote motor learning by using sensory information to reinforce weak movement patterns and discourage overactive patterns. *Inhibition* techniques reduce the abnormal influences on movement or posture that interfere the normal pattern of movement. *Key points of control* generally refer to parts of the body that are advantageous when facilitating or inhibiting the movements and postures.

Reflex Inhibiting Posture

Bobath was influenced by the animal experiments of Sherrington and Magnus who found out that identical stimulus in different

Hospital Anxiety and Depression Scale (HADS)

Tick the box beside the reply that is closest to how you have been feeling in the past week. Don't take too long over your replies: your immediate is best

D	A		D	A	
		I feel tense or 'wound up':			**I feel as if I am slowed down:**
	3	Most of the time	3		Nearly all the time
	2	A lot of the time	2		Very often
	1	From time to time, occasionally	1		Sometimes
	0	Not at all	0		Not at all
		I still enjoy the things I used to enjoy:			**I get a sort of frightened feeling like 'butterflies' in the stomach:**
0		Definitely as much		0	Not at all
1		Not quite so much		1	Occasionally
2		Only a little		2	Quite often
3		Hardly at all		3	Very often
		I get a sort of frightened feeling as if something awful is about to happen:			**I have lost interest in my appearance:**
	3	Very definitely and quite badly	3		Definitely
	2	Yes, but not too badly	2		I don't take as much care as I should
	1	A little, but it doesn't worry me	1		I may not take quite as much care
	0	Not at all	0		I take just as much care as ever
		I can laugh and see the funny side of things:			**I feel restless as I have to be on the move:**
0		As much as I always could		3	Very much indeed
1		Not quite so much now		2	Quite a lot
2		Definitely not so much now		1	Not very much
3		Not at all		0	Not at all
		Worrying thoughts go through my mind:			**I look forward with enjoyment to things:**
	3	A great deal of the time	0		As much as I ever did
	2	A lot of the time	1		Rather less than I used to
	1	From time to time, but not too often	2		Definitely less than I used to
	0	Only occasionally	3		Hardly at all
		I feel cheerful:			**I get sudden feeling of panic:**
3		Not at all		3	Very often
2		Not often		2	Often
1		Sometimes		1	Not very often
0		Most of the time		0	Not at all
		I can sit at ease and feel relaxed:			**I can enjoy good book or radio or TV program:**
	0	Definitely	0		Often
	1	Usually	1		Sometimes
	2	Not often	2		Not often
	3	Not at all	3		Very seldom

Please check you have answered all the questions
Scoring: Total score: Depression (D).......... Anxiety (A)..........
0–7 = Normal, 8–10 = Borderline abnormal (borderline case), 11–21 = Abnormal (case)

positions elicited different reactions which means different movement patterns. By basing on these experiments, Bobath has placed and held the patient in "reflex inhibiting postures" to break up the abnormal postural and movement patterns. First she used a pattern opposite to the patient's total pattern, which she modified later into an individually adapted mixed pattern of a better coordinated flexion and extension. These positions resulted in the change of activity of the whole body due to normalization of the postural tone. Inhibition is the process of intervention that reduces dysfunctional muscle tone. It breaks up the abnormal excessive flexion or extension (Bobath 1984, Quinton 1986).[18–21]

Inhibition Combined with Stimulation and Facilitation

Once a more postural tone is achieved, the patient need to learn to move in many different combinations of more normal movement patterns. Bobath later felt that a near normal active movement when felt by a patient can only learn the movement with minimal effort and the therapist task is to make it possible. The importance of postural reactions, i.e. righting and equilibrium reactions were recognized. It was inspired by the work of Magnus, Peiper, Weisz and Zador.[22–24] It is believed that all voluntary and skilled functional activities with their complex and selective patterns of coordination are performed on the basis of automatic postural reactions. During normal development, in the beginning, there is an influence of tonic reflexes which later disappear and are suppressed by the development of righting reactions. These are later overlapped and integrated into balance and voluntary movements. Thus, the facilitation of sequences of righting reactions, equilibrium reactions, supporting reactions and other automatic reactions (Kong 1991)[25] are incorporated in the approach.

Dynamic Treatment with Control from Key Points

Mrs. Bobath found a way of using "key points of control" through which abnormal patterns could be controlled or inhibited. The various body parts, mainly proximal—head, shoulders, pelvis are considered to be the key points. These key points can also be used to influence the strength and distribution of postural tone, and facilitate the normal movement pattern. From the key points, the therapist can control and guide the movement of the whole body.

Inhibitory control is used with facilitation. It is accomplished simultaneously with the least amount of physical intrusion. The therapist uses techniques that reduce the dysfunctional tone, the patient makes more efficient movement adaptations. The treatment is done by "handling" and based on the close interplay between the patient and the therapist and the therapist is guided by the patient's reaction to his handling.

During the treatment, it is necessary to reduce the therapist's control, handing it over gradually to the patient to allow him/her to control the movements. Much guided control and repetition of the required reactions may be necessary to assure their quality.

Treatment in Functional Situations

There is a need for transition of treatment to functional skills. Not all movements obtained through treatment are spontaneously carried over to activities of daily living but it is the only way to influence the quality of prehension and manipulation.

The treatment approaches incorporate systemic preparation to improve specific functions in the present and to prepare the patient to perform in future.

Brunnstrom Approach

The Brunnstrom approach is based on using the reflexes that represent normal stages of development, and be used in functional rehabilitation.

Principles of Brunnstrom Approach

1. Reflexes should be used to elicit movement when there is no movement (normal developmental sequence)
2. Proprioceptive and exteroceptive stimuli can be used therapeutically to evoke desired movement or tone changes

Treatment Principles

When no movement is possible, movement is facilitated by using primitive reflexes or associated reactions, proprioceptive facilitation, and/or exteroceptive facilitation to develop muscle tone in preparation for voluntary movement.

The responses of the patient from such facilitation combine with the patient's voluntary effort to produce a semivoluntary movement.

Proprioceptive and exteroceptive stimuli assist in eliciting the synergies.

When voluntary effort appears then the patient is asked to contract isometrically. If it becomes possible, then the patient is asked for an eccentric contraction of the same. Later, a concentric contraction of the muscle is attained voluntarily by the patient. Later the pattern is reverses between agonist and antagonist.

Facilitation by postural primitive reflexes is reduced or stopped as soon as the patient shows voluntary control.

Incorporate these activities in ADL.

Brunnstrom classified the stages of motor recovery into 6 stages as follows:[26,27]

1. Stage 1: The patient is completely flaccid, no voluntary movement is possible and patient is confined to bed
2. Stage 2: Basic limb synergies develop, no voluntary movement is possible
3. Stage 3: Basic limb synergy develops voluntarily and there is marked spasticity
4. Stage 4: Spasticity begins to decline, four movement combinations deviate from basic limb synergies and become variable, which are placing the hand behind the body, alternative pronation–supination with elbow at 90° flexion and elevation of the arm to a forward horizontal position
5. Stage 5: There is relative independence of the basic limb synergies. Spasticity is wanning, and movements can be performed as arm raising to a side horizontal position, alternative pronation–supination with the elbow extended and bringing hand over the head
6. Stage 6: There are isolated joint movements

Peto Approach

The concept of conductive education (CE) is an educational system, based on the work of Hungarian professor Andras Peto. This approach has six elements—froup, facilitation, daily routine, rhythmic intention, task series and conductor. The patient is encouraged to verbalize the activities as they perform them and focuses on function.

Margaret Rood Approach

This was proposed by Margaret S Rood in 1950. It is based on the philosophy of treatment concerned with the interactions of somatic, autonomic, psychologic factors and their interactions with motor activities. Rood has used sensory stimuli by stroking or brushing at a given speed for a given duration for activation of a phasic muscle response. Rood applied cold for visceral stimulation and somatic relaxation and applied pressure/stretch for postural muscle activation.

The main principles of Rood's theory to normalize the tone in the muscles are:

- Treatment begins at the developmental level of functioning.

- Movement is directed towards functional goals.
- Repetition is necessary for the re-education of muscular response.

What are the sensory inputs that can be used in Margaret Rood approach?

For facilitation:
- Light moving brush
- Fast brushing
- Icing

Proprioceptive facilitatory techniques in the form of:
- Heavy joint compressions
- Stretches
- Intrinsic stretch
- Secondary ending stretch
- Stretch pressure
- Resistance
- Tapping
- Vestibular stimulation
- Inversion
- Therapeutic vibrations
- Osteopressure

Inhibitory:
- Gentle shaking or rocking movements
- Slow stroking
- Slow rolling
- Light joint compressions
- Tendinous pressure
- Maintained stretch
- Rocking in developmental stages

PNF Techniques

These are the techniques based on the principle of α-neuronal activity by modifying the effects of higher centers through stimulation of proprioceptors. This approach was developed by Knott and Voss in the year 1968. General treatment in PNF includes the use of recapitulation of total patterns of developing motor behavior, spiral and diagonal patterns of movements, coupling voluntary movement with postural and righting reflexes, appropriate sensory and verbal cues, maximal resistance for maximal excitation and inhibition and repetitive activity for conditioning and training.

Principles of PNF

1. **Stretch:** Stretching is used to stimulate the activity of muscle-spindle. The length and position of muscle is the starting position and stretch is maintained throughout the movement.
2. **Pattern:** These are the movements in a straightline in a diagonal direction with a rotatory component acting as grip, e.g. flexion–adduction–lateral rotation [upper extremity]
   Note: Patterns are named according to direction of movement. Therefore, finishing position is the name of pattern, e.g. for flexion–adduction–lateral rotation pattern of upper extremity, the starting position is extension abduction–medial rotation.
3. **Timing:** In normal, rotation movement initiates the movement. Movements at distal joints completed first before movements at proximal joints.
   Change in normal timing can be altered and is called *timing for emphasis*, i.e. emphasize on contraction of particular muscle group.
4. **Grip:** The therapist's grip is the key factor to facilitate and should provide stretch, exteroception, resistance, traction or approximation.
5. **Irradiation:** Maximum resistance can be used to cause of overflow of the impulses from the stronger to a weaker pattern or from stronger muscle group in a pattern to weaker in same pattern, e.g. strong resisted dorsiflexion of ankle can be used to facilitate contraction of quadriceps.
6. **Voice:** Voice of the therapist adds to proprioceptory input. The therapist uses commands to stimulate the patient and must go hand in hand with stretch given by therapist.

7. **Vision:** The patient is encouraged whenever possible to observe movement.

Techniques of PNF

The techniques of PNF are classified according to the effect needed and achieved.
   a. Strengthening techniques
   b. Lengthening techniques

a. Strengthening techniques: These techniques are used to strengthen a pattern or a muscle group. It includes: 1. Repeated contractions, 2. Slow reversals.

1. *Repeated contractions:* These techniques can be applied with 3 variations:
   a. Normal timing
   b. Timing for emphasis
   c. Combining isotonic and isometric muscle work.

Normal timing: It consists of repeating any chosen pattern several times through full ROM with or without full resistance ensuring smooth movement.

Timing for emphasis: The irradiation principle is used in this technique. The patient's strong muscles made to contract maximally, facilitating recruitment of force in weaker group.

Combining isotonic and isometric muscle work: This method is designed to strengthen a muscle in a specific part of range of movement by combining isometric and isotonic muscle contractions of a muscle, e.g. in treating frozen shoulder.

Common patterns: Upper extremity:
Note:
   1. It is noted that always flexion at shoulder joint goes with lateral rotation and extension goes with medial rotation.
   2. Adduction at shoulder joint goes with flexion at wrist joint and abduction goes with extension at wrist joint.

Patterns:
1. Flexion–adduction–with elbow flexion
2. Flexion–abduction–with elbow extension
3. Extension–adduction–with elbow flexion
4. Extension–abduction–with elbow extension
5. Flexion–adduction–with elbow extension
6. Flexion–abduction–with elbow flexion
7. Extension–adduction–with elbow extension
8. Extension–abduction–with elbow flexion

Lower extremity:
Note:
   1. It is noted that always flexion at hip joint goes with dorsiflexion at ankle joint and extension goes with plantar flexion at ankle joint.
   2. Adduction at hip joint goes with lateral rotation and abduction goes with medial rotation.

Patterns:
1. Flexion–adduction–knee flexion
2. Flexion–adduction–knee extension
3. Flexion–abduction–knee flexion
4. Flexion–abduction–knee extension
5. Extension–adduction–knee flexion
6. Extension–adduction–knee extension
7. Extension–abduction–knee flexion
8. Extension–abduction–knee extension

2. **Slow reversals:** In this technique, the contraction of strong muscles in the antagonist pattern is used to facilitate the contraction of the weaker muscles of agonist group.

This technique is based on the Sherington's principle of successive induction, e.g. to strengthen deltoid muscle in the flexion–abduction–lateral rotation pattern, the patient is instructed to complete extension–adduction–medial rotation pattern. Then therapist ensures the patient is working maximally, on completion of the movement, the therapist, smoothly without pause, changes grip and the patient moves into flexion–abduction–lateral rotation.

The sequence is then repeated several times.

Rhythmic stabilization: This technique improves stability of the trunk, hip and shoulder girdle.

In this technique, the patient is made to attain a weight-bearing position while the therapist applies manual resistance.

The patient should hold in such a way that even on therapist resistance, there should not be any motion.

For example, when the patient is in sitting posture, the therapist shall give manual resistance to shoulders. The therapist usually applies simultaneous resistance to the anterior left shoulder and posterior right shoulder for 2–3 sec and then switches to posterior left shoulder and the anterior right shoulder.

Note: The therapist movement should be smooth and continuous.

If the same technique applied on the same side of joint, then it is called *alternating isometrics*, e.g. if the resistance is applied to anterior left shoulder and anterior right shoulder, it is called *alternating isometrics*. If the resistance is applied to anterior left shoulder and posterior right shoulder or vice versa, then it is called *rhythmic stabilization*.

These techniques are not stretching techniques but are used in strengthening of muscles by using proprioceptors.

Rhythmic initiation: This technique is best used in Parkinson's disease to overcome the effects of rigidity.

The pattern is started by the therapist moving passively followed by active–assistive, assistive resistive–resistive range of motion.

b. **Lengthening techniques:** These techniques are used for relaxation and stretch the shorten muscles. Two techniques provide the desired effect: 1. Working the hypertonic group, 2. Working the reciprocal group.

1. *Working the hypertonic group*: The therapist elicits a maximal contraction of the antagonist group ensuring that maximum number of motor units are contracting simultaneously. Following the contraction, the muscle will relax and the therapist takes advantage of this relaxation to move the part further into agonist group range. This can be achieved by two methods.
   a. *Contract-relax*: In this, the muscles will contract isotonically (antagonists)
   b. *Hold-relax*: In this, the muscles will contract isometrically (antagonists).
2. *Working the reciprocal group*: Contraction of a muscle accompanied by relaxation of antagonist group.

Effects and uses:
a. Improves co-ordination
b. Improves muscle strength
c. Strengthens the flaccid muscles.
d. Relaxation
e. Lengthens the shortened muscles.
f. Improves joint range in case of frozen shoulder.

Neck Patterns

Neck patterns in PNF are shown in Fig. 2.3.

Upper Limb Patterns

Upper limb patterns in PNF are shown in Fig. 2.4.

Lower Limb Patterns

Lower limb patterns in PNF are shown in Fig. 2.5.

Motor Relearning Programme

The Motor Relearning Programme (MRP) was developed by the Australian

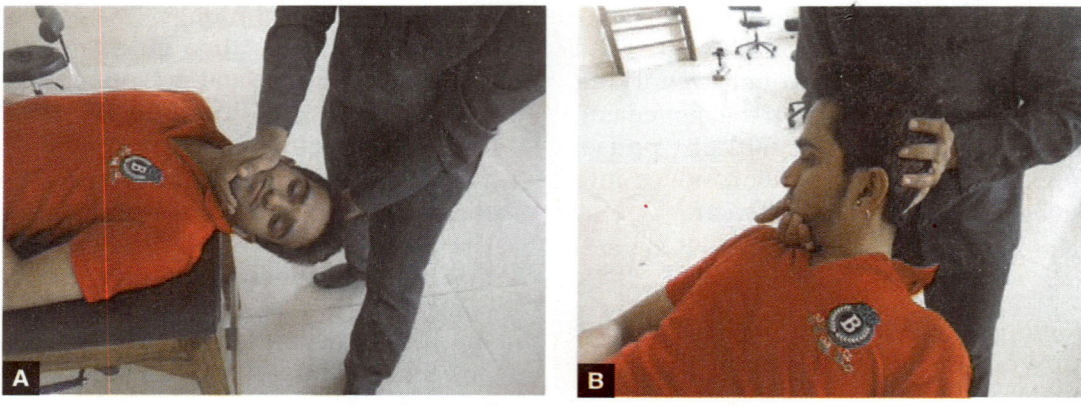

Fig. 2.3: Neck patterns in PNF

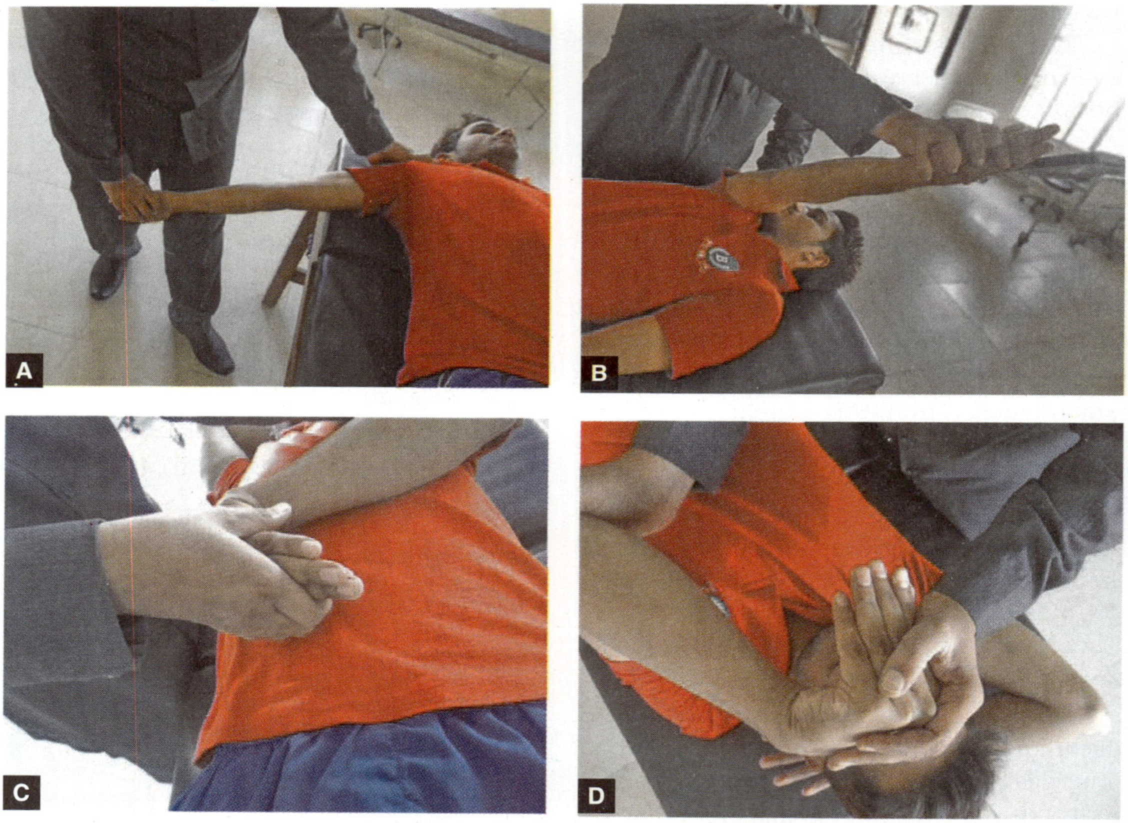

Fig. 2.4: Upper limb patterns in PNF

Physiotherapists Janet Carr and Roberta Shepherd.[28] It is a task-oriented approach to improve motor control and relearning activities of daily living.

According to this concept, development of a functional task can be achieved best when the task itself is practiced on the whole in the original environment rather than developing

Principles of Assessment and Treatment in Neurological Conditions

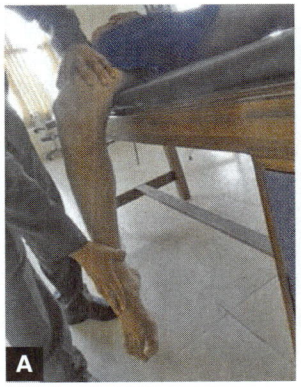

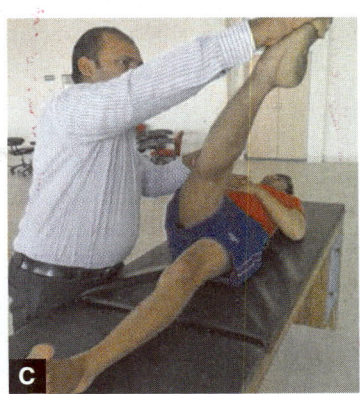

Fig. 2.5: Lower limb PNF

other neurological factors like tone, strength or coordination in a position or setting that is totally different from the actual task.

Postulates of MRP Theory

Postural adjustments are anticipatory and ongoing. Changes in muscular organization of a person occur simultaneously with the plan to move and prepare the person for performing the task.

Motor behaviors emerge as a result of context or regulatory conditions in the environment and solely depend upon the performer–environment interactions.

Postural adjustments can be learned only through the concept of task performance.

Skilled motor performances are the abilities to perform activities in different fashions in accordance with the environmental demands.

Deficits in generating appropriate models of action are the primary problem but not to be attributed to the spasticity or pathologic movement synergies.

Stereotypic movement patterns are compensatory strategies that result when movement is attempted.

Basic Guidelines for MRP

The programme is composed of guidelines for evaluating and improving 7 daily functions:

- Upper limb functions
- Orofacial functions
- Sitting up from supine
- Sitting
- Standing up and sitting down
- Standing
- Walking

Note: When training the patient with MRP, the patient must always be actively participating in the activity without resistance and be given some opportunity to make mistakes.

How to Train

There are four steps involved in Motor Relearning programme.

Step 1: Analysis of task: It involves observation of the patient's performance of a functional task.
- Compare the task with normal component
- Note any missing components or incorrect timing of components within a movement pattern.
- Check for the absence of any muscle activity that is desired for normal activity
- Check for presence of excessive or inappropriate or compensatory mechanisms in the activity
- Select the most essential components which are found missing.

Step 2: Practice of missing components:
- The missing components of the pattern practiced at peak performance for at least 30–60 minutes or more at least twice daily.
- The therapist can use verbal, visual cues or manual feedback, if the patient is not responding at first instance.

Step 3: Practice of task:
- Explain about the task in detail and the goal clearly to the patient. Motivate him.
- Practice the task repeatedly along with the verbal, visual or tactile feedback
- Progress the activity by reducing the verbal, visual and tactile cues dependency
- The progression can also be increased by increasing the complexity of the activity.

Step 4: Transfer of learning: Carry over the learning into task performance by incorporating in ADL.

Principles of Instructing Patient for MRP
- Verbal instruction is kept to a minimum. The therapist identifies the most important aspect of movement on which the patient is made to concentrate.
- Visual demonstrations are provided by the therapist.
- Manual guidance is given to give a clarification of mode of action by passively guiding the patient through the path of movement or by physically constraining inappropriate components.
- Accurate, timely feedback about the quality of performance helps the patient to learn which strategies to repeat and which one to avoid.
- Consistency of practice facilitates development of skill in task performance.

Vojta Method of Therapy

Vojta established 18 points in the body for stimulating and used the positions of reflex crawling and reflex rolling. He proposed that placing the child in these positions and stimulation of key points in the body would enhance CNS development. In this way, the child is presumed to learn normal movement patterns in place of abnormal motion.

According to Vojta, reflex locomotion is activated from three main positions—prone, supine and side lying.

Two coordination complexes in reflex locomotion: In the practical use of reflex locomotion, there are two coordinated complexes:
- Reflex creeping
- Reflex rolling

The movement sequences of reflex locomotion are retrievable at all times.

The three main positions—prone, supine and side lying and have more than 30 variations.

By combining and varying stimulation zones and resistances, as well as making changes in directions of pressure and joint angles in the starting position, therapy can be adapted to the patient's individual treatment goal and condition.

Reflex Creeping

Reflex creeping is a movement sequence that include the most fundamental components of locomotion.

Specific postural control:
- Upright posture or extension against gravity
- Goal-directed stepping movements of arms and legs
- The main position is prone lying with the head resting on the bed rotated to one side.

In newborn babies, reflex creeping can be fully activated from one zone. In adults, a combination of several pressure points is necessary.

Movement predominantly ensures in so-called cross-pattern, in which the right leg and left arm, or vice versa, move simultaneously. A leg and its contralateral arm support the body and move the trunk forwards (Fig. 2.6).

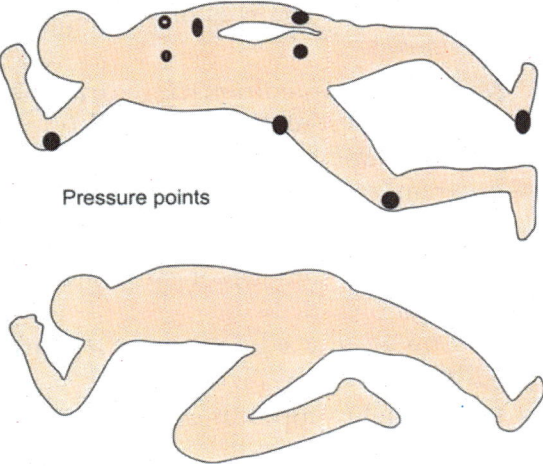

Fig. 2.6: Common pressure points in therapy

Reflex Rolling

Reflex rolling transitions from supine to side lying and leads to crawling.

Therapeutically, the reflex rolling is used in different phases of supine and side lying.

First phase: The first phase starts in supine lying with upper limbs and lower limbs extended.

Now stimulate the breast zone in the intercostals space beneath the nipple on the mammillary line, rotation to the side is achieved.

The rotation of the head is resisted by the therapist.

The fundamental reactions are:
1. Extenson of spine and flexion of hip, knee and ankle joints
2. Maintenance of lower limbs in this position against gravity
3. Preparation of upper limbs for the support function
4. Lateral eye movements and initiation of swallowing
5. Increase in depth of breathing
6. Coordinated activation of abdominal muscles.

Second phase: As now the child has obtained side lying, the second phase shall start from side lying.

The body is supported by the underlying upper and lower limbs which move in upwards and forwards against gravity (Fig. 2.7).

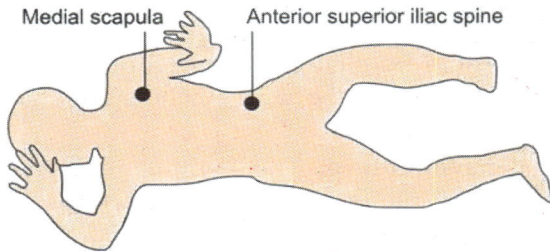

Fig. 2.7: Reflex rolling

Fundamental reactions:
- Contrary flexion and extension movements of the over and underlying arms and legs with increase in support function on the underlying shoulder progressing to the hand and underlying pelvis progression to leg.
- Extension of the spine during the entire rolling sequence.
- Maintenance of the head in side lying.

NEUROPLASTICITY

It is the capacity for continuous alteration of the neural pathways and synapses of the central nervous system in response to injury or repetitive experience.

The CNS may respond to this stimuli by reorganizing its structure, function and/or neural connections.

New neural connections may form in order to compensate for injury or loss of function or both.

Mechanisms of Neuroplasticity[29]

Axonal sprouting: The undamaged axons grow new nerve endings to reconnect damaged neuron links.

New neural pathways: Undamaged axons sprout to other undamaged nerve cells forming new neural pathways to accomplish a needed function.

Cortex changes: Use of dependent competition among neurons can alter brain network in both sensory and motor cortices.

What Happens to Neuroplasticity by Rehabilitation?

There will be changes in neuroplasticity mechanisms with the aid of rehabilitation and has a promotive effect:

Behavioral level: There will be recovery of sensory, motor or autonomic functions

Physiological: There will be normal responses seen in reflexes.

Structural: There will be axonal or dendritic strengthening

Cellular level: There will be strengthening of synapses.

How to Improve Neuroplasticity and Functional Recovery Potential?

The motor relearning is aimed at gaining functional independence through learning a specific task-oriented movement. The motor control strategies have five components:

I. **Motor programs:** It is asset of pre-set sequences to get a coordinated voluntary movement by muscle activation and is carried out without peripheral feedback.

II. **Motor planning:** It is strategical planning for a movement requiring coordination of various motor programs.

III. **Feedback:** It is the stimulation of control centers in brain by sending the information from peripheral receptors regarding the accuracy of movement and adaptations necessary.

IV. **Feed forward mechanism:** It is the strategy of the musculoskeletal system to adapt or respond in anticipation of a movement or changes of an ongoing movement.

V. **Motor skill acquisition:** It is the training of a goal-oriented problem solving strategy through development of motor programs and integration to form a motor plan and execute.

The various techniques incorporated to enhance neuroplasticity are:
- Constraint-induced movement therapy (CIMT)
- Body weight support treadmill training (BWSTT)
- Exoskeleton training.

Constraint-Induced Movement Therapy

This therapy mainly aims to improve the upper extremity function in patients with stroke and other CNS damage patients.

This therapy is developed by Edward Taub of the University of Albama at Birmingham. According to him, in stroke patients, the patient denies to use affected limb and depends completely on the unaffected limb for ADL. This causes "learned non-use" leading to further deterioration of the condition. According to Taub, the patient engaging in repetitive exercises with the affected limb, the brain grows new neural pathways thus enabling the neuroplasticity.

The focus of this technique is to combine restraint of the unaffected limb and intensive use of the affected limb. The restrain can be achieved by using slings or triangular bandages, splints, etc. The type of restraint should be selected by required level of safety and intensity of therapy. Some patients use slings or splints only to restrain the wrist and

Principles of Assessment and Treatment in Neurological Conditions

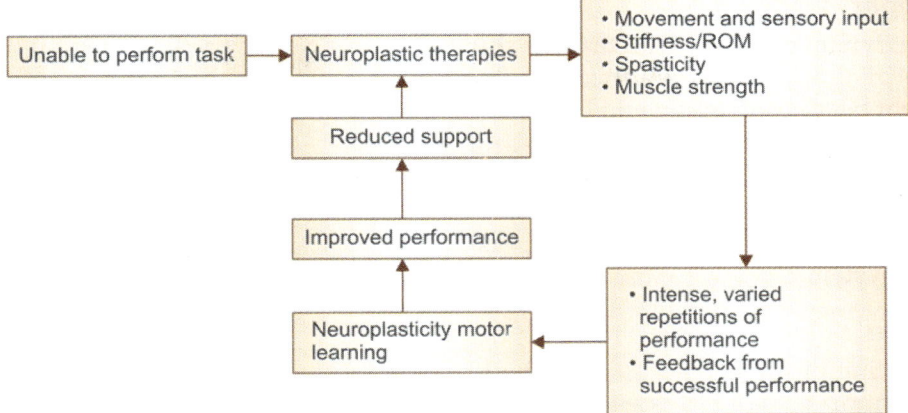

Fig. 2.8: The potential influence of neuroplastic interventions on functional performance

hand of the involved limb but allow the other joints to move in order to protect during loss of balance or falls by extension of the arm. Then the patient is encouraged to use the affected limb, this is called shaping.

Thus, the patient is encouraged to perform supervised structured tasks with the affected limb at least 6 hours a day for 10 days over a 14-day period.

Variants

Constraint-induced aphasia therapy: This technique is used for treating patients with aphasia where the patient is made to use verbal communication without using other forms of communication like gestures, sign language, wirting notes, etc.

Constraint-induced movement therapy: This therapy has proven to be effective in reducing spasticity and increasing function of the hemiplegic upper extremity in chronic hemiplegics. The various strategies used in this therapy are:

- *Monitoring*: The patient or caretaker is adviced to document their performance to target behaviors.
- *Problem solving*: The patients create solutions and identify outcomes to potential obstacles.
- *Behavioral contracting*: This involves getting patients to identify the components and methods of carrying out normal behaviors.

Body Weight Support Treadmill Training (BWSTT)

It is the technique of using overhead suspension system and harness to support patient's body weight as the patient walks on the treadmill (Fig. 2.9).

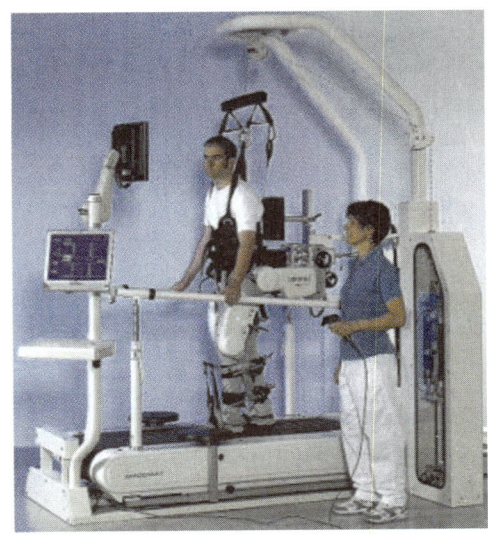

Fig. 2.9: Body weight support treadmill training

Advantages

- It is a dynamic activity with task specific repetitive activity which incorporates weight bearing, stepping and balance.
- Influences early weight-bearing and ambulation and discourages compensatory gait patterns from developing.
- Helps in conditioning the cardiovascular system.

Parameters

In the starting position, 35–50% of the body weight support to avoid the subject walking on toes, hip, knee flexion during the swing phase.

Speed of the treadmill is 0.25 m/sec and slowly the speed is increased as the patient progresses.

Time: 10–20 min daily

Use of handrail has been shown to increase single limb support and improve gait symmetry. But can be weaned in later stages.

Inclination: 8% to improve the cardiovascular conditioning.

Exoskeleton Training (Fig. 2.10)

- Recent advances in robotic technology have led to emergence of lower limb exoskeletons.
- It is the technique to provide legged mobility to individuals with lower extremity paralysis.

Fig. 2.10: Exoskeleton

- Exoskeletons facilitate over the ground walking in a reciprocating, relatively normal biomechanical positions.
- These exoskeletons enable the individuals with lower extremity paralysis to stand and walk over the ground with a weight-bearing, four-point reciprocal gait.
- Walking is achieved by the user's forward lateral weight shift to initiate a step.
- Battery powered hip and knee motors drive the legs and replace neuromuscular function.

REFERENCES

1. Van Allen MW. A guide to the performance and interpretation of the neurologic examination. Van Allen's Pictorial Manual of Neurologic Tests. 3rd ed. St Louis, MO: Mosby-Year Book Medical 1988.
2. Haslam RH. Clinical neurological examination of infants and children. Handb Clin Neurol 2013. 111:17–25. [Medline].
3. Bates B. Nervous system. In: A Guide to Physical Examination and History Taking. 8th ed. Philadelphia, PA: Lippincott, Williams & Wilkins. 2004.
4. Sharshar T, Citerio G, Andrews PJ, Chieregato A, Latronico N, Menon DK, et al. Neurological examination of critically ill patients: a pragmatic approach. Report of an ESICM expert panel. Intensive Care Med. 2014 Apr. 40(4):484–95.
5. Teasdale G, Jennett B. "Assessment of coma and impaired consciousness. A practical scale". Lancet 1974;2 (7872):81–84.

6. Jon Brillman; Scott Kahan (1 March 2005). In A Page Neurology. Lippincott Williams & Wilkins. ISBN 978-1-4051-0432-6. Retrieved 27 June 2011;pp. 4.
7. Brodal A. Neurological Anatomy in Relation to Clinical Medicine, 2nd ed., New York, Oxford University Press, 1969.
8. Medical Council of the UK. Aids to the Examination of the Peripheral Nervous System. Palo Alto, Calif., Pendragon House, 1978.
9. Monrad-Krohn GH, Refsum S. The Clinical Examination of the Nervous System, ed. 12, London, H.K. Lewis & Co., 1964.
10. Wolf J. Segmental Neurology, Baltimore, University Park Press, 1981.
11. DeJong RN. The Neurologic Examination, 4th ed. New York, Paul B. Hoeber, Inc., 1958.
12. Wartenberg R. The Examination of Reflexes: a Simplification. Chicago, Year book Publishers, 1945.
13. Mahoney F, Barthel D. "Functional evaluation: the Barthel Index". Md Med J 1965;14:61–65.
14. Granger CV, Dewis LS, Peters NC, Sherwood CC, Barrett JE. "Stroke rehabilitation: analysis of repeated Barthel index measures". Arch Phys Med Rehabil January 1979;60(1):14–7.
15. Katz S, et al. Studies of illness in the aged. The index of ADL: A standardized measure of biological and psychosocial function. JAMA 1963;185:914–9.
16. Leddy AL, Crowner BE, et al. "Functional gait assessment and balance evaluation system test: reliability, validity, sensitivity, and specificity for identifying individuals with Parkinson disease who fall." Phys Ther 2011;91(1):102–113.
17. Zigmond AS, Snaith RP. "The hospital anxiety and depression scale". Acta Psychiatrica Scandinavica 1983;67(6):361–370. doi: 10.1111/j.1600-0447.1983.tb09716.
18. Bobath B, Bobath K. The Neuro-Developmental Treatment. In: Scrutton D, et al. Management of the Motor Disorders of Children with Cerebral Palsy. Clinics in Developmental Medicine 90, Spastics International Medical Publications, Oxford, 1984;pp. 6–18.
19. Quinton MB. The Importance of the Body Image in our daily Lives and in Therapy. Schweizerischer, Bundder Therapeuten cerebraler Bewegungsstörungen, Mittteilungsblatt. 25, 1986;pp. 9–17.
20. Quinton MB. Structure of NDT Baby Treatment In: Book of Abstracts. The First World Congress of the Neuro-Developmental Treatment Concept; 1997 June;13–16; Ljubljana, Slovenia, 1997;pp. 44–45.
21. Quinton MB, Nelson CA. Concepts & Guidelines for Baby Treatment. Clinician's view, Albuquerque, New Mexico. 2002.
22. Peiper A. Cerebral Function in Infancy and Childhood. Consultants Bureau, New York, 1963; pp. 147–210.
23. Weisz B. Studies in Equilibrium Reactions. J Nerve and Ment Dis 1938;88:153–162.
24. Zador J. Les Réactions d´Equilibre chez l, Homme. Masson et Cie, Paris, 1938.
25. Köng E. Geschichte und Entwicklung des BobathKonzeptes.Der Kinderartzt 1991;22:705–710.
26. Wade Derick T, Wood Victorine A, Hewer Richard Langton. "Recovery after stroke—the first 3 months". Journal of Neurology, Neurosurgery and Psychiatry 1985;48(1):7–13.
27. Sawner K, La Vigne J. Brunnstrom's Movement Therapy in Hemiplegia: A neurophysiological approach. 2nd ed., JB Lippincott Company, Philadelphia, 1992.
28. Janet H Carr, Roberta B Shepherd. A motor relearning programme for stroke, Aspen Publishers, Business and Economics, 1987; 188 pages.
29. Livingston RB. "Brain mechanisms in conditioning and learning". Neurosciences Research Program Bulletin 1966;4(3):349–354.

3

Cerebrovascular Accidents

LEARNING OBJECTIVES

At the end of this chapter, you will be able to:
1. Describe briefly the cerebral circulation and its physiological importance
2. Identify the various etiological factors of CVA.
3. Identify the clinical features of various cerebral artery syndromes
4. Demonstrate skills in assessing the patient with cerebrovascular accidents
5. Design treatment protocols based on various approaches in rehabilitating the patient

INTRODUCTION

- Stroke is becoming an important cause of premature death and disability in developing countries like India.[1] It is the chief cause of mortality and morbidity worldwide affecting millions of people every year.
- Brain cell function requires a constant delivery of oxygen and glucose from the bloodstream. A stroke, or cerebrovascular accident (CVA) occurs when blood supply to part of the brain is disrupted, causing brain cells to die. Blood flow can be compromised by a variety of mechanisms. A stroke is due to lesion affecting the opposite side of cerebrum. Thus right lesion causes left stroke. It can cause sensory, motor, perceptual, mental deficits. It can also lead to language disorders.
- A stroke which is for less than 24 hours duration is called transient ischemic attack (TIA).

Definition

It is rapidly developing clinical signs of focal disturbances of cerebral function of presumed vascular origin and of more than 24 hours duration.

Cerebral Circulation: Circle of Willis or Circulus Arteriosus[2]

- The greater part of the brain is supplied with arterial blood by circle of Willis (Fig. 3.1).
- It is the anastomosis between the branches of vertebral arteries and internal carotid arteries.
- The vertebral arteries arise from the subclavian arteries, passes upwards through the foramina in the transverse process of cervical vertebra enter the

Cerebrovascular Accidents

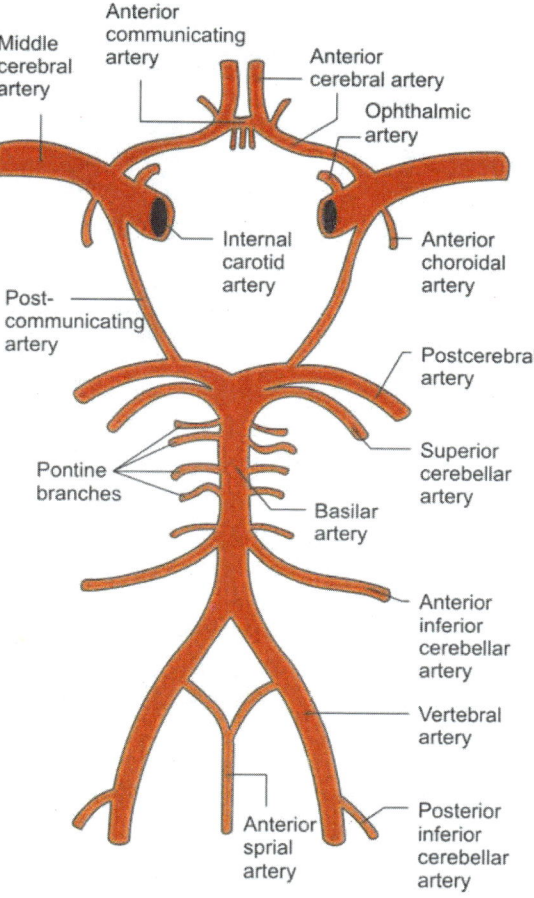

Fig. 3.1: Circle of Willis

skull through the foramen magnum, then join to form the basilar artery.
- The basilar artery gives right and left posterior cerebral arteries (Fig. 3.2).

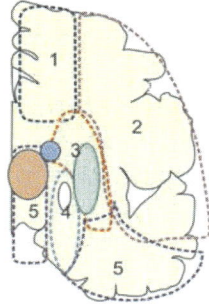

Fig. 3.2: Cerebral artery areas: 1. anterior cerebral, 2. middle cerebral, 3. penetrating branches of middle cerebral, 4. anterior choroidal, 5. posterior cerebral

- Each posterior cerebral arteries is joined with corresponding internal carotid artery by means of posterior communicating artery.
- Each internal carotid artery gives two branches anterior cerebral artery and middle cerebral artery (Fig. 3.2).
- The right and left cerebral arteries are connected by means of anterior communicating artery.
- Cerebral blood flow is maintained by a number of autoregulatory mechanisms.
- These mechanisms provide homeostatic balance, counteracting fluctuations in systolic blood pressure while maintaining a normal flow of 50 to 60 mL/100 mg of brain tissue per minute.
- The brain requires a continuous blood flow to deliver oxygen and glucose to the tissues.
- Cerebral flow represents approximately 70% of cardiac output.
- Brain has high energy requirements and very little metabolic reserves.

Incidence in India: The annual incidence rate of stroke in India is 13/100,000 population with 15.2/100,000 in males and 10.8/100,000 in females (Abraham J et al. 1972, Sunder Rao PSS 1971). In 2012–13, an overall 3126 patients in India were identified with stroke with an annual incidence rate of 140/100,000.

Incidence in western countries: Stroke is the leading cause of serious, long-term disability in United States. Each year, approximately 795,000 people suffer from stroke (US Centers for Disease Control and Prevention).

According to WHO, 15 million people suffer from stroke worldwide each year (World Health Report 2002).

Etiology
- Atherosclerosis
- Emboli
- Hemorrhage
- Hypertension
- Ischemia

Warning Signs of Stroke

- Sudden numbness or weakness of face, arm, or leg especially one side of the body
- Sudden confusion
- Trouble in speaking or understanding
- Sudden trouble in seeing in one or both eyes
- Sudden trouble in walking
- Dizziness
- Loss of balance or coordination
- Sudden, severe headache
- Sudden nausea

Ischemic Strokes

- These are result of a thrombus, or embolism leading to block to the cerebral blood flow which deprives the brain of needed oxygen and glucose, disrupts cellular metabolism, and leads to injury and death of tissues.
- *Cerebral thrombosis*: Formation of blood clot within cerebral arteries or their branches.
- Thrombi lead to ischemia or occlusion (cerebral infarction or ABI)
- *Cerebral embolus (CE)*: It is a collected matter (blood clot, plaque) formed elsewhere released into blood stream, produces ischemia or infarction (Fig. 3.3).

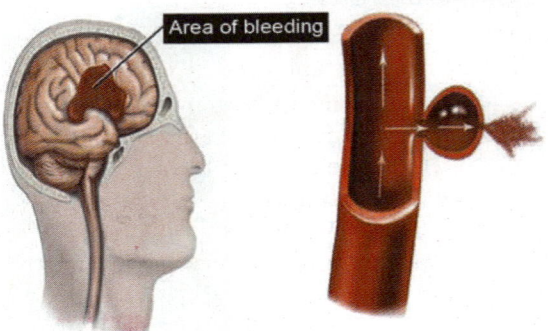

Fig. 3.4: Hemorrhagic stroke

Hemorrhagic Strokes (Fig. 3.4)

It is due to the rupture or trauma of intracerebral vessels leading to abnormal bleeding into the extravascular areas of the brain resulting increased ICP (intracranial pressure). The hemorrhage can be:

- *Intracerebral hemorrhage*: Rupture of cerebral vessel with bleeding inside the brain.
- *Primary cerebral hemorrhage*: Mostly occurs in small blood vessels weakened by atherosclerosis producing an aneurysm (abnormal dilatation).

Subarachnoid hemorrhage: Bleeding into subarachnoid space typically from berry or saccular aneurysms.

What are the risk factors to develop stroke?

- Hypertension (>160/95 mm Hg)
- Diabetes
- Elevated total cholesterol and LDL
- Decreased HDL
- Elevation of hematocrits (the percentage of total blood volume occupied by RBC)
- Cardiac disorders like rheumatic heart valvular disease, endocarditis or cardiac surgeries (CABG).

Pathophysiology

The two mechanisms causing brain damage in stroke are ischemia and hemorrhage. In ischemic stroke, there is decreased or absent

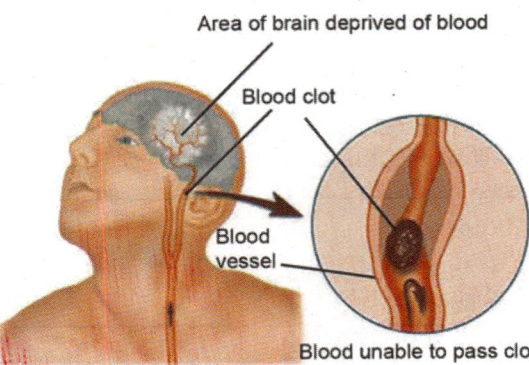

Fig. 3.3: Cerebral embolus

circulating blood deprives neurons of necessary substrates. The effects, if ischemia, are rapid as the brain is incapable of storing glucose reserves and anaerobic metabolism.

The vascular compromise leading to acute stroke is a dynamic process, the progression and extent of the damage is influenced by:
1. Rate of onset and duration of ischemia (brain can tolerate an ischemia of short duration or one with slow onset)
2. *Collateral circulation*: The effect and severity of ischemic injury is influenced by the state of collateral circulation in the area of affection.
3. *Systemic circulation status*: Constant cerebral perfusion pressure depends on the adequate systemic blood pressure. Systemic hypotension from any reason can result in global cerebral ischemia

Hematological factors: Hypercoagulable state increases the progression and extent of thrombi.

Temperature: Elevated body temperature is associated with greater cerebral ischemic injury.

Glucose metabolism: Hyperhypoglycemia can adversely influence the size of an infarct.

Ischemic Penumbra

Within an hour of hypoxic–ischemic insult, there is a core of infarction surrounded by oligemic zone (cells that are alive but metabolically less active) called ischemic penumbra (IP).

The critical time period during which the IP zone is at a risk is referred to as the *"window of oppurtunity"*, since the neurological deficits created by ischemia can be partly or completely reversed by reperfusing the ischemic area yet viable brain tissue within a critical time period may be 2–4 hours of onset of ischemia (Jones et al. 1981, Pulsinelli, 1995).

Neuronal death: The two processes by which the affected neurons die are coagulation necrosis and apoptosis.

The necrotic tissue swells rapidly mainly because of excessive intercellular and intracellular water content and lack of O_2.

The vascular lesion to the brain causes release of neurotransmitters like glutamate and aspartate by the ischemic cells which excite neurons and produce an intracellular influx of Na^+ and Ca^+ leading to irreversible cell damage.

Cerebral edema begins within a few minutes and reaches a maximum for about 4 days and mostly disappears by 3 weeks.

TYPES OF STROKE

According to management category:
- TIA (transient ischemic attack)
- Minor stroke
- Major stroke
- Deteriorating stroke

According to clinical category:
- Anterior cerebral artery syndrome
- Middle cerebral artery syndrome
- Posterior cerebral artery syndrome
- Internal carotid artery syndrome
- Vertebrobasilar artery syndrome

Transient Ischemic Attack

If the patient recovers from neurological dysfunction within 24 hours of attack, then it is termed TIA.

Minor Stroke

It is one in which there is no neurological deficit with good rehabilitation outcome.

Major Stroke

It is the one in which there is neurological deficit with bad rehabilitation outcome.

Deteriorating Stroke

This term is used to refer to the patient whose neurological status is deteriorating after admission into the hospital.

Anterior Cerebral Artery Syndrome

Structure involved	Signs and symptoms
Primary motor area, medial aspect of cortex, internal capsule	Contralateral hemiparesis, lower extremities are more involved than upper extremities, proximal muscles are more involved than distal
Primary sensory area, medial aspect of the cortex	Contralateral hemisensory loss, lower extremities are more involved than upper extremities, proximal muscles are involved than distal
Corpus callosum	Problem with imitation and bimanual tasks, apraxia
Uncertain localization	Abulia (akinetic mutism), slowness, delay, lack of spontaneity, motor inaction.
Bilateral involvement of posteromedial part of superior frontal gyrus	Urinary incontinence

Middle Cerebral Artery Syndrome

Structure involved	Signs and symptoms
Primary motor cortex and internal capsule	Contralateral hemiparesis involving mainly the upper limbs and face (UE > LE, distal > proximal)
Primary sensory cortex and internal capsule	Contralateral hemisensory loss involving mainly the upper limb and face (UE > LE, distal > proximal)
Broca's cortical area (third frontal convolution) in the dominant hemisphere	Motor speech impairment: Broca's or nonfluent aphasia with limited vocabulary and slow, hesitant speech
Wernicke's cortical area (posterior portion of the temporal gyrus) in the dominant hemisphere	Receptive speech impairment: Wernicke's or fluent aphasia with impaired auditory comprehension and fluent speech with normal rate and melody
Both third frontal convolution and posterior portion of the superior temporal gyrus	Global aphasia: Nonfluent speech with poor comprehension
Parietal sensory association cortex in the non-dominant hemisphere.	Perceptual deficits: Unilateral neglect, depth perception, spatial relations, agnosia
Premotor or parietal cortex	Limb—kinetic apraxia
Optic radiation in internal capsule	Contralateral homonymous hemianopsia
Frontal eye fields or their descending tracts	Loss of conjugate gaze to the opposite side
Parietal lobe	Ataxia of contralateral limb(s) (sensory ataxia)
Upper portion of posterior limb of internal capsule	Pure motor hemiplegia (lacunar stroke)

Lacunar Syndromes

- It is caused by small vessel disease deep in cerebral white matter.
- It is consistent with specific anatomic sites.
- Pure motor lacunar stroke is due to involvement of ventrolateral thalamus.
- Dysarthria/clumsy hand syndrome due to involvement of base of pons, internal capsule.
- Ataxic hemiparesis due to involvement of pons, genu of internal capsule, corona radiata or cerebellum.
- Sensory/motor stroke due to involvement of junction of internal capsule and thalamus.
- Dystonia/involuntary movements due to infarction of the putamen or globus pallidus.

Posterior Cerebral Artery Syndrome

Structure involved	Signs and symptoms
Primary visual cortex	Contralateral homonymous hemianopsia
Calcarine cortex	Bilateral homonymous hemianopsia
Left occipital lobe	Visual agnosia
Lesion of inferomedial portions of temporal lobe bilateral or on the dominant side only.	Memory defect
Ventral posterolateral nucleus of thalamus.	Central post-stroke (thalamic) pain
Subthalamic nucleus	Involuntary movements, intention tremor.
Cerebral peduncle of midbrain	Contralateral hemiplegia
Third cranial nerve and cerebral peduncle of midbrain	Weber's syndrome, occular nerve palsy and contralateral hemiplegia.

Internal Carotid Artery Syndrome

Internal carotid artery occlusion in its proximal segment immediately after its generation from the common carotid artery is often silent.

Course occlusion of the ICA produces massive infarction in the region of the brain supplied by MCA.

Significant edema can be seen with uncal herniation, coma and death.

Vertebrobasilar Artery Syndrome

The basilar artery is formed by the two vertebral arteries and supplies pons, inner ear and cerebellum. The complete occlusion can be fatal.

Progressive lesion often starts with occipital headache, diplopia, hemiplegia, quadriplegia and coma.

Complete basilar syndrome affecting the ventral pontine nuclei (without affection of reticular system) results in **"locked in syndrome"**.

Locked in syndrome is characterized by anarthria, quadriplegia with preserved consciousness, alertness and vertical eye movements.

Vertebrobasilar artery system occlusion can produce either ipsilateral or contralateral symptoms depending upon whether the tracts involved are crossed or uncrossed.

Drop attack: Sudden loss of tone in the lower limb muscles due to involvement of medullary pyramids and is common in vertebrobasilar artery syndrome.

Symptoms like visual loss, diplopia, homonymous hemianopia, facial numbness or weakness, tinnitus, dysarthria or dysphagia are common.

Hemiparesis due to involvement of corticospinal tract is evidenced.

Paresthesia of the face or limbs can occur with involvement of the leminiscal system.

Spinothalamic tract ischemia results in the loss of pain and temperature sensations on the opposite side of involvement.

Level of consciousness is altered due to involvement of reticular spinal tract.

PRIMARY IMPAIRMENTS IN STROKE

The primary impairments observed in stroke patients are:
- Motor deficits
- Sensory deficits
- Presence of abnormal reflexes
- Presence of associated reactions
- Speech and language disorders
- Perceptual problems

Motor Deficits

Immediately after the onset of disruption to cerebral blood flow, there will be stage of cerebral shock. This stage is manifested as *flaccidity* and *areflexia*.

It is replaced by development of *spasticity*, *hyperreflexia* and mass movement patterns called *synergies*. These symptoms are seen in the side opposite to lesion.

The duration of flaccidity may vary from days to weeks to infinite. In some cases of pure pyramidal tract lesions, the signs of hypertonicity (spasticity) can only be seen in some group of muscles, viz. elbow flexors, wrist flexors, quadriceps and calf muscles.

Brunnstrom (1966, 1970), a Swedish physical therapist, described the common sequence of recovery from stroke. A patient can plateau at any stage, but generally follow the sequence. The variability depends on the location and severity of the lesion and potential for adaptation.

There are six stages of recovery process of a patient after stroke.

Stage	Phase description
1	It is immediate after the onset of stroke characterized by flaccidity whereby no movement of the limbs on the affected side occurs.
2	Recovery begins with developing spasticity, hyperreflexia and onset of basic limb synergies
3	Spasticity is more pronounced and the limb synergies become strong.
4	Spasticity and the influence of synergy begins to decline and the patient is able to move with less restrictions. The ease of movements progresses from difficult to easy within this stage
5	Spasticity continues to decline, and there is a greater ability for the patient to move freely from the synergy pattern. Here the patient is also able to demonstrate isolated joint movements and more complex movement combinations
6	Spasticity is no longer apparent, allowing near-normal to normal movement and coordination.

Bobath further simplified the process of recovery into three stages:

Stage 1: Flaccidity
Stage 2: Spastic
Stage 3: Stage of spontaneous recovery

Spasticity: There is increased tone in the muscles. The spasticity is prevalent in antigravity muscles for reasons that are not clearly understood. One of the acceptable explanations might be that the antigravity muscles are relatively more stretched than the progravity muscles in neutral position, hence stimulating the stretch reflex which results in spasticity.

The *clasp knife mechanism* is exhibited by the spastic muscles and clonus is present.

The irregular distribution of spasticity results in the peculiar abnormal mass movement called **synergy**.

The distribution of spasticity in the upper and lower limbs are as follows:

Upper limbs	Lower limbs
Shoulder girdle depressors and retractors	Pelvic girdle retractors
Shoulder joint medial rotators and adductors	Hip joint extensors, adductors and medial rotators
Elbow joint flexors	Knee joint extensors
Wrist and finger flexors	Ankle and toe-plantar flexors and supinators
Forearm pronators	

Synergy (Fig. 3.5): These are abnormal, stereotypes, primitive mass movement pattern of the limbs associated with spasticity.

The synergy patterns can be initiated reflexively or voluntarily.

The pattern of synergy movements in upper and lower limbs is given in tabular format on the next page.

Abnormal attitude of limbs: The hemiplegic patient attains a peculiar abnormal attitude of the limb. The abnormal attitude of the limbs is not the same as synergy. The attitude is attained due to the combination of the strongest components of

Extremity	Flexion synergy	Extension synergy
Upper limb	Scapular retraction/elevation Shoulder abduction and external rotation Elbow in flexion Forearm supinated Wrist and fingers flexion	Scapular protraction Shoulder adduction and internal rotation Elbow in extension Forearm pronation Wrist and fingers flexion
Lower limb	Hip flexion, abduction and externally rotated Knee flexion Ankle dorsiflexion and inversion Toes dorsiflexion	Hip extension, adduction and internal rotation Knee extension Ankle plantar flexion and inversion Toe plantar flexion

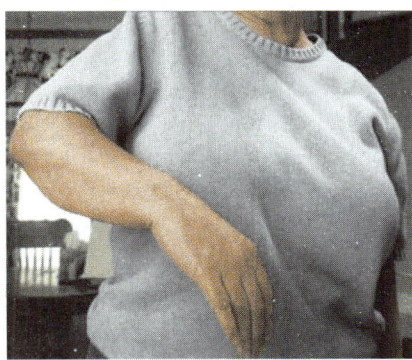

Fig. 3.5: Flexion synergy

the flexor and extensor synergy in both upper and lower limbs.

The strongest components of various synergies in upper and lower limbs are as follows:

Limb	Strongest component	
	Flexor synergy	Extensor synergy
Upper limb	Shoulder girdle retraction Elbow flexion Wrist and fingers flexion	Shoulder girdle depression Shoulder adduction and medial rotation Forearm pronation
Lower limb	Hip joint lateral rotation and inversion	Pelvic retraction Hip adduction and lateral rotation Knee extension and ankle plantar flexion

The hemiplegic patient has the chance of developing an attitude of shoulder retraction and depression, shoulder adduction and medial rotation, elbow flexion, forearm pronated and wrist, fingers in flexion.

The lower limb attitude shall be pelvic retraction, hip extension, adduction and external rotation, knee extension and ankle plantar flexion, inversion.

Sensory Deficits

The sensory deficits are experienced in patients with CVA depending upon the site and extent of lesion.

Touch, pain, temperature, joint proprioceptive sense would be affected to a variable degree depending upon the lesion.

The lesion involving the thalamus leads to contralateral hemianesthesia as the thalamus is the relay center for all sensations.

If the lesion is in the somatosensory area of cerebral cortex, then the cortical sensations are impaired like tactile localization, two-point discrimination, stereognosis and tactile extinction.

The visual disturbances commonly seen due to involvement of visual cortex, internal capsule or optic radiation is homonymous hemianopia, where there is loss of vision in the nasal half of opposite eye (normal side) and temporal half of the eye on the hemiplegic side.

The involvement of the upper part of the pons results in *crossed anesthesia* which means there is loss of sensations on the same side of lesion over the face and opposite side of trunk and limbs. Lack of conjugate gaze is also observed in these patients.

Presence of Abnormal Reflexes

During the initial stages of stroke because of cerebral shock, all the deep tendon reflexes will be absent or suppressed.

When the spasticity sets in, there will be exaggeration of clinical reflexes (deep tendon jerks).

The Babinski will be positive and shows extensor response.

The superficial reflexes, especially the abdominal and cremasteric reflexes, are diminished or absent.

Dominance of primitive reflexes is observed. The occurrence of primitive reflexes is due to lack of inhibitory regulation of higher centers over the lower level of central nervous system. The most common primitive reflexes observed are:

Asymmetrical tonic neck reflex: When the head is rotated to one side, increase in the extensor tone on face side and increase in flexor tone of limbs on occipital side.

Symmetrical tonic neck reflex: When the neck is in flexion, there will be increased flexor tone in upper limbs and extensor tone in lower limbs. When the neck is in extension, there will be increased extensor tone in upper limbs and flexor tone in lower limbs.

Symmetrical tonic labyrinthine reflex: There will be increase in the extensor tone in supine position and flexor tone in prone muscles in the limbs.

Tonic lumbar reflex: When the patient's trunk is rotated over pelvis to the hemiplegic side, there will be flexion in the hemiplegic upper limb and extension in hemiplegic lower limb.

Flexor withdrawal reflex: When a strong stimulation is given to the sole of hemiplegic foot in the form of stroking the sole with a blunt object, there will be sudden hip-knee flexion.

Positive supportive reactions: Pressure at the bottom of the foot causes co-contraction of the lower limbs making the limb rigid.

Presence of Associated Reactions

Associated reactions are tonic postural reactions in the muscles and are abnormal. There will be associated reactions in normal people too, which are known as synkinetic movements, e.g. swinging of hands during walking. But the associated reactions which are present in stroke patients are abnormal and manifest themselves as attenuation of the hemiplegic's attitude in response to movement either on the same side of the body or occurring in different parts.

Souques' phenomenon or sign: When the hemiplegic upper limb is raised to a position of above 90° flexion or abduction, there will be fan-shaped finger extension on the hemiplegic hand.

Raimiste's phenomenon or sign: When resistance is applied to abduction or adduction of the unaffected lower limb, there will be similar reaction in the affected limb.

Homolateral limb synkineses: When there is flexion of hemiplegic upper limb, there will be flexion of hemiplegic lower extremity. This is due to dependency that exists between the synergies of the involved upper and lower extremities.

Speech and Language Deficits

The language and speech disorder is due to the involvement of the dominant parietal lobe. The inability or difficulty in speech and language is known as aphasia. There are three most common types of aphasia observed in hemiplegic patients:

1. **Broca's aphasia** is also known as non-fluent aphasia. It is due to lesion in Broca's area which is on the dominant hemisphere. Patient can understand the

words but could not reply. Commonly known as motor or expressive aphasia.
2. **Wernicke's aphasia** is known as fluent aphasia. It is due to involvement of Wernicke's area. In this, the patient can speak but comprehension and perception of spoken words is lost. The patient will talk non-stop and irrelevant. It is commonly known as sensory or receptive aphasia
3. **Global aphasia:** Patient has deficit in both production and comprehension of speech, it is a very difficult condition for rehabilitation. This is due to involvement of conduction zone between the Broca's and Wernicke's area or involvement of both areas. This is commonly known as conductive or total aphasia.

Perceptual Problems

Apraxia

It is the inability to perform a learned movement though the patient has good muscle strength, coordination, intact attention and senses. It is due to the lesion on the dominant parietal lobe. The bilateral apraxia commonly involves lips and tongue also. It is of three types:
1. **Idiomotor:** Where movement cannot be carried by patient on command but can occur automatically.
2. **Ideational:** Where the purposeful movement is not possible either automatically or on command.
3. **Constructional apraxia:** It is the difficulty to imitate simple arrangements of objects or movements, e.g. the patient cannot redraw or copy a circle when asked to.

Agnosia

- Agnosia is the inability to recognize or make sense of upcoming information despite sensory capacities.
- The patient cannot identify the objects despite intact visual, auditory and tactile sensations due to lack of association pathways to arrange sensory images, memory and disposition towards action.
- **Visual agnosias:** It is the inability to recognize familiar objects despite normal function of the eyes and optic tract. The lesion is in the dominant parieto-occipital cortex.
- **Auditory agnosia:** It is the inability to recognize familiar sounds or music despite normal auditory pathways. The lesion is in the dominant temporal lobe.
- **Tactile agnosia:** It is also called *asteriognosis*. It is the inability to recognize objects by feeling them by hand with closed eyes. It is due to the lesion involving the dominant parietal lobe.
- **Autotopagnosia:** It is the inability to recognize different parts of his own body.
- **Anosognosia:** It is the the unawareness of the hemiplegic side.
- **Simultanagnosia:** It also known as Balint's syndrome. It is the inability to perceive a visual stimulus as a whole, due to lesion in the dominant occipital lobe.
- **Gerstmann's syndrome:** This is due to lesions in the inferior region of the dominant parietal cortex and is characterized by:
    - Confusion between right and left finger
    - Agraphia (difficulty in writing)
    - Acalculia (difficulty in calculation)

Anosognosia: Lack of awareness of the presence or severity of one's paralysis. Area of lesion is still in experiment.

Visuospatial relation disorders: It is the difficulty or inability in judging the distance, size, position and relationship of various parts to the whole. It is due to the lesion of the non-dominent parietal lobe.

Visuospatial impairment: It is the inability to recognize or respond to the stimuli which are present on the side contralateral to the affected non-dominant hemisphere.

Somatognosia: The patient is unable to differentiate the self real and mirror image, e.g. the patient will start combining the image in the mirror. The patient may also have difficulty in differentiating the self and other's body parts.

Visual perceptual impairment: It is the inability in the perception of the vertical and horizontal planes leading to frequent falls. The patient has topographical disorientation.

Cognitive and behavioral problems: In normal individuals, the right cerebral hemisphere is dominant for negative emotions and the left cerebral hemisphere is dominant for positive emotions (Hellige, 1993). Hence, the patients with left hemisphere damage result in depressed, low profile, anxious and negative attitudes. The patients with right hemispheric lesions result in euphoric, overconfident, impulsive attitude.

Dysphagia: It is the difficulty in swallowing. The exact relation between the dysphagia and stroke is not clear. Some researches shown that the dysphagia may be due to bilateral cerebral hemisphere or brainstem lesions.

Bladder and bowel impairment: In flaccid stages, there will be overflow incontinence. Anterior communicating artery blockage results in ischemia of paracentral lobule resulting in loss of voluntary control of micturition.

Sexual dysfunction: There will be problems with libido, orgasm, sexual desires, erection or lubrication.

SECONDARY IMPAIRMENTS OF STROKE

Psychological complications: The stroke is a great insult to the body and the patient especially who is young and only earning member in family is affected the most psychologically. The patient will suffer from chronic depression, social withdrawal, anxiety syndrome, insomnia, secondary mania, irritation disorder, abnormal expression of emotions.

Musculoskeletal complications: The musculoskeletal complications would be osteoporosis, contractures, deformities, secondary osteoarthritis. The complications are mainly due to muscular imbalance and inactivity.

The shoulder joint is prone to sublaxation due to paralysis of the rotator cuff muscles.

Pain at the shoulder can result in reflex sympathetic dystrophy.

Scalenus anticus syndrome occurs because of the abnormal posture or attitude developed by the patient. The posture which precipitates this syndrome is cervical spine in flexion, shoulder joint in medial rotation and scapula retraction. It is manifested by neck and shoulder discomfort, tingling numbness in the hand and fingers.

Circulatory disturbances: Deep vein thrombosis is most common circulatory disturbance seen in these patients. Edema in lower limbs due to inactivity of calf muscles is also common.

Cardiac and respiratory deconditioning: Although the cardiac and respiratory systems are not directly affected by stroke, the physical inactivity is the main cause of cardiac and respiratory deconditioning.

Gastrointestinal and renal system disturbances: The patients may exhibit the symptoms of constipation, renal caliculi, urinary tract infections.

GAIT IMPAIRMENTS IN STROKE PATIENTS

- The patient has unilateral weakness and spasticity with the upper extremity held in flexion and the lower extremity in extension.
- The foot is in extension so the leg is "too long", therefore, the patient will have to

circumduct or swing the leg around to step forward.

In hemiplegic patients, the anticipation of the weight-bearing excites the extensor synergy which results in plantar flexion, resulting in missing of heel strike component in gait. In flaccid stage of stroke, there will be buckling of knee joint during heel strike.

In hemiplegic patients during midstance, the forward progression of the body weight results in over stretching of calf muscles resulting in increased spasticity of plantar flexors, hence the forward shift in hemiplegics is prevented.

There is lack of heel off phase in hemiplegics due to quadriceps spasticity and plantar flexor spasticity. This results in slowness of walking.

In hemiplegics due to activation of extensor synergy, the adductors contract in place of abductors (abductors contract in normal gait) and hence there will be *Trendelenburg* sign visible.

In swing phase, the ground clearance is poor due to extensor synergy with hip, knee flexion and predominance of plantar flexion at ankle. In absence of forward rotation of pelvis, there will be circumduction of the limb resulting in classical circumduction gait.

INVESTIGATIONS DONE FOR STROKE PATIENTS

The purpose of doing investigations is to identify the causes or risk factors and level of damage to the brain due to CVA. The common investigations are:

1. **Routine and special blood investigations:** Complete blood picture along with thyroid function tests, lipid profile and blood glucose levels.
2. **Carotid duplex:** It is an ultrasound scan of the major arteries in the neck which supply blood to the brain. The atherosclerosis can be identified with this investigation.
3. **CT scan (Fig. 3.6):** The area of brain affected and extent of damage can be ruled out.

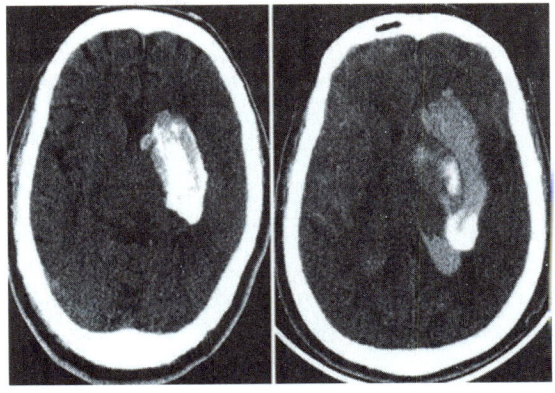

Fig. 3.6: CT scan showing the ischemic lesion of lentiform nucleus

4. **ECG:** To assess the heart function and electrical activity abnormalities of the heart. If further required, an echocardiography can be taken.
5. **MRI scan:** This may be required, if CT scan cannot reveal or give a clinical picture of symptomatology. An MRI scan may be used to look at the blood vessels in the neck to rule out any blocks (Fig. 3.7).

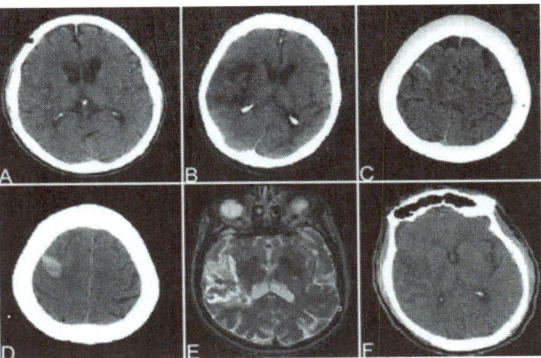

Fig. 3.7: MRI scan of ischemic stroke

6. **Cerebral arteriogram/cerebral angiogram:** A cerebral arteriogram is an X-ray or series of X-rays of the head that show the arteries of the brain. A series of X-rays are taken after a contrast medium is injected into the main arteries of the patient's head (Fig. 3.8).

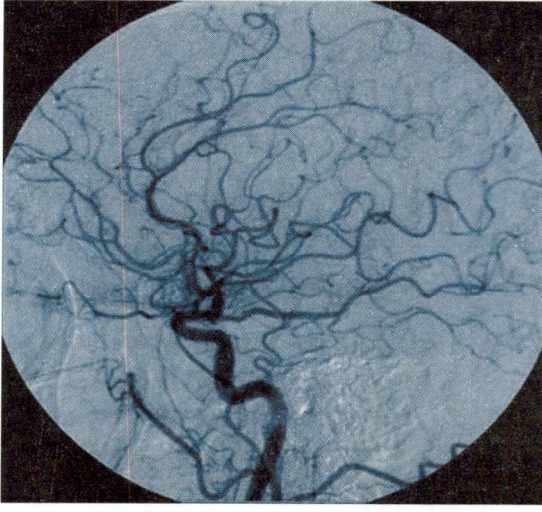

Fig. 3.8: Cerebral arteriography

PHYSIOTHERAPY ASSESSMENT OF STROKE PATIENT

The physiotherapy assessment starts with taking the demographic data, history taking from the patient or from medical records.

Observation

- Observe the general built of the patient, obese patients are difficult to rehabilitate.
- Posture and attitude of the limbs: Observe the posture, attitude of the limbs and facial symmetry.
- Check for the tropical changes of the skin. In stroke, due to muscular inactivity, the circulatory disturbances are common, leading to tropical changes in the skin.
- Observe for any swellings, scars over the upper and lower limbs.
- Observe for any contractures or deformities.
- Observe for any abnormal movements like tremors.
- Observe for the changes in muscle for wasting.

Palpation

- Palpate and compare with unaffected side for temperature.
- Palpate and grade the edema whether localized or generalized, indurated or non-indurated, pitting or non-pitting type.
- Palpate for any tender points over the affected limbs.

Examination

- Examination of higher functions like memory, intelligence, level of consciousness, behavior and topographical orientation and speech.
- Test the cranial nerves bilaterally and check for any abnormalities.

Sensory examination: Sensory examination would be very important to identify the level of lesion and type of lesion. A thorough examination of sensory system—superficial, deep and cortical senses should be evaluated. A loss of cortical sensations indicates the lesion is in the sensory cortex, loss of other sensations indicates an extensive subcortical lesion, loss of all senses indicates thalamic lesions. Intact sensory system indicates the lesion is in the motor cortex only.

Reflex examination:
- The superficial reflexes, like abdominal and cremasteric, are diminished or lost.
- The deep tendon reflexes show exaggerated response.
- The Babinski is positive or shows extensor response.

- There may be presence of abnormal primitive reflexes like ATNR, STNR, etc.

Motor examination: The motor examination in stroke patients involves assessment of tone, measurement of muscle girth, power of the muscles, voluntary control of the muscles, presence of abnormal associated movements.

Tone of the muscles: The tone of the muscles is assessed by repeated passive movement of the limbs. The patients with stroke show hypotonicity or flaccidity during the acute cerebral shock.

Later the spasticity in the muscles set in. The spasticity typically shows clasp knife mechanism and also clonus can be exhibited.

The tone of the muscles can be graded by modified Ashworth scale of spasticity.

Modified Ashworth scale of spasticity	
Grades	Description
0	No increase in muscle tone
1	Slight increase in muscle tone with catch and release at the end
1+	Slight more increased range as catch and followed by release for half of the range
2	Resistance is increased throughout the range during passive movements but part moved freely
3	Passive movements difficult with considerable increased muscle tone
4	Part remains rigid in flexion or extension

In spasticity, the voluntary control of movements should be assessed. The brain appreciates movement performed and not the muscles. The voluntary control should be assessed at all the joints and for all movements.

The voluntary control testing can also be graded by STREAM (stroke rehabilitation assessment of movement) testing. It is a performance based measurement that assesses the voluntary movement of extremities

Voluntary control testing	
Grades	Description
0	No contraction
1	Initiation of contraction or flicker contraction
2	Half range of motion in pattern
3	Full range of motion in pattern
4	Initial half range in isolation and the later half in pattern.
5	Full range of motion in isolation but goes into pattern when resistance offered
6	Full range of motion in isolation against resistance

and mobility following a stroke. It specifically measures amplitude of movement and quality of movement.

STREAM voluntary control testing	
Grades	Description
0	Unable to perform the test movement through any appreciable range (includes flicker or slight movement)
1	Able to perform only part of movement or complete the full movement with marked deviation from normal pattern
2	Able to complete the movement in a manner that is comparable to the unaffected side

Examination of Associated Reactions/Movements

Souques' phenomenon or sign: When the hemiplegic upper limb is raised to a position of above 90° flexion or abduction, there will be fan-shaped finger extension on the hemiplegic hand.

Raimiste's phenomenon or sign: When resistance is applied to abduction or adduction of the unaffected lower limb, there will be similar reaction in the affected limb.

Homolateral limb synkineses: When there is flexion of hemiplegic upper limb, there will be flexion of hemiplegic lower extremity. This is due to dependency that exists between the synergies of the involved upper and lower extremities.

Muscle girth: Muscle girth is the circumference or thickness or diameter of the limb. Girth or circumference of all the major muscles is measured. Usually in UMN lesions, there will be no wasting.

Range of motion: The range of motion at all the joints is measured and recorded by goniometry. It is important to assess the ROM to check whether any secondary musculoskeletal complications are set in.

Coordination: Tests of coordination are performed both equilibrium and non-equilibrium. Along with coordination, the synergy is also checked.

Perception and cognitive functions: Assess for any perceptual or cognitive dysfunction by performing tests for agnosia and apraxia.

GAIT ASSESSMENT

This is very important aspect of neurological examination. The gait parameters are checked and compared with normal.

Observe the symmetry of the gait, ability to walk with a narrow base, the length of the stride when walking at a normal pace, and ability to turn with a minimum of steps and without loss of equilibrium. Check for tandem walking (ability to walk on a straight line). Further for the sake of evidence-based practice, Functional Gait Assessment Scale can be used to assess the gait in stroke patients.

ASSESSMENT OF ADL

The physiotherapist's ultimate aim is to rehabilitate the patient back to his near normal or normal lifestyle. ADL assessment is very important tool of assessment to identify the functional disability.

An ADL assessment is simply an assessment to analyze a person's ability to perform his personel care and general activities, he/she needs to do on a routine basis in and around home.

A physiotherapist by assessing the ADL assessment determines the level and type of assistance a person requires in order to live an independent life as possible.

This assessment will also able the therapist to plan for the modifications that the patient may be deemed necessary for making him independent.

What are the areas to assess for ADL?

Basic ADL:
- Self-care activities
- Mobility
- Communication

Self-care:
- Feeding
- Grooming
- Dressing
- Bathing
- Toilet activities

Mobility:
- Bed mobility
- Wheelchair mobility
- Transfers from bed to wheelchair and vice versa
- Ambulation

Communication:
- Writing skills
- Speaking
- Using various communication devices like telephone, mobile, etc.

Feeding: Assess the patient how he/she is feeding, the type of set up, utensils and spoons used, ability to chew and swallow.

Grooming and hygiene: Assess how the patient is able to do or how far he needs assistance in oral care, hair dressing, bathing, shaving, etc.

Dressing: Assess the patient, her/his ability to wear the clothes, buttoning, lacing the shoes, buckling the belt, etc.

There are many scales available to assess the ADL performance of the patient like FIM (functional independent measure) can be used.

It has the following rubric measurements:
- 7—independent
- 6—modified independent
- 5—supervision/set up modification
- 4—minimal assistance
- 3—moderate assistance
- 2—maximal assistance
- 1—total assistance

Apart from this, an evidence-based assessment can also be done by Barthel index, Katz index of independence in ADL, etc.

PSYCHOLOGICAL ASSESSMENT

The patients with stroke often experience anxiety and depression. It is the major secondary complication and more attention is required. To assess, HADS (Hospital Anxiety and Depression Scale) was developed by Zigmond and Snaith in 1981.

PHYSIOTHERAPY MANAGEMENT

Various techniques were made available by various therapists in rehabilitating stroke patients. However, an integrated approach involving all the methods is found beneficial in all stroke patients. The various approaches are as follows.

Bobath Concept

It is an approach to neurological rehabilitation based on normal movement or neurodevelopmental approach. The main principle is to promote motor learning for efficient motor control in various environments, thereby improving participation and function. The Bobath approach mainly concentrate to prevent the synergic patterns and facilitate normal movement. It also promotes to learn how to control postures and movements.

The main strategies of this concept are:
- *Therapeutic handling*: Whereby the patient's movements are influenced by facilitation and inhibition techniques. *Facilitation* techniques to promote motor learning by using sensory information to reinforce weak movement patterns and discourage overactive patterns. *Inhibition* techniques which reduce the abnormal influences on movement or posture that interfere the normal pattern of movement.
- *Key points of control* generally refers to parts of the body that are advantageous when facilitating or inhibiting the movements and postures.

Brunnstrom Approach

The Brunnstrom approach is based on using the reflexes that represent normal stages of development, and be used in functional rehabilitation.

The Brunnstrom approach principles:
1. Reflexes should be used to elicit movement when there is no movement (normal developmental sequence).
2. Proprioceptive and exteroceptive stimuli also can be used therapeutically to evoke desired movement or tone changes.

Peto Approach

The concept of conductive education (CE) is an educational system, based on the work of Hungarian professor Andras Peto. This approach has six elements—group, facilitation, daily routine, rhythmic intention, task series and conductor. The patient is encouraged to verbalize the activities as they perform them and focuses on function.

Johnstone Approach

Margatet Johnstone pionerred the use of air splints in active training of the hemiplegic limb in the severly impaired stroke patients. The treatment is based on reflex inhibition with special attention to inhibiting the tonic neck reflexes through use of air splints and positioning. The optimal position for the affected upper extremity in the air splint at

Hospital Anxiety and Depression Scale (HADS)

Tick the box beside the reply that is closest to how you have been feeling in the past week. Don't take too long over your replies: your immediate is best

D	A		D	A	
		I feel tense or 'wound up':			**I feel as if I am slowed down:**
	3	Most of the time		3	Nearly all the time
	2	A lot of the time		2	Very often
	1	From time to time, occasionally		1	Sometimes
	0	Not at all		0	Not at all
		I still enjoy the things I used to enjoy:			**I get a sort of frightened feeling like 'butterflies' in the stomach:**
0		Definitely as much		0	Not at all
1		Not quite so much		1	Occasionally
2		Only a little		2	Quite often
3		Hardly at all		3	Very often
		I get a sort of frightened feeling as if something awful is about to happen:			**I have lost interest in my appearance:**
	3	Very definitely and quite badly	3		Definitely
	2	Yes, but not too badly	2		I don't take as much care as I should
	1	A little, but it doesn't worry me	1		I may not take quite as much care
	0	Not at all	0		I take just as much care as ever
		I can laugh and see the funny side of things:			**I feel restless as I have to be on the move:**
0		As much as I always could		3	Very much indeed
1		Not quite so much now		2	Quite a lot
2		Definitely not so much now		1	Not very much
3		Not at all		0	Not at all
		Worrying thoughts go through my mind:			**I look forward with enjoyment to things:**
	3	A great deal of the time	0		As much as I ever did
	2	A lot of the time	1		Rather less than I used to
	1	From time to time, but not too often	2		Definitely less than I used to
	0	Only occasionally	3		Hardly at all
		I feel cheerful:			**I get sudden feeling of panic:**
3		Not at all		3	Very often
2		Not often		2	Often
1		Sometimes		1	Not very often
0		Most of the time		0	Not at all
		I can sit at ease and feel relaxed:			**I can enjoy good book or radio or TV program:**
	0	Definitely	0		Often
	1	Usually	1		Sometimes
	2	Not often	2		Not often
	3	Not at all	3		Very seldom

Please check you have answered all the questions
Scoring: Total score: Depression (D).......... Anxiety (A)..........
0–7 = Normal, 8–10 = Borderline abnormal (borderline case), 11–21 = Abnormal (case)

40 mm Hg pressure inflated by mouth is shoulder external rotation, elbow, wrist and fingers in extension, forearm in supination and thumb in abduction. The lower limb is positioned in protracted pelvis, hip in internal rotation, hip, ankle and knee in flexion.

Motor Relearning Program

This was first described by Carr–Shepherd in the year 1987. The training is based on an understanding of kinematics and kinetics of normal movement, motor control process and motor learning. The major factors in the learning or relearning process identified by Carr and Shepherd are:
- Identification of a goal
- Inhibition of unnecessary activity
- Ability to cope with the effects of gravity and, therefore, to make balance adjustments while shifting weight
- Appropriate body alignment
- Practice
- Motivation
- Feedback

The motor relearning is aimed at gaining functional independence through learning a specific task oriented movement. The motor control strategies have five components.

1. **Motor program:** It is asset of pre-set sequences to get a coordinated voluntary movement by muscle activation and is carried out without peripheral feedback.
2. **Motor planning:** It is strategical planning for a movement requiring coordination of various motor programs.
3. **Feedback:** It is the stimulation of control centres in brain by sending the information from peripheral receptors regarding the accuracy of movement and adaptations necessary.
4. **Feed forward mechanism:** It is the strategy of the musculoskeletal system to adapt or respond in anticipation of a movement or changes of an ongoing movement.
5. **Motor skill acquisition:** It is the training of a goal-oriented problem solving strategy through development of motor programs and integration to form a motor plan and execute.

Proprioceptive Neuromuscular Facilitation

This approach was developed by Knott and Voss in the year 1968. General treatment in PNF includes the use of recapitulation of total patterns of developing motor behavior, spiral and diagonal patterns of movements, coupling voluntary movement with postural and rightning reflexes, appropriate sensory and verbal cues, maximal resistance for maximal excitation and inhibition and repetitive activity for conditioning and training.

Rood Approach

This was proposed by Margaret S Rood in 1950. It is based on the philosophy of treatment concerned with the interactions of somatic, autonomic, psychologic factors and their interactions with motor activities. Rood has used sensory stimuli by stroking or brushing at a given speed for a given duration for activation of a phasic muscle response. Rood applied cold for visceral stimulation and somatic relaxation and applied pressure/stretch for postural muscle activation.

Vojta's Approach

Vojta established 18 points in the body for stimulating and used the positions of reflex crawling and reflex rolling. He proposed that placing the child in these positions and stimulation of key points in the body would enhance CNS development. In this way the child is presumed to learn normal movement patterns in place of abnormal motion.

According to Vojta, reflex locomotion is activated from three main positions—prone, supine and side lying.

Two Coordination Complexes in Reflex Locomotion

In the practical use of reflex locomotion, there are two coordinated complexes:
- Reflex creeping
- Reflex rolling

The movement sequences of reflex locomotion are retrievable at all times.

The tree main positions—prone, supine and side lying and have more than 30 variations.

By combining and varying stimulation zones and resistances, as well as making changes in directions of pressure and joint angles in the starting position, therapy can be adapted to the patient's individual treatment goal and condition.

Temple Fay and Doman Approach

Temple Fay and Doman place emphasis on the use of more primitive patterning initially and encourage reptilian and amphibian pattern movement before working towards the more sophisticated patterns of the mature human being. They also emphasize, more positively, the need for constant repetition and the importance of stimulating the patient physically and intellectually.

Recent studies have shown that there is no single approach can be promotive to get a desired benefit as the patients, therapists and environments are different. So it is beneficial to use an integrated approach to deal with the hemiplegic patients by incorporating the principles of these approaches in rehabilitation of the patient.

Acute/Short-term Goals of Management

1. To provide psychological support and develop good rapport with the patient
2. To develop awareness of the affected side of the patient and prevent unilateral neglect.
3. To decrease the tendency of developing abnormal movement patterns.
4. To prevent the complications of the prolonged bedrest.
5. To prevent joint contractures and deformities.
6. To maintain normal joint range of motion at all the joints.
7. To encourage early weight bearing.

Plans of Physiotherapy

To achieve the above mentioned goals, the following plans are to be implemented.

1. Develop good rapport with the patient. Counsel the patient and explain the importance of physiotherapy. Provide moral, spiritual, social support to the patient. Encourage the patient to involve actively in all self-care activities.
2. Hemiplegic patients have a tendency of neglecting affected side. To prevent this, the hemiplegic side should never be towards the wall. Feeding, medications should always be given from the affected side. Ask the family and visitors to talk to him by standing towards affected side.
3. *Positioning*: To prevent patterns to set in, positioning is done and maintained with the help of sandbags, splints and pillows. The following positions are maintained.
    - *Supine*: Shoulder girdle protracted, shoulder abducted and externally rotated, elbow–extension, forearm supinated, wrist and fingers in extension.
    - *Affected LE*: Hip forward (pelvis protracted), knee on a small towel roll to prevent hyperextension, nothing against the soles of feet. For persistent plantar flexion, a splint can be used to position the foot and ankle in neutral position.
4. The air splints designed by Margaret Johnston can be used.

5. To prevent contractures and deformities, apart from positioning, stretching, mobilization and splinting is done.
6. Weight bearing activities are necessary to promote development of tone in muscles and also maintain the absorption of calcium into the bones. Thus the patient is taught and made to perform bridging exercises sitting with weight bearing on the affected arms.
7. Achieving adequate joint ROM is necessary for functional independence and ambulatory training. The joint range of motion is maintained and improved by stretching exercises. Use of appropriate orthotic supports would be useful to maintain the joint range. Strengthening exercises of weak muscles would be necessary to prevent the contractures and deformities in order to maintain full range of motion. It is mostly observed that shoulder elevation, scapular depression, scapular retraction, elbow extension, supination, hip extension, ankle dorsiflexion, great toe extension are the common movements that are impaired and result in contractures. In some cases, tightness of muscles may be preserved for functional independence. It should be kept in mind that aggressive ROM exercises are contraindicated, if the patient has unstable fractures, active heterotropic ossification and deep vein thrombosis. In case of spastic muscles, daily stretching and terminal sustained stretch is to be considered for rehabilitation for limitation of joint range of motion.
8. There is a high chances of shoulder subluxation, due to paralysis of rotator cuff muscles, to prevent this, the patient is advised to wear a shoulder sling/Bobath splint to prevent shoulder sublaxations when sitting or standing (Fig. 3.9).
9. Prevention of secondary complications (refer Head Injury Management).

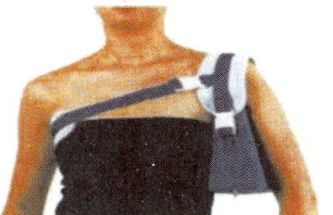

Fig. 3.9: Bobath sling

Long-term Goals of Physiotherapy

- To continue counseling the patients and motivation
- To normalize the tone or to maintain normal tone in the muscles
- To prevent contractures and deformities
- To maintain respiratory and cardiovascular proficiency
- Improve functional capabilities and train the patient to be functionally independent.
- Vocational training

Plans of Physiotherapy

1. Continue to counsel the patient regularly. Do not allow the patient to develop negative thoughts. The importance of effective communication with the patient is proven in physiotherapy care for better prognosis and improving the rapport with the patient. Have patience to listen to all the patient's complaints and problems. Try to answer and explain the doubts or queries of patient about his condition. Never give false promises or promise goals that cannot be achievable. Effective channels of communication and coordination is required among the members of the multidisciplinary team and also with patient and guardians. In a research by Barnes et al, 2012, it was believed that facilitating a successful intervention might require a health care professional to enhance their communication skills.

2. Normal tone in the muscles is very important to improve the overall functional activities of the patient. It is very important to decrease the tone in the spastic muscles and strengthen the antagonist muscles.
3. The various techniques to normalize tone are as follows:
   a. Gentle rhythmic passive movements in full range of motion (Fig. 3.10).

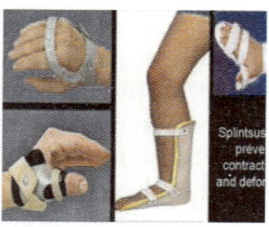

Fig. 3.11: Splints used to prevent contractures and deformities

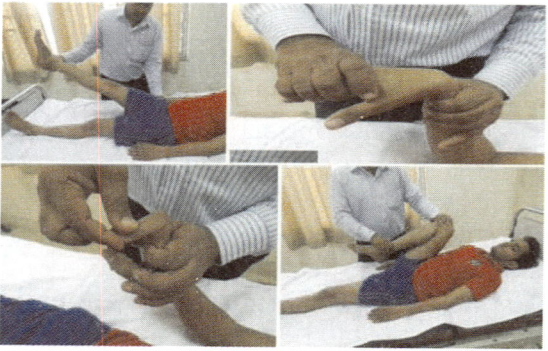

Fig. 3.10: Full range rhythmical passive moments

   b. Suspension therapy
   c. Aquatic exercises, whirlpool therapy
   d. Cryotherapy in the form of prolonged icing
   e. Positioning in reflex inhibiting postures
   f. Continuing the use of Johnston air splints.
   g. PNF techniques
   h. Faradic stimulation of antagonistic muscles can reduce the spasticity in the agonist muscles by reciprocal inhibition.
   i. Use of splints to maintain the full length of the muscles (Fig. 3.11).
   j. Biofeedback techniques to decrease spasticity in the agonist and strengthen the antagonist muscles
   k. Use of primitive reflexes like ATNR, STNR, TLR to normalize the tone.
   l. *Vibrations*: It is a selective means of stretching and are found to be facilitatory to tonic postural muscles. When a muscle is vibrated at small amplitudes of a few millimeters at 50–500 Hz, a tonic vibration reflex is elicited and the muscle slowly contracts. Hagbarth and Eklund found that vibration of spastic muscles of the Hemiparetic patients caused an initial phasic response followed by a tonic response that lasted as long as stimulus is applied. Bishop noticed that, when vibrations were applied, there were decreased spasticity in the antagonists muscles and an increase in the response of weak muscles.
4. Exercises in spastic stage to improve the functional activities and promote independence:
   a. **Exercises in lying** (Fig. 3.12):
   • Mobilize the scapula passively and promote the protraction and elevation of scapula voluntarily by the patient.
   • PNF patterns of upper limb and lower limb in full range. Incorporate the various techniques like rhythmic stabilization, overflow principles, etc. into the patterns
   • The fingers are in flexed position due to spasticity, to make the fingers open, use of primitive reflex—Souque's phenomenon is incorporated. When the hemiplegic upper limb is raised to a position of above 90° flexion or abduction, there will be fan-shaped finger extension on the hemiplegic hand.

Cerebrovascular Accidents 71

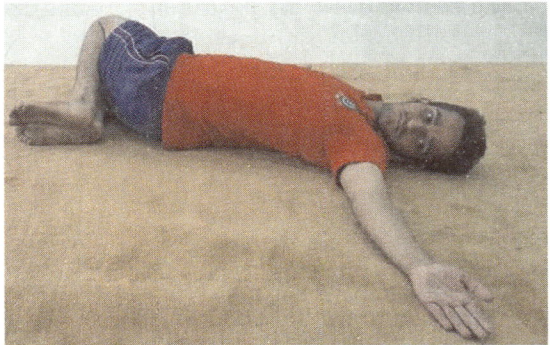

Fig. 3.12: Upper picture—bridging exercise. Lower picture—unilateral pelvic rotation exercise

- Train the patient to touch the head by maintaining external rotation and supination.
- Bridging exercises in lying and unilateral bridging on the affected limb support.
- Unilateral rotation of the pelvis: The patient is made to lie supine with hip-knee flexion. Then ask the patient to rotate the pelvis to each side. The patient tries to maintain this posture so that the spasticity of the trunk is inhibited. This exercise also helps in encouraging forward rotation of the pelvis and correct pelvic retractions.
- Incorporate the **Raimiste's phenomenon or sign:** When resistance is applied to abduction or adduction of the unaffected lower limb, there will be similar reaction in the affected limb.

- The patient is trained for dorsiflexion by using principle of irradiation in PNF. Strong contraction of quadriceps results in dorsiflexion of the foot.

b. **Exercises in sitting:**
- Make the patient sit by therapist supporting from behind. The patient is asked to abduct and extend the elbow of the hemiplegic side. The therapist supports the elbow and then the patient is asked to weight bear on the affected upper limb.
- Shoulder shrugging activities with tactile stimulation by the physiotherapist.
- Beading activities, holding the objects with affected limb, play dough activities will help to improve the hand activities.
- Supination and pronation activities of the forearm.
- The patient is made to sit on the vestibular ball to improve the balance exercises by vestibular stimulation

c. **Exercises in standing** (Figs 3.13 and 3.14):
- Static balance exercises
- Dynamic balance exercises
- Forward bending and straightening of the trunk
- Kicking activities: Kick the ball
- Weight shifting activities
- Trunk rotations
- Forward marching and walking—sideways in straight line, figure of eight and circle
- Standing with narrow base of support and then throwing of balls
- Standing and outreach activities. Ask the patient to extend the upper limb and hold an object placed on a table or in therapist hand.

5. **Gait training** (Fig. 3.15): Before starting the gait training to the patient, identify

In stroke, the most common impairments that cause hindrance to patient ability to ambulate are decreased muscle strength, impaired voluntary control, balance and equilibrium.

The gait training protocol can be framed based on ICF (international classification of functioning) model:

1. Identify the primary factors resulting in gait problems.
2. Selection of appropriate walking ability related outcome measures.
3. Development of tailored made training programmes.
4. Identification of potential environmental or personal factors that facilitate or impede individual's walking ability.

- Improve muscle strength particulary hip abductors, hip flexors, knee extensors, ankle dorsiflexors, plantar flexors, hip extensors and knee flexors.
- Develop balance and equilibrium reactions by training on tilt board, wobble board, vestibular ball exercises.
- Determine the individual status of independency and determine the type of assistance required like cane, forearm crutch, standard walker, rolling walker, rollator, etc.
- Train for forward walking first in parallel bars, then with walking aids and progress to independent walking.
- Train sideways and backward walking.
- Teaching turning techniques –90° and 180° turning.
- Once the patient is confident progress to stair climbing and stair descending first with support and independently.
- Advance the progression of gait by negotiating stairs without railing, walk on uneven surfaces, carrying the objects when walking.

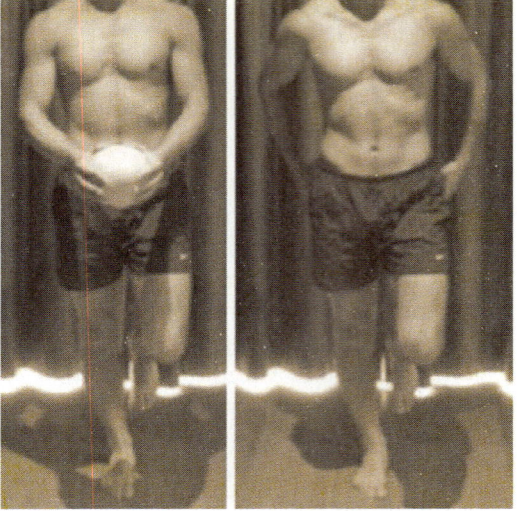

Fig. 3.13: Exercises in standing

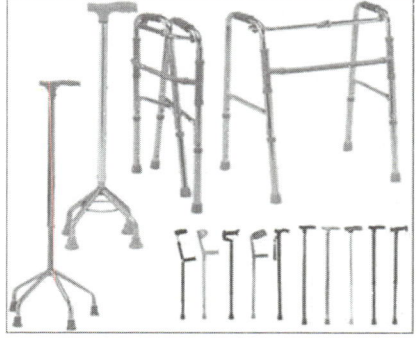

Fig. 3.14: Walking aids

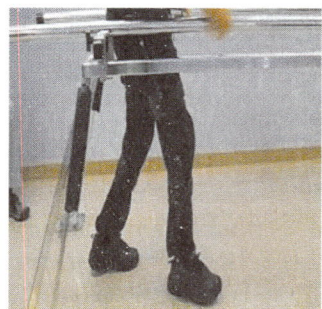

Fig. 3.15: Gait training in parallel bars

the impairments that primarily determine walking ability of the patient. This will help the therapist to develop effective gait training strategies.

Cerebrovascular Accidents

FUNCTIONAL RE-EDUCATION—TRANSFER ACTIVITIES

These are taught from and to the wheelchair and the bed, plinth, bath/lavatory seat and motorized vehicle. Transfer from a mat to wheelchair and vice versa are also taught for various treatment sessions. Sitting up from a lying position is a prerequisite for independent dressing and transfers. A thorough evaluation should be conducted to determine the most appropriate transfer technique for any individual with paralysis. Dependent transfers include sliding transfers and dependent standing pivot technique or use of Hoyer lifts. Transfers that require some active patient participation include the two man lift, sliding board transfer or assisted pivot transfer. Always remember the aim of assisted transfer is to gradually reduce the assistance required until the patient can perform the transfer independently. Floor to chair transfer training is very important for anyone who falls out of wheelchair or otherwise ends up on the floor and needs to get back to the chair (Fig. 3.16).

Principles of Training of Transfer Technique for Physiotherapists

- Thoroughly assess the patient's ability and preparations for resettlement for transfer activities
- Functional charts useful to record these activities which could be useful to check the prognosis and also helps to stimulate the patient's interest.
- Correct positioning of the chair, ensuring that the brakes are fully on and lifting the legs with the hands on plinth, placement of legs on foot rest would be the principles of safe training
- During the transfer techniques care must be taken not to knock the leg or drag them along a hard surface.
- Check that the tyres of wheelchair are not worn out and the floor should not be slippery.

To Improve Overall Functional Capabilities

To improve functional independence mat activities are taught depending upon the level of independence. Training activities done on adjustable therapy mat and are composed of sequenced activities that progress from the

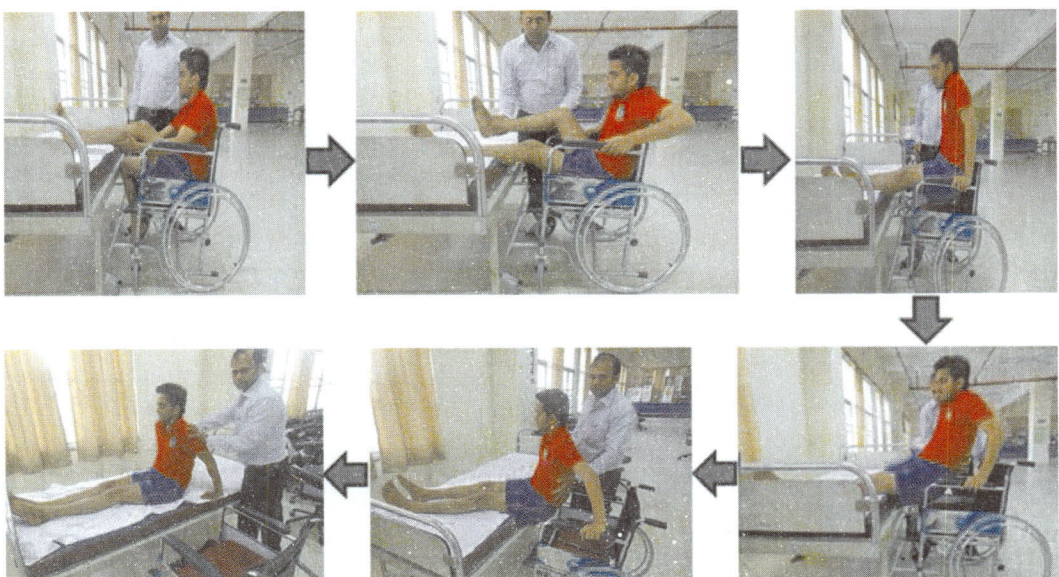

Fig. 3.16: Transfer techniques

easiest to the most difficult (Fig. 3.17). The usual progression is from bed mobility to rolling, prone lying, long sitting, short sitting, kneeling, half kneeling and standing (Figs 3.18 and 3.19). Muscles needed for individuals with lower limb paralysis to be able to move or position their legs in bed are wrist extensors, bicep, anterior deltoid, middle deltoid, and shoulder girdle stabilizers. Individuals with tetraplegia are taught to use their arms, head and neck for momentum to roll in bed with keeping the elbow straight while the shoulder is flexing across the body.

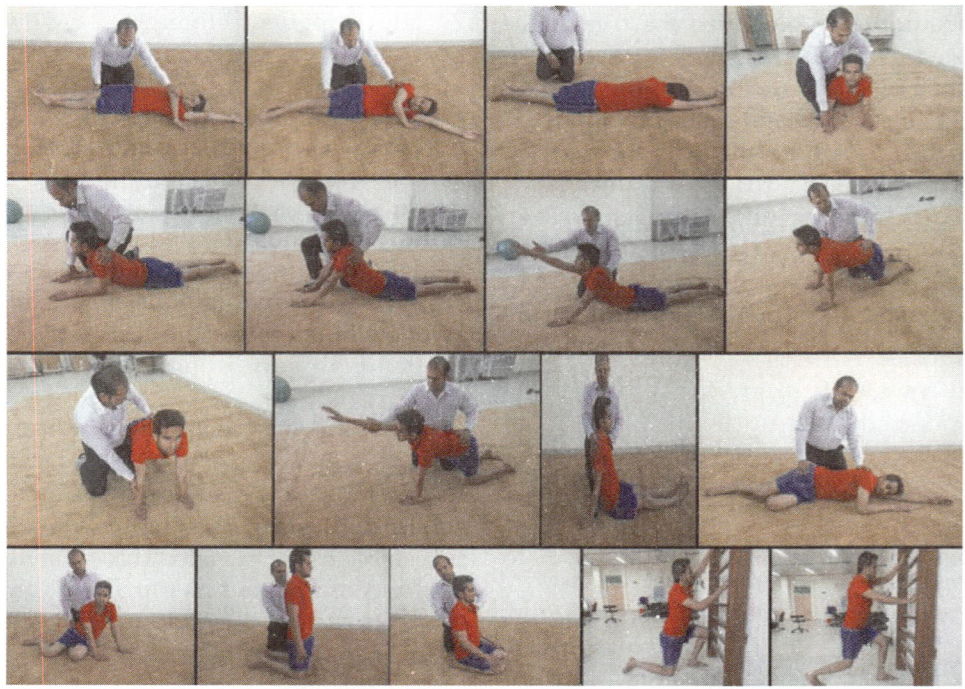

Fig. 3.17: Mat exercises

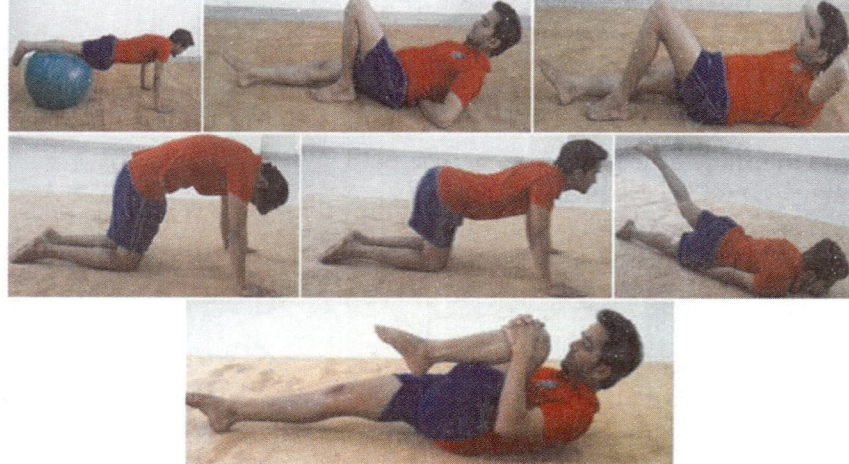

Fig. 3.18: Some activities in lying

Fig. 3.19: Activities in standing exercises

REFERENCES

1. Jeyaraj Durai Pandian, Paulin Sudhan Stroke. Epidemiology and Stroke Care Services in India, J Stroke 2013;15(3):128–134.
2. Cipolla MJ, San Rafael (CA): Cerebral circulation, Morgan & Claypool Life Sciences 2009.
3. CDC, NCHS. Underlying Cause of Death 1999–2013 on CDC WONDER Online Data-base, released 2015. Data are from the Multiple Cause of Death Files, 1999–2013, as compiled from data provided by the 57 vital statistics jurisdictions through the Vital Statistics Cooperative Program. Accessed Feb. 3, 2015.
4. Wolf PA, D'Agostino RB, Belanger AJ, Kannel WB. Probability of stroke: a risk profile from the Framingham Study. Stroke 1991;22:312–318.
5. Brown RD, Whisnant JP, Sicks RD, O' Fallon WM, Wiebers DO. Stroke incidence, prevalence, and survival: secular trends in Rochester, Minnesota, through 1989. Stroke 1996;27:373–380.
6. Xuetao Chen, Liang Zhou et al. Risk factors of stroke in Western and Asian countries: A systemic Review and Meta analysis of Prospective cohort studies, BMC Public Health 2014;14:776.
7. Garcia JH, Anderson ML. Circulatory disorders and their effects on the brain. In Davis RL, Robertson DM (Eds): Textbook of Neuropathology, 3rd edition. Williams & Wilkins, Baltimore 1997;715–822.
8. Toni D, Fiorelli M, Bastianello S, et al. Hemorrhagic transformation of brain infarct: predictability in the first five hours from stroke onset and influence on clinical outcome. Neurology 1996;46:341–345.
9. Zohrevandi B, et al. "Third Ventricle Colloid Cyst as a Cause of Sudden Drop Attacks of a 13-Year-Old Boy." Emerg (Tehran) 2015;3(4):162–164.
10. Brust JCM, Plank CR, Healton EB, Sanchez GF. The pathology of drop attacks: a case report. Neurology 1979;29:786–790.
11. Wade Derick T, Wood Victorine A, Hewer Richard Langton. "Recovery after stroke—the first 3 months". Journal of Neurology, Neurosurgery & Psychiatry 1985;48(1):7–13.

12. Souques' phenomenon Taber's Cyclopedic Medical Dictionary 2009; Issue 21:2158.
13. Sara Ahmed, Nancy E Mayo, et al. The Stroke Rehabilitation Assessment of Movement (STREAM): A Comparison With Other Measures Used to Evaluate Effects of Stroke and Rehabilitation, APTA, July 2003.
14. Mahoney F, Barthel D. "Functional evaluation: the Barthel Index". Md Med J 1965;14:61–65.
15. Granger CV, Dewis LS, Peters NC, Sherwood CC, Barrett JE. "Stroke rehabilitation: analysis of repeated Barthel index measures". Arch Phys Med Rehabil 1979;60(1):14–7.
16. Katz S, et al. Studies of illness in the aged. The index of ADL: A standardized measure of biological and psychosocial function. JAMA 1963;185:914–9.
17. Leddy, AL, Crowner, BE et al. "Functional gait assessment and balance evaluation system test: reliability, validity, sensitivity, and specificity for identifying individuals with Parkinson disease who fall." Phys Ther 2011;91(1):102–113.
18. Bobath B, Bobath K. The Neuro-Developmental Treatment. In: Scrutton, D, et al., 1984. Management of the Motor Disorders of Children with Cerebral Palsy. Clinics in Developmental Medicine 90, Spastics International Medical Publications, Oxford, 1984;6–18.
19. Janet H Carr, Roberta B. Shepherd, A motor relearning programme for stroke, Aspen Publishers, Business & Economics 1987;188.
20. Livingston RB. "Brain mechanisms in conditioning and learning". Neurosciences Research Program Bulletin 1966;4(3):349–354.
21. Barnes S, Gardiner C, et al. Enhancing patient-professional communication about end-of-life issues in life-limiting conditions: a critical review of the literature, J Pain Symptom Manage 2012;44(6):866–79.
22. Eklund G, Hagbart K-E. Normal Variability of Tonic Vibration Reflexes in Man Exp Neurol 1966;16:80.
23. Brown MC, Engberg I, Matthews PBC. The Relative Sensitivity to Vibration of Muscle Receptors of the Cat J Physiol Lond. 1967;192:773.
24. Bishop B, Machover S, Johnston R, Anderson M. A quantitative assessment of gamma-motoneuron contribution to the achilles tendon reflex in normal subjects. Archives of Physical Medicine & Rehabilitation 1968;49:145–154.
25. Bishop B. Vibratory stimulation. Part II. Vibratory stimulation as an evaluation tool. Physical Therapy 1975;55:28–34.
26. Karatas M, Cetin N, Bayramoglu M, Dilek A. "Trunk Muscle Strength in Relation to Balance and Functional Disability in Unihemispheric Stroke Patients," American Journal of Physical Medicine and Rehabilitation 2004;83(2):81–87.
27. Verheyden G, Vereeck L, Truijen S, Troch M, Herregodts I, Lafosse C, et al. "Trunk Performance after Stroke and Relationship with Balance, Gait and Functional Ability," Clinical Rehabilitation 2006;20(5):451–458.

4

Head Injuries and Comatose Patients

LEARNING OBJECTIVES

At the end of this chapter, you will be able to:
1. Describe various etiological factors and types of head injuries
2. Identify various clinical manifestations of the patients with head injuries
3. Identify the various investigations done in patients with head injuries
4. Demonstrate the skills to assess the head injury and comatose patients.
5. Design treatment protocols for rehabilitating the patients with head injuries and coma.

INTRODUCTION

A traumatic head injury is a condition in which there is an insult to the brain tissue by an external force leading to altered state of consciousness, impairment of physical, cognitive, behavioral or emotional disturbances. In most cases of head injuries, the mechanism of injury is a blunt force to the head. The head injuries in India have increased prevalence recently due to increased traffic and decreased traffic sense and safety measures.[1] Currently head injuries are the major causes of disability in young adults.

Etiology

- Road traffic accidents
- Gunshot injuries
- Riots and assaults
- Fall from a height
- Industrial accidents

TYPES OF HEAD INJURIES

Head injuries are mainly of two types (Fig. 4.1): Closed head injuries and open head injuries.

Closed Head Injuries

These are non-missile injuries and the skull vault remains intact after the injury. The brain tissue may be damaged due to inertial forces acting on the brain when the skull is struck by a powerful blow.

The brain damage in closed head injuries is due to:
- The rigidity and internal contours of the skull
- Shearing forces over the brain due to trauma

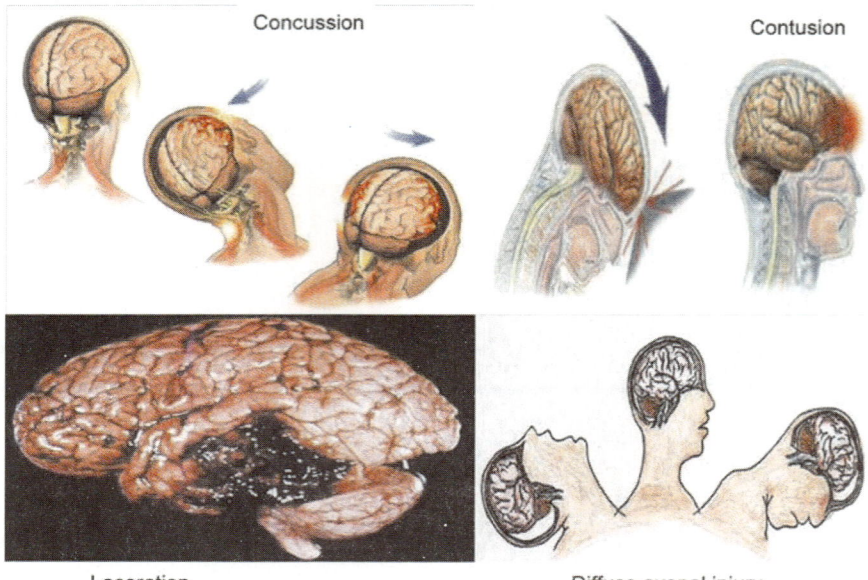

Fig. 4.1: Types of head injuries

- The quality of incompressibility of the brain tissue leading to increased pressure.

If the brain injury occurs on the same side of the blow over the skull, then it is called *coup injury* and if it arises on the opposite side, then it is called *countrecoup injury*. The countrecoup injury is due to bouncing of the brain to the other side of the skull due to the blow.

The brain injuries are further classified depending upon the tissue involved, and severity of the injury.

Concussion: It is the mildest form of injury.
- The injury is a result due to blow to the head resulting in victim's physical, cognitive and emotional disturbances.
- The severity of concussion is measured by concussion grading systems.[2,3]

The Colarado Medical Society Grading of concussion:

Grade I: Concussion consists of confusion only.

Grade II: Confusion and post-traumatic amnesia.

Grade III and IV: Loss of consciousness.

In some cases, a slightly greater injury is associated with both antegrade and retrograde amnesia. The amount of time that the amnesia is present correlates with the severity of the injury.

Patients also develop post-concussion syndrome which includes memory problems, dizziness, tiredness, sickness and depression.

Cerebral contusion: It is the bruising of the brain tissue. In majority of the cases, the contusions occur in the frontal and temporal lobes. The cerebral contusions can be extremely dangerous and life-threatening, if there is diffused cerebral edema and transtentorial herniation.

Diffuse axonal injury: It occurs usually as a result of an acceleration or deceleration motion. Axons are stretched and damaged widely due to slide over another. The prognosis vary depending on the extent of damage.

Intracranial hemorrhage: The blood vessels in the cranium can be ruptured due to injury and results in hemorrhage. The hemorrhage is considered as a focal brain injury. It occurs in a localized spot rather than causing diffuse spread.

The intracranial hemorrhage is classified into intra-axial and extra-axial hemorrhage.

Intra-axial hemorrhage is bleeding within the brain itself. It is also known as cerebral hemorrhage. The bleeding can be within the brain tissue or can be in brain's ventricles. Intra-axial hemorrhages are more dangerous and harder to treat.

Extra-axial hemorrhage: The bleeding that occurs within the skull but outside the brain tissue is termed extra-axial hemorrhage. It is further classified into:

Epidural hemorrhage also known as extra-dural hemorrhage. This hemorrhage occurs between dura mater and skull. It is mainly because of laceration of an artery due to trauma. The middle meningeal artery is commonly involved (Fig. 4.2A).

Subdural hemorrhage: It is the bleeding due to rupture of the bridging veins in the subdural space which are present between dura and arachnoid mater (Fig. 4.2B).

Subarachnoid hemorrhage: It results either from trauma or from ruptures of aneurysms. Blood starts layering into the brain along the sulci and fissures. It occurs between the arachnoid and pia meningeal layers (Fig. 4.2C).

Degree and Severity of Head Injuries

Mild head injury: The damage to the neural structures is minimal with slight or no impairment of neurological function.

The common clinical features shall be headache, dizziness, fatigue, decreased concentration power, poor memory and irritability.

The head injury is considerd to be mild, if the GCS[4-6] (Glasgow Coma Scale) score is more than 13, no focal neurological dysfunction.

Moderate head injury[4-6]: The injuries in which the GCS score lies between 8 and 13. There may be post-traumatic amnesia lasting between 1 and 24 hours.

Severe head injury[4-6]: These head injuries result in focal neurological dysfunction. They produce an obvious[4-6] disability, but the patient regains conscious status.

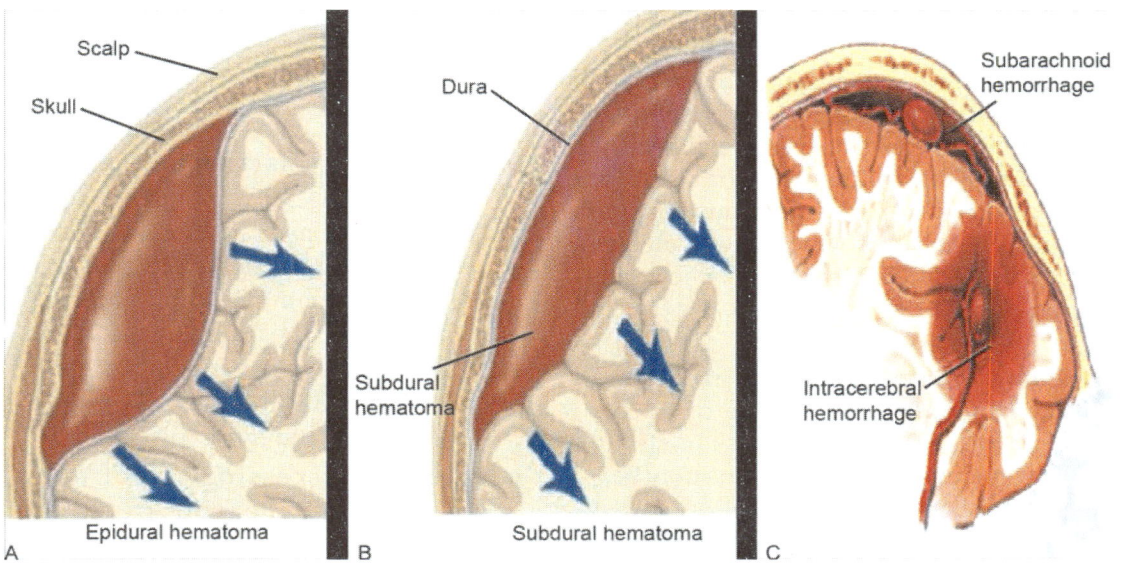

Fig. 4.2: A. Epidural hematoma, B. Subdural hematoma, C. Subarachnoid hemorrhage

Persistent vegetative state: It occurs in small percentage of head injury cases. The restoration of level of consciousness is not possible due to severity of the brain damage. These are acute brain traumas resulting in disorientation, unconsciousness, aphasia and inability to perform voluntary movements.

Open Head Injuries

In this, the skull vault is fractured, the cranial cavity is exposed to the external environment. There may be laceration of the brain tissue by the fracture fragments.

The head injuries can be *diffused* where the trauma is spread over a wide area.

Focal: The lesion is located in a small, specific area.

CLINICAL MANIFESTATIONS

The clinical features of traumatic brain injury depends upon the severity and location of injury. However, the most common symptoms are listed below.

1. Altered State of Consciousness

The head injuries can result in the alterations in the level of consciousness mainly because of the bilateral cerebral hemisphere dysfunction or direct depression of brainstem function or brainstem activating system. The altered state of consciousness need not have the direct relation with the motor loss (Plum and Posner).[7,8] The altered consciousness need not only be the coma (unconsciousness) but can also result in other forms. The various altered states of unconsciousness exhibited by the patients of TBI are:

a. **Coma:** It is the complete arrest of all cerebral functions, a state of unconsciousness unresponsiveness to stimuli, however, strong the stimulus may be.
b. **Stupor:** This is a state of unresponsiveness but responds to painful stimulus through some bodily movements.
c. **Obtundity:** This is the state of over sleepiness and disinterest in the surrounding environment.
d. **Delirium:** This is the state of disorientation, where the stimulus is misinterpretated and often noted when patient comes out of unconscious state.
e. **Clouding of consciousness:** This is a state of confusion, distractibility, memory loss and slowed response to stimulus.

2. Cognitive Dysfunction

The head injury patients can have difficulties in the following cognitive functions. These problems can be temporary or permanent. The cognitive dysfunction is generally due to a focal or generalized lesions.

a. **Attention deficits:** The patients have difficulty in paying attention. The patient can be easily distracted.
b. **Memory loss:** The patient will have a temporary or permanent memory loss. The memory loss be loss of specific memories or inability to store new memories. The memory loss is termed amnesia. The head injury patients will exhibit three forms of amnesia:
   i. *Antegrade amnesia*: It is the inability to store new memory after the injury in future. This retards the patients in learning and remembering new things.
   ii. *Retrograde amnesia*: It is the inability of the patient to recollect events or specific memories that took place just before the trauma.
   iii. *Post-traumatic amnesia*: It is the loss of memory in the time lapse between the injury and the recovery.
c. **Speed of processing/executing the functions:** Patients with frontal lobe trauma in head injuries, will have problems in speed of processing and

executing the functions. These patients lack the ability in:
  i. Choosing a goal
  ii. Developing the plans to reach the goal
  iii. Executing the plans in line with the goal
  iv. Evaluation of the strategy used.
d. **Emotional changes:** If there is orbitofrontal area involvement due to trauma, the patient will have euphoria, intolerance, irritability, inappropriate sexual behavior and unacceptable social behavior. Involuntary expression of emotions in the form of crying or laughing may be present.
e. **Behavioral changes:** The patient may show deviation from the normal behavior and may exhibit the symptoms of depression, impulsiveness or hyperactivity.

3. Speech and Language Deficits

Depending upon the area of lesion, the patient will demonstrate the difficulties in speech which could be:
a. **Broca's aphasia:** The patients have difficulty in speaking, the speech is broken, unclear and syntax of language may be lost. It is also called non-fluent or motor aphasia.
b. **Wernicke's aphasia:** The patient will speak continue sentences. The language may be clear and syntax is present. The patient will not understand the spoken words. These patients usually will not stop talking or talks continuously without a pause. The speech will be non-essential. This is also known as fluent or sensory aphasia.
c. **Global aphasia:** In this type of aphasia, the patient neither would be able to understand the spoken words or able to speak clearly. It is due to extensive brain damage.
d. **Dysarthria:** The patient with lesions in basal ganglia will demonstrate this type of speech problem. In this, the patient will have difficulty in articulation of speech.
e. **Prosodic dysfunction:** In this type of speech disorder, the patient shall have difficulties with intonation or inflection. (Variation of spoken pitch that is not used to distinguish words; instead it is used for a range of functions such as indicating the attitudes and emotions of the speaker.)

4. Sensory Deficits

The sensory cortex involvement in the trauma results in faulty or impaired sensory interpretation. The patient may exhibit sensory impairment in general and special senses. The patient may also develop paresthesia or abnormal sensations. The patient may complain of ringing of bells in the ears (tinnitus) or a persistent bitter taste or an unpleasant smell. Common problems of sensory dysfunction would be but not limited to:
a. Difficulties with interpretation of touch, temperature, movement, limb position and fine discrimination.
b. Difficulty in the integration or patterning of sensory impressions into psychologically meaningful data.
c. Visual disturbances like partial or total loss of vision, weakness of eye muscles and double vision (diplopia), blurred vision, problems in judging the distance, involuntary eye movements (nystagmus) intolerance of light (photophobia).
d. Auditory disturbances like decreased capability to hear or loss of hearing, ringing in the ears (tinnitus) increased sensitivity to sounds.
e. Olfactory impairments like loss or diminished sense of smell (anosmia).
f. Gustatory disability like loss or diminished sense of taste or feeling of abnormal tastes for a prolonged period of time.

5. Motor Deficits

The motor deficits are common and depend upon the site and degree of injury to brain tissue. The patient may exhibit decerebrate rigidity or decorticate rigidity (Fig. 4.3).

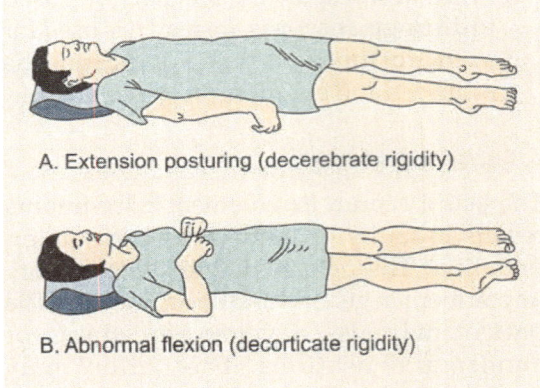

Fig. 4.3: Types of rigidity

Decorticate rigidity leads to adduction and flexion of upper extremities and internal rotation of the lower extremities with plantar flexion of feet.

Decerebrate rigidity leads to extension and lateral rotation of the upper extremities and plantar flexion of the feet.

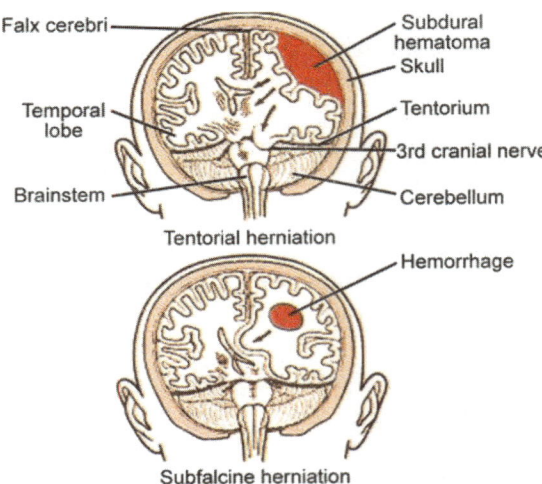

Fig. 4.4: Types of herniation

Motor deficits could be in the form of hemiplegia, monoplegia with either presence of spasticity or flaccidity.

There will be gradual progress from flaccidity to rigidity and can be unilateral or bilateral.

ASSESSMENT OF PATIENTS WITH HEAD INJURIES

The patients with head injuries need a careful and detailed examination. There is no fixed protocol for assessing a head injury patients because of its diverse clinical presentation. Often the patient will be uncooperative during the examination. The assessment of head injury may not be completed in a single session, a periodic examination would be required.

1. Take the demographic data of the patient either by direct interview or from the medical records.
2. Take a detailed history of the patient. It should include but not limited to:
   - Type of injury
   - Location
   - First aid and technique of carrying the patient.
   - Duration of unconsciousness, present of lucid interval
   - External bleeding and leak of CSF through the nose.

If the patient is in unconsciousness state, then these details are obtained from the medical records.

The therapist should also take the past history of any trauma to head.

Observation

- Observe the general built of the patient
- Observe for the abnormal posture (decerebrate or decorticate)
- If the patient is on artificial ventilation, observe and note the parameters.

- Observe the various tubings like catheter, Ryle's tube, etc.
- Observe for discoloration of skin, raccoon's eye (bluish discoloration around the eyes or ecchymosis in the periorbital area) (Fig. 4.5).

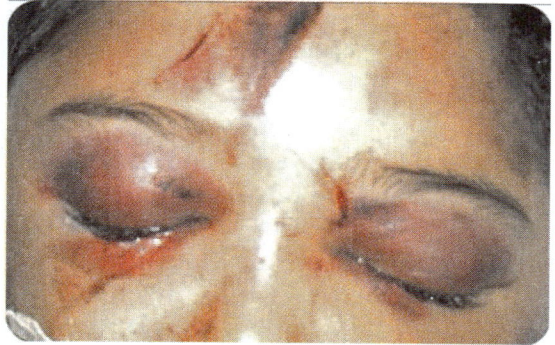

Fig. 4.5: Raccoon's eye

- Observe for site and extent of scar tissue.
- Observe for bruises anywhere else in the body
- Observe for swelling elsewhere to identify the associated injuries.

Examination

Higher functions: Examine the higher functions—memory, intelligence, level of consciousness, behavior, orientation and speech.

The level of consciousness is assessed by GCS (Table 4.1).

Interpretation

Generally, brain injury is classified as:[4-6]
- Severe, with GCS <8–9
- Moderate, GCS 8 or 9–12.
- Minor, GCS ≥13.

Examination of cranial nerves: Examine the cranial nerves in craniological order and report. Most commonly affected nerves in head injuries are facial, abducent, vestibulocochlear and optic nerves.

Sensory examination: Sensory examination would be very important to identify the level of lesion and type of lesion. A thorough examination of sensory system—superficial, deep and cortical senses should be evaluated.

Reflexes: Deep and superficial reflexes are evaluated. Apart from these clinical reflexes, primitive reflexes may be set in like ATNR, STNR, etc.

Motor Examination

Tone: The tone is examined by simple passive movement. The tone is assessed by Modified Ashworth scale of spasticity (Table 4.3).

Power/voluntary control: The muscle strength is graded by manual muscle testing. If there is spasticity, then voluntary control is graded.

Range of motion: All the joints are measured for active and passive range of motion by goniometry.

Table 4.1: Glasgow Coma Scale						
	1	2	3	4	5	6
Eye	Do not open eyes	Open eyes in response to painful stimuli	Opens eyes in response to voice	Opens eyes spontaneously	–	–
Verbal	Makes no sounds	Incomprehensible sounds	Utters inappropriate words	Confused, disoriented speech	Oriented, converses normally	–
Motor	Makes no movements	Extension to painful stimuli	Abnormal flexion to painful stimuli	Flexion/withdrawal to painful stimuli	Localizes painful stimuli	Performs voluntary action on command

Table 4.2: Rancho Los Amigos Level of cognitive function

Level	Response	Description
Level I	No response	Patient does not respond to external stimuli and appears asleep
Level II	Generalized response	Patient reacts to external stimuli in nonspecific, inconsistent, and non-purposeful manner with stereotypic and limited responses
Level III	Localized response	Patient responds specifically and inconsistently with delays to stimuli, but may follow simple commands for motor action
Level IV	Confused, agitated response	Patient exhibits bizarre, non-purposeful, incoherent or inappropriate behaviors, has no short-term recall, attention is short and nonselective
Level V	Confused, inappropriate, non-agitated response	Stimuli: Simple commands are followed consistently, memory and selective attention are impaired, and new information is not retained
Level VI	Confused, appropriate response	Patient gives context appropriate, goal-directed responses, dependent upon external input for direction. There is carry-over for relearned, but not for new tasks, and recent memory problems persist
Level VII	Automatic, appropriate response	Patient behaves appropriately in familiar settings, performs daily routines automatically, and shows carry-over for new learning at lower than normal rates. Patient initiates social interactions, but judgment remains impaired
Level VIII	Purposeful, appropriate response	Patient oriented and responds to the environment but abstract reasoning abilities are decreased relative to premorbid levels

Table 4.3: Modified Ashworth scale of spasticity

0	No increase in muscle tone
1	Slight increase in muscle tone, manifested by a catch and release or by minimal resistance at the end of the range of motion when the affected part(s) is moved in flexion or extension
1+	Slight increase in muscle tone, manifested by a catch, followed by minimal resistance throughout the remainder (less than half) of the ROM
2	More marked increase in muscle tone through most of the ROM, but affected part(s) easily moved
3	Considerable increase in muscle tone, passive movement difficult
4	Affected part(s) rigid in flexion or extension

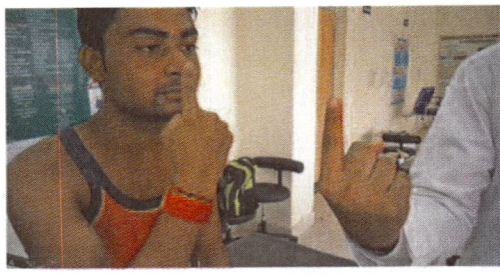 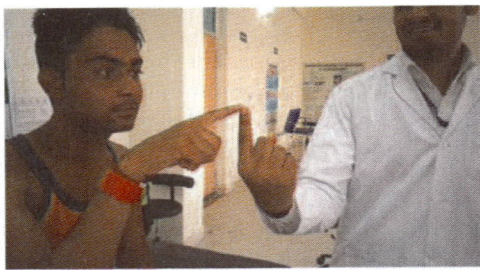

Fig. 4.6: Finger nose test for assessing co-ordination

Co-ordination: The co-ordination is tested by equilibrium and non-equilibrium tests.

Balance: Balance is examined by Berg Balance Scale.[9]

Synergy: The synergy is assessed and type is noted. Common synergies are shown in Table 4.4.

Table 4.4: Common synergies

Extremity	Flexion synergy	Extension synergy
Upper limb	Scapular retraction/elevation	Scapular protraction
	Shoulder abduction and external rotation	Shoulder adduction and internal rotation
	Elbow in flexion	Forearm pronation
	Forearm supinated	Wrist and fingers flexion
	Wrist and finger flexion	
Lower limb	Hip flexion, abduction and externally rotated	Hip extension, adduction and internal rotation
	Knee flexion	Knee extension
	Ankle dorsiflexion and inversion	Ankle plantar flexion and inversion
	Toes dorsiflexion	Toe plantar flexion

Gait Assessment

The gait parameters are evaluated and observed for changes. If a patient is functionally mobile, a brief gait assessment and if required wheelchair assessment is carried out.

Functional Assessment

Check for the functional capabilities of the patient which shall guide for modifications to be done in rehabilitation. Depending upon the assessment done, mat mobility, transfers, self-care activities like grooming, feeding, dressing and other activities of daily living, leisure activities should be assessed. The *Rappaport disability scale* can be used to assess the functional disability status.

Arousability Awareness and Responsivity

1. Eye Opening

0: Spontaneous: When the patient's eyes open up with the sleep/wake rhythms indicating active arousal mechanisms. This does not assume that the patient is aware.

1: To speech and/or sensory stimulation: When the eyes move in response to any verbal stimulation, whether the patient is spoken to or shouted at. This is not necessarily a command to open the eyes. Eyes can also open in response to a mild touch or pressure.

2: To pain: When the eyes open as a result of the patient feeling pain.

3: None: When the eyes will not open for anything—even painful stimulus.

2. Communication Ability

0: Oriented: This is when the patient is aware their surroundings. In this state, the patient can tell you basic facts about his/her location and other details of his/her life.

1: Confused: This is when the patient's attention can be held and he/she can answer questions. When answering questions, the answers may be delayed and/or indicate a level of disorientation or confusion.

2: Inappropriate: The patient is able to talk with intelligible articulation but nothing meaningful is said. Patient's speech is typically random or exclamatory. Having sustainable conversations with the patient is not possible.

3: Incomprehensible: Patient is able to make sounds such as groaning or moaning but is not able to make recognizable words. Conversations with the patient are impossible.

4: None: The patient displays no signs of communication or sounds whatsoever.

3. Motor Response

0: Obeying: The patient obeys commands such as "move your fingers". This also includes other commands such as "blink your eyes" or "move your lips". Grasping, reflexes, and other complicated movements should not be used.

1: Localizing: When the patient moves his/her limb (even a little bit) to move away from painful stimulus occurring on more than one point on that limb, there must be a deliberate motor act to move away from, or remove, the source of stimulation. This is very similar to withdrawing.

2: Withdrawing: When the patient moves away from a stimulus and exhibits more than a reflex response.

3: Flexing: When the patient flexes at the elbow and attempts to withdraw in a result of feeling a painful stimulus.

4: Extending: When the patient extends his/her limb after feeling a painful stimulus.

5: None: When the patient exhibits no response to stimulus whatsoever.

4. Cognitive Ability for Self-care Activities

I. Feeding:

0: Complete: When the patient continuously shows awareness about how to feed and the patient can convey the information that he/she knows when feeding should occur.

1: Partial: When the patient can sometimes show awareness that he/she knows how to feed and/or conveys information that he/she knows when feeding should occur.

2: Minimal: When the patient rarely shows awareness about how to feed and/or rarely shows that he/she knows when this is to occur. The patient can communicate desire to feed with certain signs, sounds, or activities.

3: None: Shows no awareness of how to feed or when to feed. The patient cannot convey any information by signs, sounds, or activity.

II. Toileting:

0: Complete: When the patient continuously shows awareness that he/she knows how to use the toilet and conveys information that he/she knows when this should occur.

1: Partial: When the patient can sometimes show awareness that he/she knows how to use the toilet and/or can convey information that he/she knows when the act should occur.

2: Minimal: When the patient rarely shows awareness that he/she knows how to use the toilet and/or rarely shows that he/she knows when this is to occur.

3: None: Shows no awareness of how to use the toilet or when he/she should go. The patient cannot convey any information by signs, sounds, or activity.

III. Grooming:

0: Complete: When the patient continuously shows awareness that he/she knows how and when to groom.

1: Partial: When the patient can sometimes show awareness that he/she knows how to groom and/or convey information that he/she knows when grooming should occur.

2: Minimal: When the patient rarely shows awareness about how to groom and/or rarely shows that he/she knows when this is to occur based on certain signs, sounds, or activities.

3: None: Shows no awareness of how to groom or when to groom. The patient cannot convey any information by signs, sounds, or activity.

Dependence on Others

Level of Functioning

0: Completely independent: The patient is able to live as he/she wishes without any restrictions regarding physical, mental, emotional, or social situations.

1: Independent in special environment: The patient is capable of living as he/she wishes, as long as certain requirements are met (such as mechanical aids).

2: Mildly dependent: The patient is able to care for most of her/his own needs but she/he needs a little help due to physical, mental, emotional, or social problems.

3: Moderately dependent: The patient can partially take care of himself/herself. In some cases, the patient may need another person there at times.

4: Markedly dependent: The patient needs help with all major activities and the help of another person at all times.

5: Totally dependent: The patient is not able to care for anything by himself/herself and requires 24 hours nursing care.

Psychosocial Adaptability

Employability

0: Not restricted: The patient can compete with others in a large variety of jobs that incorporate existing skills. The patient can also initiate, plan, execute, and assume responsibilities associated with homemaking. In addition, he/she can also carry out and complete most age relevant school assignments.

1: Selective jobs, competitive: The patient can compete with others in a limited variety of jobs that incorporate existing skills because of some types of limitation. He/she can also initiate, plan, execute, and assume responsibilities of some homemaking tasks. It is also possible for her/him to carry out and complete some, but not all age relevant school assignments.

2: Sheltered workshop, non-competitive: The patient cannot compete with others in any variety of jobs that incorporate existing skills because of moderate or severe limitation. He/she cannot, without major assistance, initiate, plan, execute, and assume responsibilities associated with homemaking.

In addition, the patient cannot carry out and complete age relevant school assignments without assistance.

3: Not employable: The patient is completely unemployable because of extreme limitations. He/she is completely unable to initiate, plan, execute, and assume responsibilities associated with homemaking. In addition, the patient cannot carry out and complete any age relevant school assignments.

- Score 0: Normal
- Score 1: Mild
- Score 2 to 3.5: Partial
- Score 4 to 6: Moderate
- Score 7 to 11: Moderately severe
- Score 12 to 16: Severe
- Score 17 to 21: Extremely severe
- Score 22 to 24: Vegetative state
- Score 25 to 29: Extreme vegetative state (or, if the person has a score of 29, possible death)

Respiratory assessment: It is very important to assess the respiratory functions of the patient. Check the breathing pattern, use of accessory muscles of respiration, VO_{2max}, auscultate for breath sounds. On percussion, there may be dimished sounds.

Autonomic dysfunction: Check the patient for postural hypotension, pulse in lying and standing and sitting postures and check whether the patient has any dizziness or any symptom related to postural hypotension.

PHYSIOTHERAPY MANAGEMENT

The physiotherapy management depends upon the status of the patient. The management goals depends upon the level of alertness, consciousness and ability to comprehend the commands.

Management of unconsciousness patient:
- To stimulate arousal response in the patient

- To maintain good bronchial hygiene
- To prevent secondary complications
- To maintain normal joint range of motion in all the joints

Management of conscious patient:
- To provide psychological support and develop good rapport with the patient
- To normalize the tone and improve muscle strength
- To improve the voluntary control
- To train in transfer techniques and aid in early ambulation
- To train for balance and equilibrium
- To enhance motor relearning and training
- To improve overall functional capabilities
- To prevent all possible secondary complications

Plans of Physiotherapy for Unconscious Patient

Stimulation of Arousal Response in Patients in Unconsciousness State

Stimulation of arousal response in patients in unconscious state is described in Table 4.5.

To Maintain Good Bronchial Hygiene

Prolonged unconsciousness, depressed respiratory centers depression or diaphragmatic dysfunction is associated with dyspnoea can decrease the life span. Ineffective cough can lead to respiratory infections like pneumonia (Senent et al, 2011).[10] The condition can become worse, if there is shortening and stiffness of chest muscles and reduced lung compliance.

The main role of physiotherapists in respiratory management is to closely monitor for any symptom of respiratory insufficiency and to prevent accumulation of secretions in the lungs.

Respiratory adjuncts, such as PEP, should not be indicated, instead manual assisted coughing should be used.

Regular postural drainage and deep breathing exercises are to be incorporated. Chest secretions may be mobilized by manual techniques like percussions, vibrations or mechanical assistive devices like mechanical exsufflator or a positive pressure device. *Head low position must be avoided in patients with head injuries due to risk of increasing intracranial pressure.*

Stretching of pectoral muscles, sternocleidomastoid could be helpful to improve chest mobility.

Neurophysiological stimulation (NPF) of respiration:[12] It is believed that cutaneous and proprioceptive stimulation reflexly increases the depth of breathing.

The perioral technique is thought to relate to the suckling reflex, and may facilitate slow as well as deep breathing. NPF also proven to improve coughing, swallowing and abdominal contractions.

Perioral technique: Moderate finger pressure is maintained inwards and downwards just above the lip as long as the patient is required to deep breathe. The effect may continue for some minutes afterwards.

Intercostal stretch: Apply pressure downwards towards the toes on the upper border of the rib at end, expiration, unilateral or bilateral but not on the floating ribs.

Cocontraction of abdominal muscles: Apply pressure laterally over the lower ribs and pelvis at right angles to patient, alternating right and left sides and maintain the pressure for up to 2 min or until desired effect is achieved.

Vertebral pressure: Finger pressure against thoracic vertebrae between T2 and T10.

Points to remember before starting chest physiotherapy in unconscious patients:
- Monitor physiological responses such as heart rate, BP, respiratory rate and oxygen saturation at all the times.
- The physiotherapist should be aware of effects of positioning and mobility of the

Table 4.5: Principles of management

Rationale	Principles	Techniques

• Affect the RAS and increase arousal and attention[11]
• Prevent environmental (sensory) deprivation, which has been shown to retard recovery
• Improve the quantity and quality of responses toward purposeful activity
• Provide opportunities for the patient to respond to the environment in an adaptive way
	• Do not harm the patient
	• Before starting any stimulation, check resting vital signs (heart rate, blood pressure, and respiratory rate)
	• Avoid or minimize stimulation programs with comatose patients that have a ventriculostomy when increased intracranial pressure (ICP) and/or cerebral perfusion pressure (CPP) are still issues; monitor ICP and CPP during and after treatment, if necessary
	• Control the environment to eliminate as many distractions as possible
	• Make sure the patient is as comfortable as possible
	• Organize the stimuli, present them in an orderly manner, and involve only 1 or 2 modalities of senses at a time
	• It is important to control how much and how often to provide stimulation, because patients can get accommodated the stimulation
	• Provide ample time for the patient to respond (because of slow information processing). 1 or 2 minutes between the administration of different stimuli is useful as an initial guide until the length of response delay is established
	• Keep sessions relatively brief—patients can usually tolerate up to 15–30 minutes
	• Conduct sessions frequently, allowing patients to respond several times daily
		• *Visual stimulation*
		  ◆ Provide a visually stimulating environment at the bedside, such as colorful, familiar objects, family photographs (labeled), and TV 10–15 minutes at a time
		  ◆ Provide normal visual orientation, by positioning patient upright in bed, in the wheelchair, etc. This also helps decrease complications of prolonged bedrest, such as pressure sores, breathing problems, osteoporosis, and muscle contractures
		  ◆ Eliminate distraction to allow patient to focus on visual stimuli, such as a familiar face, object, photos, and on a mirror
		  ◆ Attempt visual tracking after focusing is established, i.e. getting the patient to follow a stimulus with his/her eyes at it moves. Tracking usually begins in the center or midline
		• *Auditory stimulation*
		  ◆ Provide regular auditory stimulation at the patient's bedside. All hospital staff should be encouraged to speak to the patient as they work in the room or directly with the patient. An information sheet can be posted in the room with information about the patient's likes and dislikes
		  ◆ Permit only one person to speak at a time
		  ◆ Use radio, TV, tape recording of a familiar voice, etc. for 10–15 minutes at intervals throughout the day
		  ◆ Direct work for focusing and localizing sound and look for patient's response when you change the location of a sound, e.g. call the patient's name, clap your hands, ring a bell, rattle, whistle, etc. 5–10 seconds at a time
		  ◆ Avoid stimulation that evokes a startled response. This type of stimulation is counterproductive
		• *Touch stimulation*: Tactile input can be facilitatory (encourage a desired response) or inhibitory (discourage/interfere with a desired response). For example, pain and light touch to the skin tend to produce an inhibitory response, while maintained touch, pressure to the oral area, and slow

(Contd.)

Rationale	Principles	Techniques

	• Select meaningful stimuli, such as voice of family and friends, favorite music, cologne, etc. Stimuli that have emotional significance to the patient are usually more likely to elicit responses
	• Verbally reinforce responses to increase the likelihood of obtaining responses in later sessions
	• Try stimulating all the senses, and vary the stimuli in nature and intensity to maximize the possibility of increasing arousal
	• Avoid overstimulation
	• Include participation by family and significant others in the coma stimulation program
		stroking of the spine tend to produce a facilitatory response. The face, and especially the lips and mouth area, are the most sensitive
		  ♦ Use a variety of textures, such as personal clothing, blankets, stuffed animals, lotions, etc.
		  ♦ Use a variety of temperatures, such as warm and cold clothes or metal spoons dipped for 30 seconds in hot or cold water
		  ♦ Vary the degree of pressure—firm pressure is usually less threatening or irritating to the patient than light touch. Examples include grasping a muscle and maintaining the pressure for 3–5 seconds, stretching a tendon and maintaining the stretch for a few seconds, and rubbing the sternum
		  ♦ Use unpleasant stimuli, such as a pinprick, with caution. Avoid ice to face or body, as it may trigger a sympathetic nervous system response, i.e. increased blood pressure, heart rate, and salivation and decreased gastrointestinal activity
		• *Movement stimulation*
		  ♦ Use range of motion exercises, changes in body position such as a single or repetitive roll, a tilt table to bring the patient to a more upright position, and movement activities on a therapy mat
		  ♦ Watch for early physical protective reactions or delayed balance reactions during these activities
		• *Position stimulation*
		  ♦ Slow changes in position tend to be inhibitory, while faster movement patterns tend to facilitate arousal
		  ♦ Monitor the patient's blood pressure (and ICP if appropriate) during this stimulation
		  ♦ Use position changes that are meaningful and familiar, such as rolling, rocking in a chair or on a mat, and moving from lying down to sitting
		  ♦ Avoid spinning, which may trigger seizures, and mechanical input, such as raising and lowering the hospital bed, which has a little functional meaning and produces limited response

(Contd.)

Table 4.5: Principles of management (Contd.)

Rationale	Principles	Techniques

		• *Smell stimulation*
		 ◆ Use after shave, cologne, perfume, favored extracts, coffee grinds, shampoo, and favorite foods
		 ◆ Provide the stimuli for no more than 10 seconds
		 ◆ Avoid touching the skin with the scent, because patient may accommodate the scent and be less responsive to it
		 ◆ Use garlic and mustard as noxious stimuli
		 ◆ Avoid vinegar and ammonia because they irritate the trigeminal nerve
		 However, there may not be a response to smell stimulation because:
		 ◆ The olfactory nerve is the most commonly injured cranial nerve in TBI
		 ◆ Many TBI patients have tracheostomies, which eliminate the exchange of air through the nostrils and therefore inhibit the sense of smell
		 ◆ Patients have nasogastric tubes in place, which block one nostril and, therefore, decrease the sense of smell
		 • *Taste and oral stimulation*
		 ◆ Provide taste stimulation, unless patient is prone to aspiration—use a cotton swab dipped in a sweet, salty, or sour solution, but avoid sweet tastes, if the patient has difficulty managing oral secretions since sweet tastes increase salivation
		 ◆ Provide oral stimulation during routine mouth care, unless patient demonstrates a bite reflex:
		   * Use a sponge-tipped or glycerin swab or a soft toothbrush to diminish hypersensitivity and abnormal oral/facial reflexes
		   * Use a flavored cleansing agent, such as mint or lemon, to increase oral stimulation during routine mouth care
		 ◆ Provide stimulation to the lips and area around the mouth. If patient demonstrates defensiveness to touch, such as pursing lips, closing mouth, or pulling away from the stimulus, gently continue with stimulation techniques to decrease defensive reactions and increase level of awareness. Do not attempt feeding of patients in coma.

To Prevent Secondary Complications

Musculoskeletal complications	• Muscle weakness and atrophy Decrease of muscle strength • 10–15% loss of strength/weakness • Decrease of muscle endurance • Loss of muscle mass—atrophy • Contracture • Because of the immobilization, normal muscle tissues are replaced by collagen—leads to contracture *Changes in joints* • Reduction in normal biomechanical characters in ligament • Immobilization can cause fibro-fatty infiltration in joints that can mature into strong adhesion within the joints and might destroy cartilage • There is a cross-link of collagen between the bony ends *Disuse osteoporosis* • Bone morphology and density depend on forces that act upon the bone, such as the direct pulling action of tendon and weight bearing • Immobilization leads to bone mass loss and hypercalciuria • Trabecular bone (spongy bone) loss is more than the cortical bone (compact bone) loss • Disuse osteoporosis can lead to fractures of spine vertebrae, femur, and distal radius.
Cardiovascular complications	• Increased heart rate and decreased cardiac reserve • Heart rate increases generally more than 80 beats/min because of immobilization • Resting pulse rate speeds up one beat each minute every 2 days • The increased heart rate results in less diastolic filling time • Orthostatic hypotension • Occur when the cardiovascular system does not adapt normally to an upright posture • It occurs after 3 weeks of bedrest, because of the excessive pooling in the lower extremities and a decrease in circulating blood volume. • This always along with a rapid heart rate • Venous thromboembolism • Most deep vein thrombosis (DVT) occurs in lower leg • This is due to primarily to venous stasis and to a lesser degree increase in blood coagulability. Stasis occurs in legs following decreased contraction of the gastrocnemius and soleus.
Respiratory/pulmonary complications	• Decreased ventilation • Because of the immobilization, there is a reduction of tidal volume • Loss of strength in respiratory muscle leads to reduction in chest expansion, later it may end up in respiratory failure • Atelectasis

(Contd.)

Head Injuries and Comatose Patients

Other complications	• Immobilization can result in marked impairment in clearing the secretion • Result in accumulation of secretion in bronchial tree • Hypercalcemia—increased calcium in blood • Renal stones—because of the hypercalciuria, urinary stasis often causes stones to form in kidney and bladder • Anorexia—loss of appetite due to the loss of physical activity • Constipation—result from decreased peristalsis and the less intake of fiber and fluid
Integumentary complications/pressure sores (Fig. 4.7)	• Pressure sores or decubitus ulcers are localized areas of cellular necrosis • They usually found over bony prominance • Pressure sores are categorised into 4 stages based on severity Grade 1: Reddened skin which persists for more than 30 minutes after pressure has been relieved Grade 2: Superficial skin damage. May present as a blister or as an abrasion Grade 3: Full thickness skin loss not extending to bone or muscle. This grade is not usually painful. Grade 4: Full thickness skin loss with extensive tissue damage through muscle and bone.

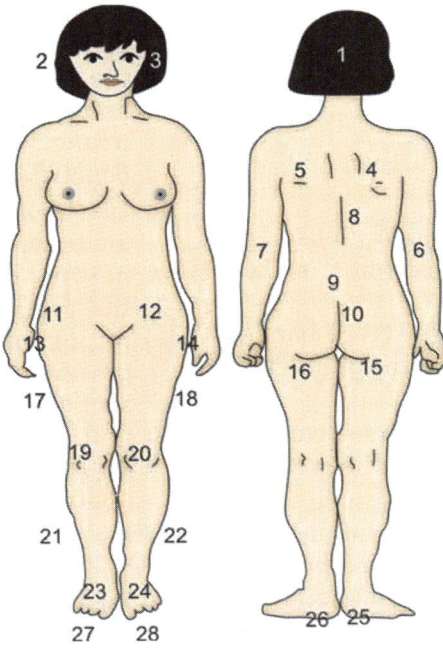

Fig. 4.7: Common areas to develop pressure sores: 1. back to head, 2. right ear, 3. left ear, 4. right scapula, 5. left scapula, 6. right elbow, 7. left elbow, 8. vertebrae (upper-mid), 9. sacrum, 16. left ischial tuberosity, 17. right thigh, 18. left thigh, 19. right knee, 20. left knee, 21. right lower leg, 22. left lower leg, 23. right ankle (inner or outer), 24. left ankle (inner or outer), 25. right heel, 26. right toe/s, 27. Left toe/s

patient on various monitoring devices and their reading.
- The physiotherapist should always deal with the patient as if he/she was conscious and wake, even if the patient appears not to be.
- Frequency and intensity of treatment sessions will be determined by the patient condition, but should generally be at least twice a day.
- Treatment should be carried out at least 1½ hr after feeding time.
- The physical therapist must be aware of patient's medication, lab investigations, patient treatment by other health care team, patient's/family concerns.
- The therapist should be aware of all the equipment used in the ICU.

Management of complications: *To avoid contractures and deformities concentrate on the following*:
1. Passive ROM of all joints and prolonged stretching. Each movement is repeated at least 8–10 times for every 3–4 hours.
2. Use of splints and proper positioning and frequent change of positioning would be helpful.

To prevent circulatory complications: To prevent DVT and swelling, concentrate on the following:
- Passive movements, elastic crepe bandaging, compression unit, and limb elevation.
- Faradic stimulation of muscles would also help in proper venous drainage.
- Gentle massage techniques can be very helpful in relieving edema, stimulating the muscles and also arousal responses in the patients by tactile stimulation.

To prevent respiratory complications:
1. Pre-treat with bronchodilator, if the patient has severe bronchospasm (20 min before treatment)
2. Modified postural drainage positions, usually with the head of the bed flat unless patient has an increase in intracranial pressure above 30 mmHg, then head of the bed should be elevated to 30°.
3. If there are no other contraindications, then the following should be done by two therapists or one therapist with an assistant.
4. Turn the patient to both sides and manually hyperventilate the patient using "ambu" and hyperoxygenate using 10–15 L oxygen; if the patient cannot be taken off ventilator, set the ventilator FIO_2 200%.
5. Use pulmonary hygiene techniques to mobilize secretions such as vibration, percussion, rib springing and shaking.
6. Endotracheal suctioning to clear retained secretions using sterile techniques.
7. The best position for relaxation, decreased dyspnea and improved ventilation and oxygenation are with the head of the bed elevated to 30° and lying on well-aerated lung. The prone lying is also proven to be beneficial.
8. Use tracheal tickle technique to elicit a cough, if not successful, then use nasopharyngeal suctioning to clear the retained secretions. It is very important to hyperoxygenate the patient with 10–15 L O_2 prior to suctioning to avoid complications.
9. If the patient has a tracheostomy, then manually hyperventilate and oxygenate the patient before suctioning.

To maintain good skin integrity and prevent pressure sores: Pressure sores are deep wounds due to constant and continuous pressure on the skin overlying the bony prominences in the body. It is the very common, painful complication of patients who are on the prolonged bedrest. The pressure sores can aggravate the existing disability. It should be noted that the development of pressure sores is solely due to negligence of the caregivers and health providers but not accidental. Hence, the health care providers and caretakers should take all necessary precautions to prevent the pressure sores.

Precautions to prevent pressure sores:
1. Closely monitor all the areas of risk daily.
2. Frequent change in the position of the patient.
3. Small rubbing friction massage with a moisturizer for a small duration (15 sec) over the areas of risk shall improve the skin elasticity and pliability.
4. Use of water mattress or air bed.
5. Proper bed making technique, avoid creases on the bed cover.

Even after taking necessary precautions, if the pressure sores appear to develop, then curative measures to be taken:
1. At redness stage, stop the massage, start changing the positions frequently and take use of adaptive devises
2. At blister stage frequent dressing to be done. Hydrocolloid dressing would be helpful.

If an ulcer has already formed, debridement of dead tissues, and dressing of the wound is done. UVR therapy, ultrasound has been found useful to promote healing.

Deep vein thrombosis is another major complication of immobility. It can be prevented by leg lifts, ankle pumping movements, toe curls, stretching and taking plenty of fluids would be helpful.

To Maintain Normal Joint Range of Motion in All the Joints

Achieving adequate joint ROM is necessary for functional independence and ambulatory training. The joint range of motion is maintained and improved by stretching exercises. Use of appropriate orthotic supports would be useful to maintain the joint range. Strengthening exercises of weak muscles would be necessary to prevent the contractures and deformities in order to maintain full range of motion. It is mostly observed that shoulder elevation, scapular depression, scapular retraction, elbow extension, supination, hip extension, ankle dorsiflexion and great toe extension are the common movements that are impaired and result in contractures. In some cases, tightness of muscles may be preserved for functional independence. It should be kept in mind that aggressive ROM exercises are contraindicated, if the patient has unstable fractures, active heterotropic ossification and deep vein thrombosis may occur. In case of spastic muscles, daily stretching and terminal sustained stretch is to be considered for rehabilitation for limitation of joint range of motion.

Physiotherapy Plans in Patients in Conscious State

To Provide Psychological Support and Develop Good Rapport with the patient

Develop good rapport with the patient. Counsel the patient and explain the importance of physiotherapy. Provide moral, spiritual, social support to the patient. Encourage the patient to involve actively in all self-care activities.

To Normalize the Tone and Maintain Normal Muscle Properties

Hypotonicity:
- Ice brisk stroking over the muscle
- Hacking over the muscle
- Weight-bearing over a joint where the muscle is crossing
- Vibrations
- ENMS is found useful in developing the tone
- Suspension therapy and aquatic exercises
- Functional re-education exercises
- Muscle energy techniques.
- EMG biofeedback techniques

Hypertonicity:
- Cryotherapy in the form of prolonged icing—ice pack
- PNF techniques
- Passive stretching exercises
- Orthotic supports to maintain the length of the muscle

To Improve the Voluntary Control

A voluntary movement is a purposeful, goal-oriented movement often learned and improved with practice. The organization of voluntary movement is as follows:
- Purpose/motivation to do a movement
- Order of movement (each set of movement prepares body for next set)
- Direction
- Rate of movement
- Strength
- Timing: One movement relative to another
- Duration of each movement:
    - Phasic (transient, discrete movements)
    - Static (e.g. stabilize joints)
- Initiate (start) movement
- Postural adjustments (+ feedback)
- Medial limb movement (+ feedback)
- Distal limb (+ feedback)
- Adjustments at every point
- Stop movement

To develop a control on the voluntary movement, normalizing tone, improving the strength in the muscles, correct pattern of movement, coordination and normal timing are taught.

PNF techniques, strengthening exercises, biofeedback, neurodevelopmental approaches would benefit the patient in acquiring a voluntary control.

The therapist should also try to minimize the effects of primitive reflexes, as they hinder the voluntary control. Reflex inhibiting postures, desensitizing techniques would be beneficial to decrease the impact of primitive reflexes.

Functional re-education exercises, vestibular ball exercises would also promote control over the voluntary movement.

Repetition of activities and functional electrical stimulation, kinetic exercises would accelerate the voluntary control over the movement.

Reversing tasks in some patients help in developing increasing control, e.g. placing the spoon onto the table would help in developing the control over taking the spoon to the mouth by improving the motor control of biceps during eccentric concentration.

The activities are modified frequently by inducing real environment settings to develop adaptation strategies.

To Aid in Early Ambulation and Transfer Activities

These are taught from and to the wheelchair and the bed, plinth, bath/lavatory seat and motorized vehicle. Transfer from a mat to wheelchair and vice versa are also taught for various treatment sessions. Sitting up from a lying position is a prerequisite for independent dressing and transfers. A thorough evaluation should be conducted to determine the most appropriate transfer technique for any individual with paralysis. Dependent transfers include sliding transfers and dependent standing pivot technique or use of Hoyer lifts. Transfers that require some active patient participation include the two men lift, sliding board transfer or assisted pivot transfer. Always remember, the aim of assisted transfer is to gradually reduce the assistance required until the patient can perform the transfer independently. Floor to chair transfer training is very important for anyone who falls out of wheelchair or otherwise ends up on the floor and needs to get back to the chair.

Principles of training of transfer technique for physiotherapists:
- Thoroughly assess the patient's ability and preparations for resettlement for transfer activities.

- Functional charts useful to record these activities which could be useful to check the prognosis and also help to stimulate the patient's interest.
- Correct positioning of the chair, ensuring that the brakes are fully on and lifting the legs with the hands on plinth, placement of legs on foot rest would be the principles of safe training.
- During the transfer techniques, care must be taken not to knock the leg or drag them along a hard surface.
- Check that the tyres of wheelchair are not worn out and the floor should not be slippery.

Gait training (Fig. 4.8): Before starting the gait training to the patient, identify the impairments that primarily determine walking ability of the patient. This will help the therapist to develop effective gait training strategies.

The gait training protocol can be framed based on ICF (International Classification of Functioning) model:
1. Identify the primary factors resulting in gait problems.
2. Selection of appropriate walking ability related outcome measures.
3. Development of tailored made training programs.
4. Identification of potential environmental or personal factors that facilitate or impede individual's walking abiity.
    - Improve muscle strength particulary hip abductors, hip flexors, knee extensors, ankle dorsiflexors, plantar flexors, hip extensors and knee flexors.
    - Develop balance and equilibrium reactions by training of tilt board, wobble board, vestibular ball exercises.
    - Determine the individual status of independency and determine the type of assistance required like cane, forearm crutch, standard walker, rolling walker, rollator, etc. (Fig. 4.9).
    - Train for forward walking first in parallel bars then with walking aids and progress to independent walking.
    - Train sideways and backward walking.

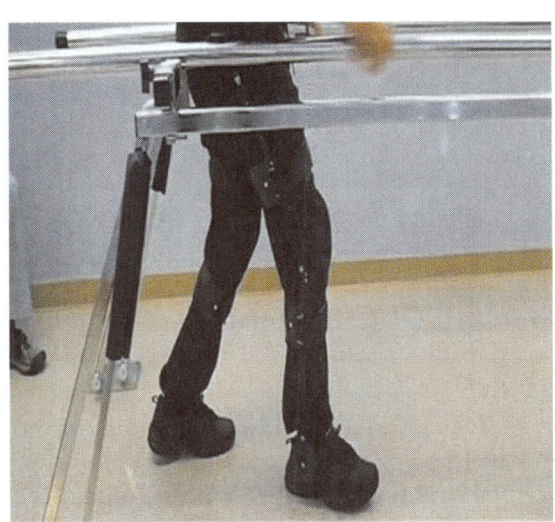

Fig. 4.8: Gait training

In head injury patients, the most common impairments that cause hindrance to patient ability to ambulate are decreased muscle strength, impaired voluntary control, balance and equilibrium.

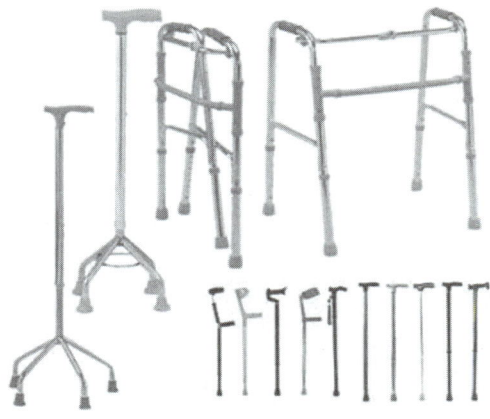

Fig. 4.9: Walking aids

- Teaching turning techniques—90° and 180° turning.
- Once the patient is confident, progress to stair climbing and stair descending first with support and independently.
- Advance the progression of gait by negotiating stairs without railing, walk on uneven surfaces, carrying the objects when walking.

To Train for Balance and Equilibrium

Thoroughly assess the balance by Berg Balance Scale and identify the strategies that are impaired. Here are some exercises to improve balance and equilibrium (Figs 4.10 to 4.13).

One foot standing training: This exercise shall improve ankle strategies and proprioception.

Fig. 4.12: Balance training

Fig. 4.10: Balance training

Fig. 4.11: Balance with out-reach activities

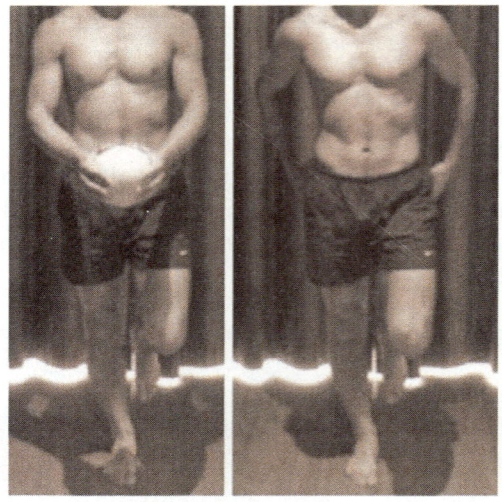

Fig. 4.13: Balance training in standing

Procedure: Ask the patient to stand on a firm surface ideally a mat. The feet should be apart at a minimum distance of shoulder width.

Now ask the patient to lift one leg off the ground and maintain balance on the other limb and maintain for 10 counts. Repeat the exercise by changing the leg.

One foot standing balance with hip 90° flexion: This exercise shall improve the ankle strategies and proprioception.

Procedure: Ask the patient to stand on a firm surface ideally a mat. The feet should be apart at a minimum distance of shoulder width.

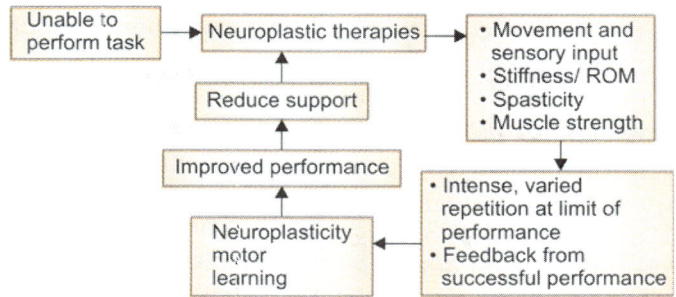

Fig. 4.14: The potential influence of neuroplastic interventions of functional performance

One foot standing balance by holding weights in a diagonal fashion: This exercise is intended to improve ankle strategies.

Procedure: Ask the patient to stand with moderate weight dumb-bell or weight cuff in hand in a diagonal fashion.

Now ask the patient to lift one lower limb.

One foot standing balance with forward bending: This exercise shall improve and train hip strategies.

Procedure: Make the patient in stride standing position.

Ask the patient to lift one lower limb and bend forward.

Balance board exercises: These exercises improve hip strategies.

Procedure:

1. Ask the patient to stand on balance board (stride standing)
2. Ask the patient to maintain the board off the ground on all the sides

Exercises to improve stepping strategies:

Exercise 1

1. Ask the patient to stand with feet apart facing a stepper or stool.
2. Ask the patient to step onto the stool or stepper with one foot
3. Now ask the patient to maintain balance while stepping up with the lead foot then brings up trailing foot.
4. Ask the client to repeat with changing the foot

Exercise 2

1. Ask the patient to stand with feet close to each other in front of the therapist.
2. Now the therapist laces his/her hands on the patient shoulders.
3. Ask the patient to lean forward into the therapist support as far as balance allows.
4. Now the therapist will quickly removes the hand which makes the patient to put a step forward.

To Enhance Motor Relearning and Training

The motor relearning is aimed at gaining functional independence through learning a specific task oriented movement. The motor control strategies have five components:

1. **Motor program:** It is asset of pre-set sequences to get a coordinated voluntary movement by muscle activation and is carried out without peripheral feedback.
2. **Motor planning:** It is strategical planning for a movement requiring coordination of various motor programs.
3. **Feedback:** It is the stimulation of control centers in brain by sending the information from peripheral receptors regarding the accuracy of movement and adaptations necessary.

4. **Feed forward mechanism:** It is the strategy of the musculoskeletal system to adapt or respond in anticipation of a movement or changes of an ongoing movement.
5. **Motor skill acquisition**: It is the training of a goal-oriented problem-solving strategy through development of motor programs and integration to form a motor plan and execute.

The various techniques incorporated to enhance neuroplasticity are:
- Constraint-induced movement therapy (CIMT)
- Body weight support treadmill training (BWSTT)
- Exoskeleton training

To Improve Overall Functional Capabilities

To improve functional independence, mat activities are taught depending upon the level of independence. Training activities done on adjustable therapy mat and are composed of sequenced activities that progress from the easiest to the most difficult. The usual progression is from bed mobility to rolling, prone lying, long sitting, short sitting, kneeling, half kneeling and standing. Muscles needed for individuals with lower limb paralysis to be able to move or position their legs in bed are wrist extensors, biceps, anterior deltoid, middle deltoid, and shoulder girdle stabilizers. Individuals with tetraplegia are taught to use their arms, head and neck for momentum to roll in bed with keeping the elbow straight while the shoulder is flexing across the body.

REFERENCES

1. Gururaj G. Epidemiology of traumatic brain injuries: Indian Scenario, Neurol Res. 2002;24(1): 24–28.
2. Esselman PC, and Uomoto JM . Classification of the spectrum of mild traumatic brain injury. Brain Injury, 1995;9:417–424.
3. Hunt T, Asplund C. Concussion assessment and management. Clinical Journal of Sports Medicine, 2020;29:5–17.
4. Teasdale G, Jennett B. "Assessment of coma and impaired consciousness. A practical scale". Lancet, 1974;2(7872):81–84.
5. World Health Organization. International statistical classification of diseases and related health problems (10th ed.). Geneva, Switzerland: World Health Organization 1992.
6. Malec JF, Brown AW, Leibson CL, Flaada JT, Mandrekar JN, Diehl NN, Perkins PK. The Mayo Classification System for Traumatic Brain Injury Severity. Journal of Neurotrauma. 2007;24(9): 1417–24.
7. Plum F. Coma and related global disturbances of the human conscious state. In: Jones, E and Peters, P, (Eds.) Cerebral Cortex, Vol. 9, Plenum Press, 1991;359–425.
8. Plum and Posner. Diagnosis of stupor and coma, 4th edition, Oxford University Press, 2007.
9. Blum, Lisa; Korner–Bitensky, Nicol. "Usefulness of the Berg Balance Scale in Stroke Rehabilitation: A Systematic Review". Physical Therapy, May 2008;88(5):559–566.
10. Senent C, Golmard JL, Salachas F, Chiner E, Morelot-Panzini C, Meninger V, Lamouroux C, Similowski T. Gonzalez-Bermejo J. A comparison of assisted cough techniques in stable patients with severe respiratory insufficiency due to amyotrophic lateral sclerosis. Amyotrophic Lateral Sclerosis, 2011;12:26–32.

5

Brain and Spinal Cord Tumors

LEARNING OBJECTIVES

At the end of this chapter, you will be able to:
1. Differentiate benign and malignant tumors.
2. Identify various tumors of CNS.
3. List out the various etiological factors of CNS oncology.
4. Identify the various clinical manifestations of the tumors of CNS.
5. Demonstrate the skills to assess the patients with CNS tumors.
6. Design the treatment protocols for rehabilitating these patients.

INTRODUCTION

- An abnormal growth of tissue resulting from uncontrolled, progressive multiplication of cells and serving no physiological function is called a tumor.[1]
- Neoplasia literally means new growth. It is an abnormal mass of tissue, the growth of which exceeds and is uncoordinated with that of the normal tissues and persists in the same excessive manner after the cessations of the stimuli which evoke the changes.
- Neoplasms behave as parasites and compete with normal cells and tissues for their metabolic needs. The study of neoplasm is referred to as "oncos", tumor and "logy", study of, that is oncology.
- Primary brain cancers account for 2% of all the cancers and is a leading cause of cancer-related death in younger patients even in developed countries.
- Primary brain tumors account for 20% of the cancers in children.
- Brain and spinal cord tumors are found in the tissue inside the skull or the bony spinal column which make-up the central nervous system.
- There are about 120 types of brain and spinal cord tumors. They are nomenclated based on the type of cell they are originated from, e.g. glioma from glial cells or based on the location, e.g. meningioma in the meninges of the brain or spinal cord.

Differences between malignant and benign tumors are shown in Table 5.1.

GRADING OF TUMORS[2]

- **Grade I:** The tissue is benign. The cells look nearly like normal brain cells, and they grow slowly.

Table 5.1: Difference between malignant and benign tumors

Malignant	Benign
• Fast growing	• Slow growing
• Non-capsulated	• Capsulated
• Invasive and infiltrate	• Non-invasive
• Metastasize	• Do not metastasize
• Poorly differentiated	• Well-differentiated
• Suffix "carcinoma" or "sarcoma"	• Suffix "oma", e.g. fibroma

- **Grade II:** The tissue is malignant. The cells look less like normal cells than do the cells in a Grade I tumor.
- **Grade III:** The malignant tissue has cells that look very different from normal cells. The abnormal cells are actively growing.
- **Grade IV:** The malignant tissue has cells that look most abnormal and tend to grow quickly.

TYPES OF BRAIN TUMORS[3,4]

Primary Brain Tumors (Figs 5.1 and 5.2)

These are tumors that result from the transformation and abnormal growth of brain cells that originate in the brain. Most of the primary tumors are benign and are excised easily but some may be malignant leading to metastasis. The common types of primary tumor are as shown in Table 5.2.

Secondary Brain Tumors

Secondary brain tumors are also known as metastatic tumors where the cancer cells from

Table 5.2: Common types of primary tumors

Astrocytoma	Arises from the astrocytes. Mostly arises from the glial cells of cerebrum Grade I and II: Low grade glioma Grade III: Anaplastic astrocytoma Grade IV: Glioblastoma or malignant astrocytic glioma
Meningioma	It is the tumor that arises in the meninges. It is usually benign and grows slowly
Oligodendroglioma	It arises from the cells oligodendrocytes which are responsible for myelinating the nerve fibers
Medulloblastoma	This is common in children and arises from the cells of cerebellum
Schwannoma	The tumor arises from the Schwann cells. The most commonly affected is the vestibulocochlear nerve. This tumor is also known as acoustic neuroma
Craniopharyngioma	The tumor arises at the base of the brain near the pituitary gland and is common in children
Hemangioblastoma	Rare tumors that develop from the endothelium of blood vessels supplying the brain. They are benign and grow slowly
Pituitary tumors	These are benign tumors and are generally called pituitary adenomas
Germ cell tumor of the brain	The tumor arises from the germ cells and is common in people younger than 30 years. The most common tumor is germinoma.
Pineal region tumors	These tumors are rare and arises in or close to the pineal body which is located between cerebrum and cerebellum. The most common tumors are germinomas, teratomas, pineocytomas and pineoblastomas

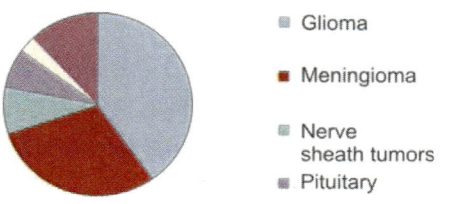

Fig. 5.1: Primary brain tumors

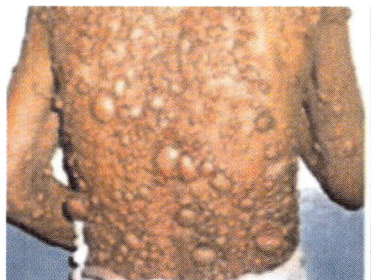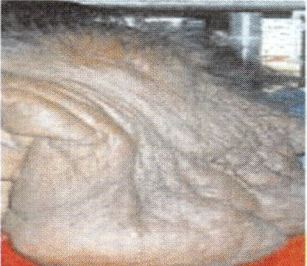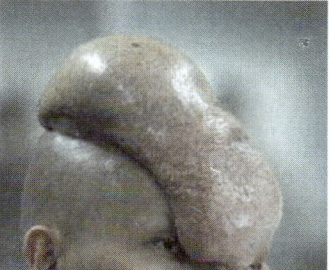

Fig. 5.2: Primary tumors of nervous system

other parts of the body, such as lung, breast, etc. can spread to the brain. The secondary brain tumors are most common than primary tumors in brain, almost about 25% of tumors are found to metastasize to the brain.

Spinal tumors: These are tumors that are located in the spinal cord. Depending upon the location, the spinal cord tumors are classified as

1. *Extradural tumors*: These tumors are located outside the dura mater of the spinal cord and are most common spinal cord tumors.
2. *Intradural tumors*: Within the dura mater and may be intramedullary (inside the spinal cord) and extrameduallry (inside the dura mater but outside the spinal cord).

Etiology

The common risk factors for developing brain tumors are:

1. **Age:** Brain tumors are more common in children and older adults, although they can arise in any age.
2. **Gender:** Men are more likely to develop brain tumors than women, meningiomas are more common in women.
3. **Occupational hazards:** Exposure to solvents, pesticides, oil products, rubber or vinyl chloride may increase the risk of developing a brain tumor.
4. **Hereditary factors:** 5% of brain tumors are linked with genetic factors including Fraumeni syndrome, neurofibromatosis, etc.
5. **Exposure to infections:** Infection with the Epstein-Barr virus increases the risk of CNS lymphoma.
6. **Electromagnetic fields:** Electromagnetic fields like energy from power lines or from cell phone have no direct effect on development of tumors but WHO recommends limiting cell phone use.
7. **Ethnicity and race:** US, European people have more risk of developing gliomas.
8. **Ionizing radiations:** Previous treatments to the head or brain with ionizing radiation, including X-rays can cause a brain tumor.
9. **N-nitroso compounds:** Some studies of diet and vitamin supplementation seem to indicate that dietary N-nitroso compounds may raise the risk of development of child and adult brain tumors. Cigarette smoke, cured meats, cosmetics can cause accumulation of N-nitroso compounds in the body.
10. **Exposure to nerve agents:** Some gulf war veterans have an increased risk of a brain tumor from exposure to nerve agents. Nerve agents are a class of phosphorus-containing organic chemicals that disrupt the mechanism by which neural messages are transferred to the organs.

Clinical Features

The symptoms of brain and spinal cord tumors depend upon the tumor type, location, size and rate of growth.

The symptoms usually develop slowly and worsen as the tumor grows.

The common symptoms of brain tumors in adults are:

- **Headache and seizures:** Headaches are the most common symptom of a brain tumor. Headaches may progressively worsen, become more frequent or constant and reoccur at irregular intervals. The headache worsens on coughing, changing posture or walking.
- Seizures can occur which can be demonstrated as convulsions, loss of consciousness or bladder control. Seizures that are demonstrated in adults for the first time without an history of previous illness or accident is a key sign of brain tumors.
- **Difficulty with language:** The patient may develop aphasia or dysarthria depending upon the location of the tumor.
- **Mood changes and change of personality:** Personality, behavior and cognitive changes can include psychotic episodes and can lead to mood swings.
- **Changes in vision, hearing, and sensation:** Vision or hearing problems include blurred or double vision, partial or total loss of vision, tinnitus in the ears and abnormal eye movements (nystagmus).
- **Motor deficits:** There may be motor deficits leading to the weakness or paralysis of muscles. Development of abnormal movement patterns (synergy) may be seen.
- **Difficulty with coordination control:** There may be loss of coordination of voluntary movements.
- **Loss of balance:** There may be balance impairment along with dizziness, trouble with walking, clumsiness or loss of normal control of equilibrium.
- Signs of raised intracranial pressure are seen.
- Sensory impairments.
- Other symptoms include endocrine disorders or abnormal hormonal regulation, difficulty in swallowing, facial paralysis, sagging eyelids, fatigue, weakened sense of smell or disturbed sleep or sleep pattern changes.
- Alteration of consciousness state.

Symptoms depend on where the tumor is located:

- **Frontal lobe:** Muscle weakness, contusion, etc.
- **Temporal lobe:** Seizures, types of aphasia, etc.
- **Occipital lobe:** Visual problems
- **Parietal lobe:** Loss of sensation

Spinal Cord Tumor Symptoms

The spinal cord tumors may include motor, and sensory problems. Pain, numbness and loss of motor control are the common symptoms.

Back pain radiating to hips, legs, feet and arms and is progressive burning or aching quality.

There may be numbness or sensory changes including decreased skin sensitivity to temperature and progressive numbness or a loss of sensation depending upon the level and location of lesion.

Diagnosis

The diagnosis of the tumors of brain and spinal cord are done by the findings from:

1. Detailed history of clinical manifestations and familial history and exposure to carcinogens, occupational history is taken.
2. **Neurological clinical examination:** A detailed clinical examination that includes higher functions examination, cranial nerve examination, examination of sensory system, reflex testing and motor examination.

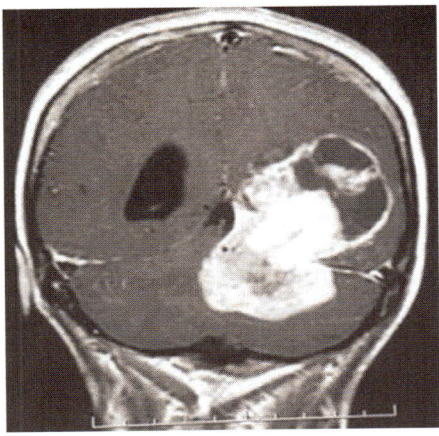

Fig. 5.3: MRI scan

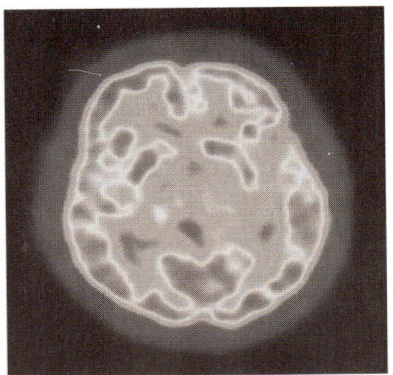

Fig. 5.5: PET scan

3. **MRI scan** (Fig. 5.3): MRI scans are used to measure the tumor's size. A special dye called contrast medium is given before the scan to create a clearer picture. The MRI scan is obtained for brain, spinal cord or both depending upon the clinical examination results and area of suspecting tumor.

4. **CT scan** (Fig. 5.4): A CT (computed tomography) scan creates a three-dimen-sional picture of the inside of the body with an X-ray machine. A CT scan is advised when an MRI is not advisable, e.g. patients having pacemaker. A contrast dye CT scan can also be done, if required.

5. **PET scan** (Fig. 5.5): Positron emission tomography is a way to create pictures of organs and tissues inside the body. A small amount of radioactive sugar substance is injected into the patient's body. As the cancer cells tend to absorb more radioactive substance, a scanner detects this substance to produce images.

6. **Cerebral arteriogram/cerebral angiogram:** A cerebral arteriogram is an X-ray or series of X-rays of the head that shows the arteries of the brain. A series of X-rays are taken after a contrast medium is injected into the main arteries of the patient's head.

7. **Lumbar puncture** (Fig. 5.6): A lumbar puncture is a procedure in which the CSF is drained to look out for tumor cells.

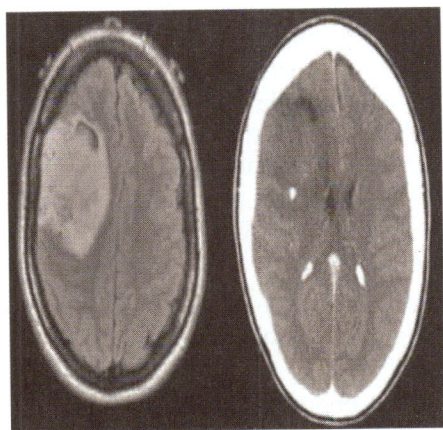

Fig. 5.4: CT scan

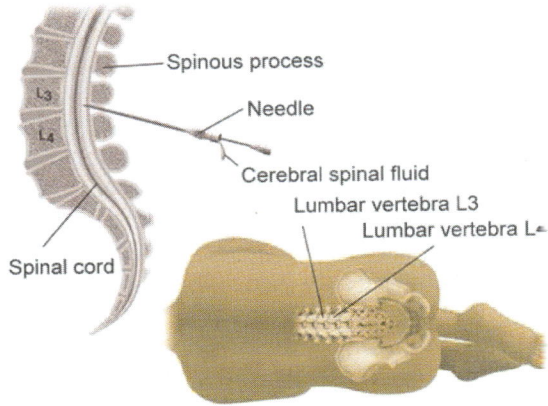

Fig. 5.6: Lumbar puncture

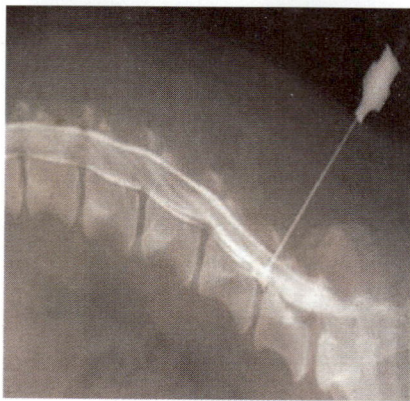

Fig. 5.7: Myelogram

8. **Myelogram** (Fig. 5.7): Some specific tumors can spread to the spinal fluid or other parts of the brain or spinal cord. A myelogram uses a dye injected into the CSF and X-ray is taken.
9. **Biopsy:** Once detected, depending on where the tumor is located, a biopsy officially is used to diagnose cancer.

Management

The management of brain and spinal cord tumors widely depends upon the following factors:
 a. Size, type and grade of tumor
 b. Extent of damage on vital parts by pressure
 c. Metastasis
 d. Possible side effect
 e. Patient's overall health condition

The most common treatment options available are surgery, radiation therapy and chemotherapy. Apart from treatment involving to stop, eliminate the tumor, care must be given to relieve patient's symptoms and side effects from the therapy. The approach is called *palliative* or *supportive care*.

The palliative care focuses on reducing the symptoms, improving the quality of life and supporting the patient and their family members.

Corticosteroids, anticonvulsants, pain killers, physiotherapy, psychological counseling, nursing are the important aspects of palliative care.

For a low grade tumors, surgery may be the only treatment needed which usually involves complete excision of tumor.

For a higher grade tumors, surgery followed by radiation and chemotherapy may be required.

The main issue in successful treatment of brain tumors is blood–brain barrier which does not allow some chemotherapeutic agents entering those structures through the bloodstream.

Sometimes it becomes difficult to excise a tumor from brain or spinal cord due to its delicate nature of the tissue. Sometimes the radiation therapy can also damage the healthy tissue.

Surgery

The main aim of the surgery is to remove the tumor and some surrounding healthy tissue during an operation.

Common surgeries performed are craniotomy, cortical mapping image-guided surgeries.

Radiation Therapy[6]

It is the use of high-energy X-rays or other particles to destroy the cancer cells. It helps in slowing or stopping the growth of the tumor.

The radiation therapy usually given after the surgery along with the chemotherapy.

The most common type of radiation therapy is called external-beam radiation therapy in which the radiation is given from a machine outside the body.

The radiation therapy can also be given by using internal implants called internal radiation therapy or *brachytherapy* (Fig. 5.8).

The protocol usually designed as a specific number of treatments given over a set period of time.

External-beam radiation therapy can be directed at the tumor in the following ways as shown in Table 5.3.

Brain and Spinal Cord Tumors

Table 5.3: Radiation therapy

Conventional radiation therapy	Treatment location is determined by anatomical landmarks and X-rays
Three-dimensional conformal radiation therapy (3D-CRT)	The treatment location is determined by CT scan or MRI scan and a 3D model of the tumor is created on a computer and is used to aim the radiation beams directly onto the tumor
Intensity modulated radiation therapy (IMRT)	It is similar to 3D-CRT, and can deliver higher doses of radiation while giving less to the surrounding areas. The radiation beams are broken up to smaller beams and intensity of these smaller beams can be changed
Proton therapy	Here the protons are used in the radiation instead of X-rays. These are used when less radiation is required, e.g. base of the skull tumors, tumors near to optic nerve
Stereotactic radiosurgery	It is the use of a single, high dose of radiation given directly onto the tumor and healthy tissue is not involved. There are different types of stereotactic radiosurgery equipment: 1. A modified linear accelerator is a machine that creates high energy radiation, 2. A gamma knife that concentrates highly focused beams of gamma radiation on the tumor, 3. A cyber knife is a robotic device used in radiation therapy to guide radiation to the tumor target.
Fractionated stereotactic radiation therapy	The radiation is delivered with stereotactic precision but divided daily small doses called fractions given over several weeks. This technique is used in the tumors at sensitive areas like optic nerves or brainstem.

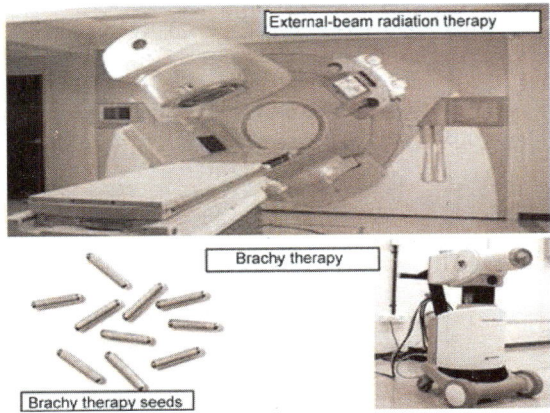

Fig. 5.8: Brachytherapy

Adverse effects of radiation therapy:
- Fatigue
- Skin reactions
- Hair loss
- Acidity and indigestion
- Neurologic deficits
- Memory deficits
- Hormonal deficiencies
- Cognitive deficits

Chemotherapy[8]

It is the use of certain drugs called chemotherapeutic agents that can have the capacity to destroy the cancer cells or slow down the ability to grow and divide.

Systemic chemotherapy gets into the bloodstream and reaches the cancer cells. Usually administered by oral or intravenous route.

Some chemotherapeutic agents are Gliadel wafers, temozolomide, procarbazine, lomustine and vincristine.

Adverse effects:
- Fatigue
- Risk of infection
- Nausea and vomiting
- Hair loss
- Loss of appetite and diarrhea
- Hearing loss and kidney damage

Targeted Therapy

In addition to the standard chemotherapy, targeted therapy is a treatment that targets the tumor's specific genes, proteins or the tissue environment that promotes growth and survival of tumor cells.

Anti-angiogenesis therapy is one type of targeted therapy used for brain tumors. The drug used is bevacizumab.

Alternating Electric Field Therapy

It is a noninvasive portable device which is given by placing electrodes that produce an electric field on the outside of a patient's head.

Physiotherapy assessment and management is same in line of head injuries or spinal cord injuries depending upon the deficits that are present.

REFERENCES

1. Dolliner M, Ko AH, Rosenbaum EH, et al. Understanding cancer. Ko AH, Dollinger M and Rosenbaum E (2008). Everyone's Guide to Cancer Therapy. How Cancer is Diagnosed. Treated and Managed Day to Day (5th Ed.), Kansas City: Andrew McMeel Publishing, 1: pp. 3–16.
2. National Cancer Institute. "Fact Sheet: Tumor Grade: Questions and Answers. Bethesda" National Cancer Institute, 2004.
3. Sobin LH, Gospodarowcz MK, Wittekind C (Eds.). TNM Classification of Malignant Tumours (7th Ed.), Wiley Blackwell, 2009.
4. Gurney JG, Smith MA, Bunin GR. "CNS and Miscellaneous Intracranial and Intraspinal Neoplasms" (PDF). SEER Pediatric Monograph, National Cancer Institute, pp. 51–57.
5. Allen Perkins et al., Primary Brain Tumours in Adults: Diagnosis and Treatment. Am. Fam. Physician, 2016;93(3):211–217B.
6. Stewart AJ, et al. "Radiobiological concepts for brachytherapy". In Devlin P Brachytherapy. Applications and Techniques, Philadelphia: LWW, 2007.
7. Mazeron JJ, Ardiet JM, Haie-Meder C, Kovacs GR, Levendag P, Peiffert D, Polo A, Roviosa A, Stmad V. "GEC-ESTRO recommendations for brachytherapy to head neck squamous cell carcinomas". Radiotherapy and Oncology 2009;91(2):150–56.
8. Charles B Wilson, et al. Current trends in the Chemotherapy of brain tumors with special reference to Glioblastomas, Journal of Neurosurgery, 1969;31(6): pp. 589–603.

6
Infections of Nervous System

LEARNING OBJECTIVES

At the end of this chapter, you will be able to:
1. List out common infections of nervous system.
2. List out the various etiological factors of various infections of nervous system.
3. Identify the various clinical manifestations of the infections to nervous system.
4. Demonstrate the skills to assess the patients with CNS infections.
5. Design the treatment protocols for rehabilitating these patients.

INTRODUCTION

Infections of the nervous system are among the most devastating diseases of the nervous system with a heavy death toll.

MENINGITIS

Meninges (Fig. 6.1) are the coverings of the brain and spinal cord—pia, arachnoid and dura mater. Meningitis is the disease caused by the inflammation of the meninges. The infection can be blood-borne or from local spread. The infection can be caused by the

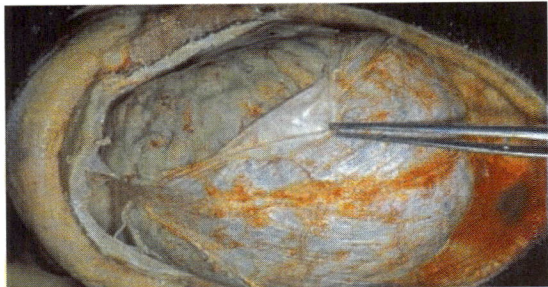

Fig. 6.1: Meninges of brain

viruses, bacteria, parasites and less commonly by certain drugs.

Types of Meningitis[1,2]

Pyogenic meningitis: Meningococcal meningitis is the most common form of meningitis and often occurs in epidemics.

H. influenzae and *E. coli* are the causative organisms in children, streptococcal and staphylococcal infections are common in adults. *Neisseria meningitidis* and *Streptococcus pneumoniae* can cause life-threatening infections.

Tuberculous meningitis: It is insidious infection caused by the *Mycobacterium tuberculosis* and is very common in the countries where tuberculosis is endemic and also in patients whose immunity is compromised like AIDS patients.

Viral meningitis: Enteroviruses, herpes simplex virus, varicella zoster virus, mumps virus, and HIV are the common viruses which cause meningitis.

Mollaret's meningitis is a chronic recurrent form of herpes meningitis.

Fungal meningitis: Cryptococcal meningitis caused by *Cryptococcus neoformans* is a common type of fungal meningitis. *Coccidioides immitis*, *Histoplasma capsulatum*, and *Candida* are the other species which can cause meningitis.

Parasitic meningitis: *Angiostrongylus cantonensis*, *Gnathostoma spinigerum*, Schistosoma, and toxocariasis are the common parasites that can cause meningitis.

Non-infectious meningitis: Meningitis can also occur as a result of non-infectious causes like spread of cancer to the meninges, NSAIDs, antibiotics and intravenous immunoglobulins. Certain inflammatory conditions like sarcoidosis, SLE, vasculitis, migrane (rare) would also lead to meningitis.

Bacterial meningitis is contagious and the bacteria are spread through exchange of respiratory and throat secretions like kissing.

Viral meningitis spreads when you are in close contact with the affected patient.

Clinical Manifestations (Fig. 6.2)

1. In most cases, it is quick onset except for the tuberculosis type which is insidious.
2. Fever, malaise and lethargy.
3. Headache which is persistant and is usually associated with vomiting.
4. The patient may have clouding of consciousness.
5. The patient is drowsy, irritable and often delirious.
6. Neck rigidity is the striking feature of meningitis. There is marked stiffness of the neck.
7. Photophobia and phonophobia.
8. Convulsions may be present.
9. Nuchal rigidity.
10. In **meningococcal meningitis**, there will be development of rashes which are numerous, small, irregular purple or red spots (petechiae) on the trunk, lower extremities, mucous membranes, conjunctiva, and (occasionally) the palms of the hands or soles of the feet.

Special Tests[3-6]

- Kernig's sign: This is the resistance met with on attempting to straightened flexed hip-knee.
- Brudzinski's sign: Flexion of the neck causes involuntary flexion of the knee and hip.
- Jolt accentuation maneuver: Headache becomes worst when a person is rapidly made to rotate the head horizontally.

Diagnosis

- CSF analysis
- Blood tests—C-reactive protein

Complete blood picture, culture and sensitivity.

Complications of Meningitis

- One of the most common problems resulting from meningitis is **hearing loss**.
- Babyish behavior
- Forgetting recently learned skills
- Reverting to bed-wetting
- Brain damage
- Epilepsy
- Changes in eye sight
- Skin rashes (Fig. 6.3)

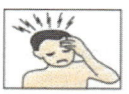

Fig. 6.2: Clinical manifestation of meningitis

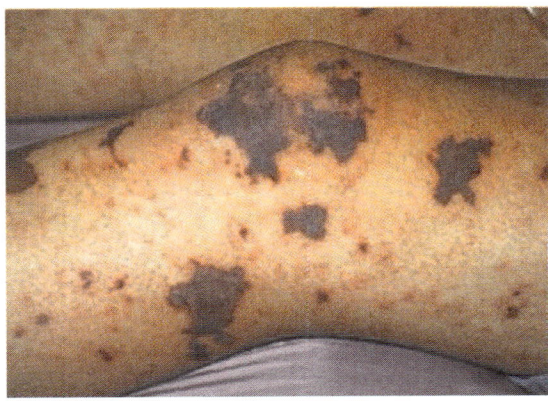

Fig. 6.3: Petechiae

CSF Analysis of Various Meningitis

CSF analysis of various meningitis is shown in Table 6.1.

Medical Management[7]

1. **Drugs:** Antibiotic therapy in pyogenic meningitis: Empiric antibiotics should be started immediately. Sulphonamides, penicillin, and chloramphenicol are the choice of antibiotics. Penicillin is sometimes given by the intrathecal route.
   Chemotherapy in tuberculous meningitis: Streptomycin, para-aminosalicylic acid, and isoniazid are given until the sensitivity of the organism is known, after the sensitivity tests, the two most effective ones are administered for at least 6 months. In serious cases, intrathecal administration may be advised. Pyridoxine is used to prevent the neurotoxic effects of isoniazid.[7,8]

   Anticonvulsants may be necessary especially in children.
   Steroid therapy: Corticosteroids, usually dexamethasone, are found beneficial especially in reduction of hearing loss.
   Supportive therapy in viral meningitis: Acyclovir would be effective, but most of the viral meningitis cases are not amenable to specific treatments.
   Antifungal therapy for fungal meningitis: Long courses of high dose antifungal therapy, like amphotericin B and flucytosine, are required.

2. **Good nursing care** is very essential which should be skillful. The patient is nursed in a subdued light and quietness is maintained. Barrier nursing is necessary to prevent spread of infections.

3. **Symptomatic treatment:** Bedrest, administration of analgesics, cough suppressants, antipyretics like paracetamol are given for symptomatic relief.

ENCEPHALITIS

It is an acute inflammation of the cells in the brain due to infection. When associated with meningitis, it is known as meningoencephalitis.

Causes

- **Viral:** It is caused by the rabies virus, HPV infection, poliovirus and measles virus. It could be a direct effect of acute infection or as a sequelae of a latent infection. Arbovirus, Bunyavirus, arena virus, and reovirus are also found to be

Table 6.1: CSF analysis of various meningitis

Type of meningitis	Glucose	Protein	Cells
Acute bacterial	Low	High	PMNs, often >300/mm³
Acute viral	Normal	Normal or high	Mononuclear, <300/mm³
Tuberculous	Low	High	Mononuclear and PMNs, <300/mm³
Fungal	Low	High	<300/mm³
Malignant	Low	High	Usually mononuclear

the other possible viruses resulting in viral encephalitis.
- **Bacterial:** Syphilis bacterium is the causative, for the secondary encephalitis. Other bacterial pathogens, like *Mycoplasma*, can also cause encephalitis.
- **Protozoa:** Toxoplasmosis, Plasmodium, Amoeba can also cause encephalitis.

Types

- **Infective encephalitis:** Usually caused by infection or latent by various bacteria, viruses or some protozoa.
- **Limbic encephalitis:** It is a system onset indicated by decrease in cognitive functions especially memory.
- **Autoimmune encephalitis:** Anti-N-methyl-D-aspartate encephalitis, and Rasmussen's encephalitis are examples of autoimmune encephalitis. These are caused by immune-mediated changes to the treatment path.
- **Encephalitis lethargic:** It is identified by high fever, headache, delayed physical respone and lethargy. The patients also exhibit upper body weakness, muscular pain, and tremors. The cause is unknown.

The groups most at risk of encephalitis are:
- Older adults
- Children under the age of 1
- People with weak immune systems

Signs and Symptoms

1. Fever
2. Headache
3. Vomiting
4. Stiff neck
5. Lethargy and weakness
6. High grade fever (103°F or higher)
7. Confusion and drowsiness
8. Hallucinations
9. Slowness of movements
10. Coma
11. Seizures
12. Irritability
13. Photophobia
14. Bulging fontanelle in children
15. Poor appetite

Diagnosis

- CSF analysis
  - Increased WBC
  - Increased protein
  - Increased glucose
- Brain CT scan or MRI
- Electroencephalography(EEG): Produces abnormal signals
- Complete blood picture
- Brain biopsy

Management

Supportive Treatment

- Rest
- Mechanical ventilation (to help with breathing)
- Lukewarm sponge baths
- Fluids replacement (sometimes through an IV)

Medications

1. Antiviral medications, e.g. acyclovir
2. Antibiotics: Streptomycin, penicillin
3. Steroid therapy
4. Anticonvulsants
5. Acetaminophen to reduce high grade fever

PHYSIOTHERAPY ASSESSMENT OF MENINGITIS AND ENCEPHALITIS

1. Take the demographic data and note the chief complaints which usually include headache, nausea, vomiting, delirium, febrile illness and coma sometimes.
2. Take the history of spread of infection, history of vaccination, history of convulsions/seizures any other medical

complications or disease preceding. Also note the medications being used along with family history and personal history.
3. Observe the general built of the patient, in case of children observe for the signs of malnutrition.
4. Observe for abnormal attitude attained and observe for any abnormal movements.
5. Observe for any unhealed wounds, asymmetrical swellings over the lymph nodes, breathing pattern of the patient.
6. Palpate for tenderness and grade it. Check for the temperature asymmetry, if any.
7. *Examine for vital signs*: Temperature, pulse, respiratory rate and blood pressure. Observe the pattern by taking notes from the TPR charts from nurses notes, if any.
8. *Examine the higher functions*:
   a. Consciousness: Usually altered, patient feels drowsy.
   b. Memory-cognitive functions, usually memory will be lost.
   c. Orientation by place, person and time is lost and usually the patient is in delirium and confusion.
   d. *Speech*: Altered, may be dysarthria, aphasia or mutism present.
   e. *Cranial nerve examination*: Check for all the cranial nerves for their functions and note the variation, if found any. In some cases, cranial nerve palsy of some nerves may be present. Usually vision, hearing and taste are impaired.
   f. Examine the sensory system for superficial, deep and cortical senses which may be impaired depending upon the severity of infection in case of encephalitis. In meningitis, the sensory loss is uncommon.
   g. Reflexes show abnormal responses. DTR are exaggerated and primitive reflexes may be prevalent.
   h. *Tone*: Check for the tone of the muscle by doing repeated passive movements.
   i. *Muscle power*: Check for the muscle power, group MMT could be helpful, usually the power is decreased.
   j. *Range of motion*: The ROM may be affected due to abnormal muscle tone, especially there is rigidity in the neck.
   k. Examine the breathing pattern, chest expansion, and depth of excursion. On auscultation and percussion, there will be features of secretions in the chest.
   l. *Bladder and bowel assessment*: There will be neurogenic bladder symptoms.
   m. *Special test*: Kernig's sign, Brudzinski sign, Jolt accentuation maneuver—positive.
   n. Check for the ADL assessment and identify the incapabilities and disability to perform the ADL.

PHYSIOTHERAPY MANAGEMENT

Aims

a. To provide psychological support and develop good rapport with the patient and caretakers
b. In case of comatose patients, to stimulate an arousal response
c. To prevent accumulation of secretions in the lungs
d. To maintain normal range of motion and prevent distorted postures
e. To improve coordination and balance
f. To normalize the tone in the muscles
g. To promote the sensory integration
h. To prevent the secondary complications
i. To improve the overall functional status of the patient.

Plans of Management

To Provide Psychological Support and Develop Good Rapport with the Patient and Caretakers

- Developing rapport with the patient is very important as any goal will be difficult to achieve without the cooperation. The patient has to be motivated well enough to gain the confidence. The goals set for the patient must be challenging at the same time achievable.
- False appreciation must be avoided.
- Initially maximum support and feedback must be given.
- Never give false hope to caretakers or to the patient.
- Explain the role of caretaker and teach the home exercises so that it can be carried at home. Remember, in case of children, always the therapy should be play therapy. Try to include games or play items into the therapy or else the kid will not show interest in the treatment.

In Case of Comatose Patients, to Stimulate an Arousal Response

Various theories have been put forward which follow the hypothesis that a multisensory balance stimulation given to the patient in an appropriate manner will have a stimulating effect on reticular formation. The various stimulations include:[6]

a. Verbal or auditory
b. Visual
c. Olfactory
d. Gustatory
e. Tactile
f. Vestibular

To Prevent Accumulation of Secretions in the Lungs

Pulmonary compromise is a leading cause of morbidity in patients with meningitis and encephalitis. Respiratory complications are very common. Accumulation of secretions leading to respiratory infections is also common problem encountered by patients due to immobility and decreased thoracic mobility or prolonged bedrest. It is also evident in patients with pre-existing chronic respiratory disorders, lowered resistance to infection. The main aim of chest care in these patients is to prevent internal secretions causing obstruction, improve ventilation to the lungs and improve vital capacity.

Deep breathing exercises, frequent change of positions would assist in clearing bronchial secretions and decrease paradoxical movement of the rib cage. (Doughlas et al 1977).[9] Nebulizations, postural drainage with manual techniques like clapping, vibrations to remove the accumulated secretions would be useful. Spirometric and chest mobility exercises are to be encouraged to improve ventilator capabilities. Coughing and huffing techniques are taught to the patient and he/she is encouraged for effective coughing.

To Maintain Normal Range of Motion and Prevent Distorted Postures

- Achieving adequate joint ROM is necessary for functional independence and ambulatory training. The joint range of motion is maintained and improved by stretching exercises. Use of appropriate orthotic supports would be useful to maintain the joint range. Strengthening exercises of weak muscles would be necessary to prevent the contractures and deformities in order to maintain full range of motion. In rare cases, overstretching is encouraged like hip external rotation in order to help the ADL activities like putting on shoes and socks. It should be kept in mind that aggressive ROM exercises are contraindicated.
- In case of spastic muscles, daily stretching and terminal sustained stretch is to be considered for rehabilitation for limitation of joint range of motion.

To Normalize the Tone in the Muscles

- To deal with hypertonicity or rigidity, the following principles of management are applied.
- Positioning the patient: Correct positioning is an important aspect of management of spasticity.
- Check for the postures that exaggerate the spasticity or promote primitive reflexes which trigger the spasticity and avoid them, e.g. tonic labyrinthine reflex, and asymmetrical tonic reflex.
- Children can be placed in prone lying for short periods which will help to improve strength in upper trunk and improve head control. A small roll under the chest or wedge can be used to assist with the position. Lying the child on their side lying is also a useful position.
- During nighttime, it is very essential to maintain position. When putting the child to bed at night do not leave them on their back, supine lying provokes the development of extensor spasticity, so side line is preferable. Curl them forwards "hugging" a pillow, if needed, to maintain the position. Always avoid any stimulus to the back of the head which provokes extensor posturing. Alter right and left side lying at least every 4 hr.
- Passive movements which are rhythmical and full range should be applied at regular intervals at least every 2 hr.
- Splinting and supportive pads and pillows can be used.
- Other measures:
  - Cold in the form of prolonged icing (ice packs) would help to inhibit the spasticity
  - Surged faradic currents can have an inhibitory effect on spasticity (Seib et al) (Alfieri).
  - Prolonged application (above the duration of 45 min) of TENS also reduces the spasticity (Potissk et al)
  - Static stretching, EMG biofeedback, and vibrations would have beneficial effects.

To deal with hypotonicity:
- Massage therapy in the form of pressure manipulations is often known for its best circulatory benefits. A good circulation to the muscles has found to be beneficial to improve the muscle tone.
- Weight-bearing and weight-training exercises would improve the muscle tone. Ice brisk stroking over the muscle bellies would improve the muscle tone. Along with this, pediatric ENMS would be beneficial.

To Improve Coordination and Balance

- Frenkel's exercise, peg and socket board exercise, rhythmic stabilization techniques, balance training devices.
- Music-based multitask exercise program involving a wide range of movements and challenged balance control system, weight shifting, walk-turn sequences would be beneficial for training balance.
- Balance training using a virtual-reality system-researchers has found that it is an effective method to train the balance in older patients.
- Static and dynamic balance exercises.
- Coordination exercises including equilibrium and non-equilibrium would be incorporated to improve the coordination.

To Promote the Sensory Integration and Sensory Retraining

Sensory integration techniques would be beneficial for:
- Daily activities and intellectual, social and emotional development of the patients.

- Sand and water play, vestibular exercises like swings, rocking toys, scooter boards, music in motion, gliders, see saws, therapy balls.
- Aromatherapy, massage also helps in sensory integration.
- Wilbarger brushing protocols:[10-12] It is a surgical brush which is used to brush the child at regular intervals.
- Oral toys, like whistles, blow toys, textured teething rings or spoons, help the children in developing oro-motor activities.

Techniques to Retrain the Sensations
Table 6.2

Other Techniques

1. Feel an object and find the same matching object from a sand tray.
2. Ask the patient to close the eyes. Place two objects of different weights and ask the patient to recognize the difference.
3. Write with the finger an alphabet or a number over the dermatome of the patient and ask the patient to feel and see and perceive. Now ask the patient to close the eyes and recognize the same.

To Improve the Overall Functional Status of the Patient

- Train for fine motor skills. Teach the child and their parents how to ideally make the child sit in a chair and appropriate sizes, supportive pillows, cushions, protective straps, etc.
- If the child is wheelchair bound, train for the wheelchair mobilization and transfer techniques.
- Handling of pens and pencils is usually difficult in these children, so train them to use flattened pencils with rubber bands bound around to make the pencil thicker.
- Train the kid with modified spoons, drinking cups for feeding activities.

Table 6.2: Sensory re-education exercises

Training for differentiation of different textures	Use cotton, sandpaper, satin, velcrow rubber, velvet, wool. Train 2–4 times for 10 min in a quiet environment daily
Steriognosis	Hide objects with different shapes like marbles, currency coins, erasers in sand. Ask the patient to close eyes and try to find the objects with the hand feel and recognize. Note: The objects should be familiar to the patient Frequency: 2 times for 10 min daily
Pressure	Apply pressure with fingers of different intensity over an area once with patient's eyes opened and patient's eyes closed. Ask the patient to feel the pressure and identify Frequency: 2 times for 10 min daily
Vibrations	Ask the patient to close the eyes and apply vibrations over a dermatome of the patient. The vibrations first should be of a powerful vibrator which is used for massage, later by mobile phone vibrations, tuning fork vibrations and vibrations as the patient progresses. Ask the patient to feel the vibrations once with eyes opened and then with eyes closed Frequency: 2–3 times for 5 min daily
Temperature	Take two bowls of water of hot and cold temperatures and ask the patient to dip the hands or a test tubes can be used instead. Ask the patient to feel and recognize. Frequency: 2–3 times for 10 min daily
Joint position sense	Position the joint in normal and affected at the same time, ask the patient first to feel the position and then repeat only on the affected limb. Ask the patient to analyze and feel Frequency: 2–3 times for 10 min daily
Joint kinesthetic sense	Move the limbs once with eyes opened and then eyes closed. Ask the patient to feel and recognize Frequency: 2–3 times for 5 min daily

POLIOMYELITIS

Poliomyelitis is an infectious disease affecting the spinal cord and is epidemic and endemic in nature.

Polio (grey) and myelon (marrow) are derived from Greek. It was first described by Michael Underwood in 1789. The polio first outbreak in Europe were reported in 19th century and in USA in 1843.

Poliovirus

It is a member of enteroviruses which are transient inhabitants of the GIT and are stable at acid pH.[13] There are three types of polio stereotypes (P1, P2, and P3) are identified. Immunity to one stereotype does not produce immunity to other stereotypes. The poliovirus is rapidly inactivated by heat, formaldehyde, chlorine and UVR.
- P1: Brunhilde
- P2: Lanchi
- P3: Leon

Epidemiology of Polio
- Found 10 million cases worldwide.
- Common in warmer, temperate countries.

Mode of transmission

Its spreads by orofecal route in countries where the hygiene is poor and by droplet infection where the sanitation is good. Transmission is by contamination of food and also by droplet infection.

Incubation period

It varies from 3 to 30 days while 7–14 days is common interval between infection and clinical illness. The incubation period decreases on provocation like injection pricks, massage, etc.

Pathogenesis (Fig. 6.4)

Clinical Manifestations

Preparalytic stage: It is a prodromal stage and the clinical presentation would be same as any other viral infection. It may last from a few hours to a few days ranging from 1 to 3 days on average.

1. Children and young adults are most commonly affected.
2. The onset is sudden with fever, headache and febrile illness.
3. Stiffness of neck with pain in the back are the early symptoms.
4. Sore throat, cough, diarrhea.
5. The disease may not progress further and abortive attacks often suspected and diagnosed in epidemic attacks.

Stage of paralysis:
- One or more limbs and/or trunk muscles are paralysed. Usually 3–6 weeks duration from the onset of symptoms.
- Child prefers sleeping position and cannot tolerate any change in the position.
- Commonly affected muscles are of lower limb, shoulder girdle and respiratory muscles.
- Deep tendon reflexes are lost.
- Wasting of muscles is a prominent and significant feature.
- The clinical picture would be a typical LMN lesion type.

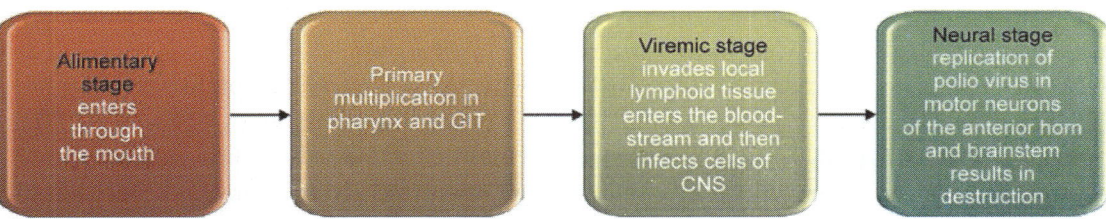

Fig. 6.4: Pathogenesis

- Pain in the affected limbs usually present with tenderness
- Respiratory failure occurs, if there is involvement of diaphragm and intercostals
- Pharyngeal paralysis leads to difficulty in swallowing, choking and asphyxia.

Convalescent stage: This is the stage where there is true or actual paralysis. The duration of this stage can be for 3 months.

The clinical manifestations of this stage depend on extent and distribution of the lesion and vary from case-to-case.

The effects of these lesions can be described as follows:
- Spinal: There will be asymmetrical flaccid paralysis (LMN lesion type) due to effect on anterior horn cells of spinal cord. The extent of paralysis varies and lower limb muscles are more involved than upper limb.
- The sensations are normal as the posterior horn cells are undamaged.
- Contractures and deformities are common due to gross muscular imbalance.

Common Deformities

Bulbar: Difficulty or inability to swallow which is an important sign of bulbar paralysis due to pharyngeal muscle paralysis. (Deglutition is impaired.) It leads to choking and asphyxia. Laryngeal paralysis causes inability to cough leading to accumulation of secretions.

The patient may have difficulty in speech due to paralysis of palate. The early signs of respiratory failure would be breathlessness, suffocation, slight cyanosis and use of accessory muscles of respiration is frequent.

Gag reflex is absent.

Spinobulbar and bulbospinal: In this type of polio, there will be signs and symptoms of both spinal and bulbar types. If the spinal symptoms are predominant and bulbar symptoms are less predominant, then it is called spinobulbar type and if bulbar symptoms predominate, then it is called bulbospinal type of polio.

Post-encephalitic: It is a sequel of encephalitis. It is usually associated with bulbar paralysis. Higher functions may be disturbed and patient may be in coma. There may be symptoms of facial paralysis. Along with neurological symptoms, systemic symptoms like headache, vomiting, neck stiffness may be seen.

Urinary retention may be present.

Stage of recovery: This is the stage of late convalescent and extends up to 2 years post-polio onset of symptoms.

Stage of residual paralysis: It is the stage where there is no further recovery and extends after 2 years of onset.

The limbs look bluish. The muscles become wasted and deformities will set in. There may be shortening of affected limb and is floppy.

The paralysis ranges between mild insignificant local weakness to almost gross paralysis of the trunk and limb musculature giving rise to severe disability and functional dependency.

This net resultant neuromuscular status at the end of this stage is known as post-polio residual palsy.

Post-polio syndrome: It may occur after 30–35 years after original attack of polio.

There will be return of the systemic symptoms like pain, fatigue, muscle weakness and functional impairment.

Medical management

In acute stage

- Immediate and complete bedrest.
- Minimal handling of the child. The affected child should be isolated.
- The other children in the family should be given booster dose of vaccination against polio.

- Diet rich in protein is a good option for nutrition.
- Splinting and correct positioning.
- Moist heat to the tender muscles by sister Kenny's hot packs. An important instruction to the parents is "**not to massage**".

Patients who have respiratory paralysis:
- Administration of analgesics, antispasmodics for a symptomatic relief.
- If respiratory paralysis, then artificial respiration is given.

Patients with bulbar paralysis: These patients will have difficulty in feeding. The patients should be nursed in semiprone position. Change the position of the patient to side line from one side to other every few hours and the foot end of the bed is raised to make an angle of 15°. A mechanical sucker may be used to remove pharyngeal secretions. Feeding should be carried by esophageal catheter preferably passed through the nose.

In chronic stage

- Prevent deformities by physiotherapy and splintage.
- To stabilize the flail joints, tenodesis, arthrodesis surgeries may be necessary.
- Tendon transfers are performed to avoid bony deformities.
- Delayed orthopedic surgeries beyond 10 years may be necessary such as osteotomy, arthrodesis and arthroplasty.

Interview with the patient: The therapist in order to develop the good rapport with the patient, first should talk to the patient and listen patiently the complaints.
- Take note of general demographics (name, age, sex, address, etc.).
- Take complete medical/surgical history by directly interviewing the patient or from parents or from medical records.
- Take a detailed vaccination history and enquire about the other family members.
- Take the growth and development history of patient. The therapist should be sensitive to the impact of the disease on the psychosocial development of the child.
- Take the past and current work abilities, social habits, level of fitness, leisure activities, cultural beliefs and behaviors, and living environment of the patient.

Observation

- Observe for the general built of the patient.
- Observe for seating, standing, walking postures of the patient.
- Check for any swellings.
- Check for muscle wasting, contractures and deformities.
- Observe and note if any tropical changes over the skin.
- Observe and note any orthotic supports or medical accessories.

Palpation

- Check for the temperature changes over the skin.
- Palpate and grade the tenderness of the muscles—mild/moderate/severe/absent.
- Palpate and describe the edema, if present.

Examination

- Note the vital signs.
- Evaluate for the cranial nerves and higher functions.
- Measure the joint range of motion both active and passive.
- Assess the power of the key muscles and grade them.

- Identify and examine the structural deformities of neck, back, chest and limb.
- Check and compare the reflexes—superficial and deep tendon reflexes.
- Check for any sensory impairments to rule out other causes of presenting illness.
- Evaluate the aerobic capacity and endurance of the patient.
- Perform detailed respiratory examination: Auscultation, percussion, breathing pattern, extent of excursion, depth of excursion, etc.

Anthropometry Measurements

- Take the limb lengths and limb girths of the patient.
- Assess BMI (body mass index) of the patient.
- Assess the ADL capacity of the patient and evaluate the need of supports and their effectivenss for its intended purpose.
- Examine gait, balance and equilibrium of the patient. Video tape analysis would be helpful to assess the gait (Table 6.3).

PHYSIOTHERAPY MANAGEMENT

The physiotherapy management of poliomyelitis depends upon the stage the patient is brought for physiotherapy intervention.

Systemic Stage

- Complete bedrest
- Isolation of the patient
- Symptomatic medical management
- No active or passive physiotherapy except for the positioning
- The child is disturbed as little as possible

Preparalytic Stage

- Complete bedrest
- Patient is nursed in isolation

Positioning of the patient:
- Shoulder abduction and slight flexion
- Elbow—extension
- Wrist neutral
- Hand: MP: 80° flexion; PIP: 45° flexion; Hip joint: Neutral; Knee: Slight flexion; Ankle: Neutral

The position can be maintained by pillows, sandbags, and foot boards.

Handling of the patient: The parents of the child should be taught how to correctly handle the kid causing less discomfort to the kid and preventing further damage to the neural structures. Advice the patient's parents not to lift the child by one hand.

Hot water fermentation or Sister Kenny's moist heat pack could be useful to reduce spasm and pains.

Passive movements which are gentle, full range and rhythmical should be given at least 2–3 times per day. The technique of giving passive movements should be taught to the parents.

Convalescent Stage

During the active paralytic stage, patient is disturbed as little as possible as handling increases the discomfort and tenderness of muscles.

Splints: Proper splinting should be done to prevent the deformities. Below knee splint or an L splint would be beneficial to prevent knee flexion and equines deformity.

In case of trunk and abdominal patchy paralysis, there is risk of hernia in these patients, hence the patient is advised to wear abdominal corset.

Positioning of the patient: Advice the patient to place in prone positions to avoid hip flexion contractures.

- Shoulder abduction and slight flexion,
- Elbow—extension
- Wrist neutral
- Hand: MP: 80° flexion; PIP: 45° flexion;

Table 6.3: Abnormal gait patterns in poliomyelitis

Gait	Description	Picture
Extensor lurch	This is seen in case of hip extensor paralysis. In stance phase of affected limb, patient is not able to keep the trunk erect. He hyperextends the trunk so that the center of gravity of body passes posterior to hip joint and hip is extended passively (gluteus maximus gait)	
Abductor lurch	During unilateral stance, pelvis on the unsupported side tends to dip down because of effect of gravity and abductors of the stance leg act to maintain the pelvis stable. In case of paralysis of abductor on one side, during unilateral stance of the affected lower limb, pelvis dips down on the unsupported side. To maintain the center of gravity body over the supported lower limb, the patient has to do lateral flexion on the same side (Trendelenburg's gait).	
Extensor—abductor lurch	If both groups of muscle are paralyzed.	
Hand to knee gait	For compensating for hip extensor paralysis, patient has to do extensor lurch. But if knee is flail with extensor lurch, line of gravity will fall posterior to knee and knee will buckle. So patient stabilizes the knee in flexion or hyperextension lurches forward, holds the knee with hand so that he/she does not buckle and also trunk is supported.	
Gait with internal rotation at hip and extensor lurch	In case of flail knee, patient walks with tight IT band which causes internal rotation at hip. This changes the plane of flexion–extension at knee movement sagittal to coronal. In sagittal plane, no movement is possible now if the ligaments of knee joint are intact. So the knee is stable. The patient does extensor lurch to compensate for hip extensor weakness.	

Hip joint: Neutral; Knee: Slight flexion; Ankle: Neutral.

The position can be maintained by pillows, sandbags, and foot boards.

Shoulder rolls under the axilla helps to prevent subluxation of the shoulder

Change the position of the patient every 2–4 hours.

Chest physiotherapy is given in the form of humidification, postural drainage to keep the chest clear from secretions.

- **Stretching:** Paralysis of a muscle results in unopposed action of antagonist and leading to tightness, and contractures. The tight and contracture structures are given slow, sustained stretching.
- **Stimulation of paralytic muscles:** If the patient can tolerate, stimulate the muscles by giving the IG stimulation to maintain the normal muscle properties.
- Ice brisk stroking, hacking over the muscles shall stimulate the muscle and facilitate improvement in the strength of the muscles affected.
- Passive movements are continued.
- Hot fermentations, moist heat therapy would be beneficial in relieving the muscle spasms.

Stage of Recovery

Strengthening of the paralyzed muscles is the key aim of the therapist.

Various techniques to strengthen the muscles may be used.

1. Sensory integration: Various sensory stimuli are used to stimulate a motor response.
2. Progressive resisted exercises with the help of springs and pulleys.
3. Hydrotherapy and suspension therapy.
4. PNF techniques by using irradiation principle, successive induction and timing for emphasis would be beneficial to strengthen the muscles.

Play therapy: Involving the children in outdoor games helps to improve the strentgth of the muscles. The game selected should be specific with respect to the muscle and also intresting for the kid.

Mat exercises: These exercises can be given to bring about overall improvement in the strength, coordination, balance, functional abilities, etc.

Weight-bearing exercises should be encouraged to promote normal growth of the bones.

Stage of Residual Paralysis

When the patient reaches the stage of post-polio residual palsy, assess the patient first for final outcome of polio and its impact.

The treatment is designed based on the assessment and usually a combination of stretching, strengthening and calliperization.

Serial plaster casting, taping, bracing could be used for correcting deformities.

Depending upon the PPRP status, calipers are given to accommodate the deficit and deformity.

If required at this stage, to correct the deformities, various surgeries are done.

NEUROSYPHILIS

Neurosyphilis is a sexually transmitted disease that affects the nervous system by *Treponema pallidum.* It has been becoming uncommon disease after the introduction of modern broad-spectrum antibiotics like penicillin.

Clinical Types

a. Asymptomatic neurosyphilis
b. Meninovascular neurosyphilis
c. Tabes dorsalis
d. General paralysis of insane

Asymptomatic Neurosyphilis

1. It is manifestated by a reactive non-treponemal CSF serology (VDRL test).
2. There will be no signs and symptoms of focal neurological disturbances.
3. The CSF usually reveals elevated protein levels, lymphocytic pleocytosis and glucose levels.

Meningovascular Syphilis

1. It occurs in the second stage of syphilis and effects the meninges and vessels of the brain.

2. Syphilitic endateritis causes infarction clinically similar to stroke.
3. The luminal narrowing predisposes to cerebrovascular thrombosis, ischemia and infarction.
4. The most common artery involved is middle cerebral artery.
5. The clinical manifestations include headaches, vertigo, insomnia and personality disorders.
6. If the base of the brain is involved, cranial nerve palsies can be evident.
7. There may be formation of leptomeningeal granulomas which are avascular in nature called *gumma*.

Tabes Dorsalis

1. It is a slowly progressive parenchymatous degenerative disease involving the posterior columns and posterior roots of the spinal cord.
2. Symptoms include loss of pain sensation, loss of peripheral reflexes, impairment of vibration and joint position sense and ataxia which is progressive.
3. Bladder incontinence and loss of sexual function are common.
4. 15% of patients experience spisodes of excruciating epigastric pain with associated nausea and vomiting.
5. There are three stages of tabes dorsalis.
   a. Preataxia
   b. Ataxia
   c. Paralysis
6. There will be wide based gait with slapping.
7. Charcot joints and trophic ulcers develop in later stages.
8. The pupils are bilaterally small and fail to constrict but demonstrate normal constriction to accommodation.

General Paralysis of Insane

1. It is commonly referred as dementia paralytica.
2. It occurs approximately 20–30 years after the initial exposure to *Treponema pallidum*.
3. It represents a chronic progressive frontotemporal meningoencephalitis with resultant ongoing loss of cortical functions.
4. There will be insidious onset of psychiatric symptoms of general paresis.
5. These include loss of interest in work, memory lapse, irritability, unusual giddiness, apathy, social withdrawl.
6. Later stages, the symptoms of schizophrenia, euphoric mania, paranoia, toxic psychosi.
7. After approximately 5 years of onset, there will be convulsions.
8. Abnormal gait, paresthesias, lightning pains of extremities, loss of proprioception, Romberg sign positive.

Investigations

- CSF analysis
- Serological tests
- VDRL test
- Treponemal tests
- Non-treponemal tests
- Radiological examination
- Electrodiagnostic tests

Medical Management

Penicillin G, administered parenterally, is the preferred drug for treating persons in all stages of syphilis.

Principles of Physiotherapy Assessment

The principles of assessment follow the routine neurological assessment with emphasis on the following:

1. Take the history of onset of general or constitutional symptoms like headache, fever, fatigue, weakness, and dizziness.
2. Check with the patient about the symptoms like pain, redness in eyes, loss/double vision, photophobia, ringing of bells in ears, loss of hearing.

3. Check the musculoskeletal symptoms like neck pain, stiffness, muscle weakness.
4. Enquire the patient about the neurological symptoms like headache, dizziness, muscle weakness, confusion, loss of consciousness, seizures, difficulty in speaking, etc.
5. Obtain the clear personality changes, if any, from the family members of the patient.
6. Assess the cranial nerves to identify the cranial nerve palsies.
7. Perform a clear and accurate sensory examination. Usually there will be loss of superficial and cortical sensations.
8. Motor examination reveals the muscle weakness. In case of meningococcal neurosyphilis, there will be motor dysfunction like stroke–spasticity, synergies, loss of voluntary control, etc.
9. Examine the reflexes—hyperreflexia.
10. Nuchal rigidity testing: Assess for meningeal inflammation by following tests:
    - Chin to chest (Brudzinski's sign)—stiffness/pain with flexion of neck, flexion of hips and knees in response to neck flexion.
    - Jolt accentuation maneuver—worsening of headache when the patient rotates head rapidly from side-to-side.
11. Assess for neurological bladder symptoms.
12. ADL assessment which shows diminished capabilities.
13. Coordination and balance is impaired.
14. On examination of gait, there will be abnormal gait pattern usually ataxic gait is seen.

Physiotherapy Management

Aims

- To educate the patient about sexually transmitted diseases.
- To improve the muscle power.
- To improve cardiovascular and respiratory endurance.
- To improve balance and equilibrium.
- To improve coordination.
- To prevent contractures and deformities.
- Early ambulation.
- Improve the functional capacity of the patient.

Plans

- Educate the patient on the safety measures and precautions to prevent sexually transmitted diseases.
- Advise the patient to avoid multiple sex partners.
- All strengthening exercises are trained depending upon his muscle power.
- Proper positioning to prevent contractures and deformities.
- Balance and equilibrium exercises.
- Frenkel's exercises in various positions.
- Deep breathing exercises like pranayama, VMT, spirometry to improve ventilatory effort.
- Aerobics to improve the cardiovascular endurance.
- Cycling, swimming exercises promotes endurance.
- Early ambulation, if needed suitable supports or splints are advised.

REFERENCES

1. Communicable Disease Report Weekly. Bacteraemia, England and Wales: Laboratory reports 2001 and 2002;13(3):4–5.
2. Thigpen MC, Whitney CG, Messonnier NE, Zell ER, Lynfield R, Hadler et al. Bacterial meningitis in the United States, 1998–2007. N Engl J Med. 2011;364(21):2016–25.

3. Kernig VM. "Ein Krankheitssymptom der acuten Meningitis". St Petersb Med Wochensch. 1882;7:398.
4. O'Connor, Simon; Talley, Nicholas Joseph (2001). Clinical Examination: A Systematic Guide to Physical Diagnosis. Cambridge, MA: Blackwell Publishers. p. 363.
5. Thomas KE, Hasbun R, Jekel J, Quagliarello VJ. "The diagnostic accuracy of Kernig's sign, Brudzinski's sign, and nuchal rigidity in adults with suspected meningitis". Clin Infect Dis 2002;35 (1):46–52.
6. Brudzinski J. "Un signe nouveau sur les membres inférieurs dans les méningites chez les enfants (signe de la nuque)". Arch Med Enf 1909;12:745–52.
7. Tunkel AR, Hartman BJ, Kaplan SL, Kaufman BA, Roos KL, Scheld WM, et al. Practice guidelines for the management of bacterial meningitis. Clin Infect Dis. 2004;39(9):1267–84.
8. Barichello T, Generoso JS, Simoes LR, et al. Vitamin B_6 prevents cognitive impairment in experimental neumococcal meningitis. Exp Biol Med (Maywood). 2014;239(10):1360–5.
9. Douglas WW, Rehder K, Beynen FM, Sessler AD, Marsh HM. Improved oxygenation in patients with acute respiratory failure: the prone position. Am Rev Respir Dis 1977;115:559–566.
10. Foss A, Swinth Y, McGruder J, Tomlin, G. Sensory modulation dysfunction and the Wilbarger Protocol: An Evidence Review. OT Practice, 8(12), CE1–CE8.
11. Wilbarger P and Wilbarger J. Sensory defensiveness in children aged 2–12: An intervention guide for parents and other caretakers. Santa Barbara, CA: Avanti Educational Programs, 1991.
12. Wilbarger J, Wilbarger P. Wilbarger approach to treating sensory defensiveness and clinical application of the sensory diet. Sections in alternative and complementary programs for intervention. In AC Bundy, EA Murray and S Lane (Eds.), Sensory integration: Theory and Practice (2nd ed.) 2002. Philadelphia, PA: FA Davis.
13. Ryan KJ, Ray CG (Eds.): Sherris Medical Microbiology (4th ed.) 2004. ISBN: 0-8385-8529-9.

7

Ataxia

LEARNING OBJECTIVES

In this chapter, the student will learn and gain skills on:
a. Definition of ataxia
b. Etiopathogenesis of ataxia
c. Assess and diagnose a case of ataxia
d. Understand the investigations done to diagnose ataxia
e. Detailed physiotherapy management of the cerebellar ataxia.

INTRODUCTION

Ataxia refers to a group of neurological disorders in which motor behavior appears uncoordinated (Fig. 7.1). Walking, speaking, swallowing, writing, reading and other fine motor activities become abnormal in these patients. Ataxia may result from abnormalities in different parts of the nervous system or different parts of the body. Greek root = *a taxis* = "without order".

Ataxia, literally means "lack of order", is a neurological sign and symptom that consists of gross lack of coordination of muscle movements.

Loss of coordination, commonly associated with:
- Tremor, impaired rapid alternating movements
- Gait impairment
- Nystagmus

Definition: It is a jerky, irrhythmic, inaccurate, purposeless movement.

ETIOPATHOLOGY

Damage, degeneration or loss of nerve cells in the part of the brain that controls muscle coordination (cerebellum), results in loss of coordination or ataxia. Cerebellum comprises two ping pong ball-sized portions of folded tissue situated at the base of our brain near the brainstem. The right side of our

Fig. 7.1: Ataxia

cerebellum controls coordination on the right side of our body; the left side of our cerebellum controls coordination on the left side of our body.

Diseases that damage the spinal cord and peripheral nerves that connect our cerebellum to muscles may also cause ataxia. Ataxia causes include:

- **Head trauma:** Damage to brain or spinal cord from a blow to the head, such as a car accident, can cause sudden onset ataxia, also known as acute cerebellar ataxia.
- **Stroke:** When the blood supply to a part of your brain is interrupted or severely reduced, depriving brain tissue of oxygen and nutrients, brain cells begin to die.
- **Transient ischemic attack (TIA):** Caused by a temporary decrease in blood supply to part of our brain, most TIAs last only a few minutes. Loss of coordination and other signs and symptoms of a TIA are temporary.
- **Cerebral palsy:** This is a general term for a group of disorders caused by damage to a child's brain during early development—before, during or shortly after birth—that affects the child's ability to coordinate body movements.
- **Multiple sclerosis (MS):** MS is a chronic, potentially debilitating disease that affects our central nervous system, which comprises brain and spinal cord.
- **Chickenpox:** Ataxia can be an uncommon complication of chickenpox and other viral infections. It may appear in the healing stages of the infection and last for days or weeks. Normally, the ataxia resolves completely over time.
- **Paraneoplastic syndromes:** These are rare, degenerative disorders triggered by our immune system's response to a cancerous tumor (neoplasm), most commonly from lung, ovarian, breast or lymphatic cancer. Ataxia may appear months or years before the cancer is diagnosed.
- **Tumor:** A growth on the brain, cancerous or noncancerous (benign), can damage the cerebellum.
- **Toxic reaction:**[3–5] Ataxia is a potential side effect of certain medications, such as barbiturates, such as phenobarbital, and sedatives, such as benzodiazepines. Alcohol and drug intoxication; heavy metal poisoning—from lead or mercury and solvent poisoning—from paint thinner, can cause ataxia.

For some adults who develop sporadic ataxia, no specific acquired or genetic cause can be found. This is known as sporadic degenerative ataxia, which can take a number of forms, including multiple system atrophy (MSA), which is a progressive, degenerative disorder.[3–5]

Hereditary Ataxias

Some types of ataxia and some conditions that cause ataxia are hereditary.

Autosomal Dominant Ataxias

These include:
- Spinocerebellar ataxias.
- Episodic ataxia.

Autosomal Recessive Ataxias

These include:
- **Friedreich's ataxia:** This is also called *sensory ataxia*.
- **Ataxia telangiectasia:** This rare, progressive childhood disease causes degeneration in the brain and other body systems.
- **Congenital cerebellar ataxia:** This type refers to ataxia that results from damage to the cerebellum at birth.
- **Wilson's disease:** People with this condition accumulate copper in their brains, livers and other organs, which can cause neurological problems, including ataxia.[6,7]

CLINICAL FEATURES

- **Dysmetria:** It is the difficulty to judge the direction, distance, extent, force and timing of the limb movement. Two types:
  i. *Hypometria*—deficient extent of movement (the patient shoots the limb well in front of the target)
  ii. *Hypermetria*—excessive extent of movement (the patient shoots the limb beyond the target)
- **Dyssynergia:** It is the disturbance of smooth muscular coordination, resulting in uncoordination and abrupt movements.
- **Rebound phenomenon:** Dysfunction in agonist–antagonist relationship due to loss of cerebellar component of a stretch reflex. When a patient is made to contract the muscle isometrically and suddenly removes the resistance force, the limb moves abruptly lacking the control.
- **Dysdiadokinesia:** Difficulty in performing rapid alternative movements. If the patient is unable to perform rapid alternating movements, then it is called *Dysdiadokinesia*. Adiadochokinesia is the inability to reverse the direction of the movement.
- **Tremor:** Oscillatory movement about a joint.
  i. Intention tremor (kinetic tremor)—occurs during the movements.
  ii. Postural tremor may be evident by back and forth oscillatory movements of the body while maintaining a standing posture.
  iii. Postural tremors also may be observed as up and down oscillatory movements of a limb when it is held in antigravity position.
  iv. Titubation—refers to rhythmic oscillations of the head (side-to-side, forward and backward or may have rotatory component).
- **Hypotonia:** Reduced tone or diminished resistance to the passive movement.
- **Dysarthria:**
  - Disorder of speech articulation
  - Speech is slow and slurred—*scanning speech (one word at one time)*.
  - Pattern is very slow, hesitant with prolonged syllables and inappropriate pauses.
- **Occular disturbances:** *Nystagmus* is a form of involuntary or seesaw rhythmical eye movement. The most common type of nystagmus is called *gaze paretic nystagmus* in which there is inability to hold eccentric position of gaze resulting in need to make repetitive saccades to look eccentrically.
  Other type of nystagmus are down beat nystagmus and rebound phenomenon nystagmus.
  Voluntary gaze is accomplished by a series of jerky movements which is called *saccadic dysmetria*.
  Ocular flutter in which there is occasional burst of very rapid to and fro flutter-like horizontal oscillation around the point of fixation.
- **Gait changes :** Broad base gait or ataxic gait:
  - Upright stand stability is often poor and arms may be held away from the body to improve balance (high guard position).
  - Stepping patterns are irregular in direction and distance.
  - Initiation of forward progression of a lower extremity may start slowly, and the extremity may unexpectedly be flung rapidly and forcefully forward and audibly hit on the floor.
- **Cognitive and mood problems:** In addition to motor dysfunction, patients with cerebellar degeneration may have cognitive and emotional difficulties.

INVESTIGATIONS AND DIAGNOSTIC APPROACH

A diagnostic approach to ataxia may include:
- Neurological and medical history, including drug and toxin exposures.
- Family history of neurological problems.
- Neurological and medical examinations.
- Blood tests to rule out specific deficiencies and toxins.
- Urine screen for mercury exposure.
- Brain imaging: Magnetic resonance image (MRI) or computed tomogram (CT) (Fig. 7.2).
- Possible neuroimaging of the spinal cord.
- Electrophysiologic testing (electromyography and nerve conduction velocity testing), if there are signs or symptoms of peripheral nerve dysfunction.

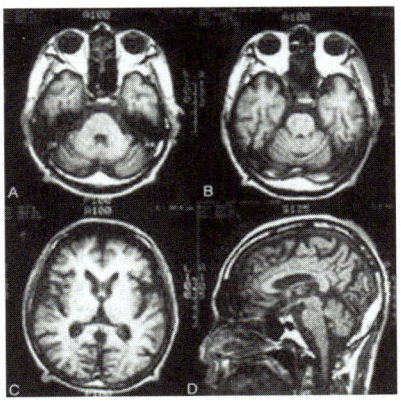

Fig. 7.2: Brain imaging

Berg Balance Scale (Table 7.1)

Description: 14-item scale designed to measure balance of the older adult in a clinical setting.

Category	Directions	Select best response 0, 1, 2, 3 or 4	Score
1. Sitting to standing	Stand up without using hands	4 = Able to stand without hands independently; 3 = Able with hands independently; 2 = Able with hands >1 try; 1 = Needs assistance to stand/stabilize; 0 = Needs moderate or maximum assistance	
2. Standing unsupported	Stand for 2 minutes without holding on	4 = Able to stand safely for 2 minutes; 3 = Able to stand 2 min without support; 2 = Able to stand 50 sec unsupported; 1 = Needs >1 try to stand 30 sec unsupported; 0 = Unable to stand 30 sec unsupported	
3. Sitting with no back support. Feet support on ground	Sit with arms folded for 2 minutes	4 = Able to sit safely 2 min; 3 = Able to sit 2 min under support; 2 = Able to sit 30 sec; 1 = Able to sit 10 sec; 0 = Unable to sit without support 10 sec	
4. Standing to sitting	Sit down	4 = Sits safely with minimum use of hands; 3 = Controls descent with hands; 2 = Uses back of legs on chair; 1 = Sits independent but uncontrolled; 0 = Needs assistance	
5. Pivot transfers (chair-to-chair or chair-to-bed)	Transfer 1 way to seat with arms, and 1 way to seat without arms	4 = Transfers safe with minor use of hands; 3 = Transfers safe needs hands; 2 = Transfers with verbal cues or support; 1 = Needs 1 assistance; 0 = Needs 2 assistance or support	

(Contd.)

Table 7.1: Berg balance scale (Contd.)

Category	Directions	Select best response 0, 1, 2, 3 or 4	Score
6. Standing unsupported eyes closed	Close eyes and stand still for 10 seconds	4 = Stands 10 sec safe; 3 = Stands 10 sec with support; 2 = Stands 3 sec; 1 = Unable to keep eyes closed 3 sec but stands safe; 0 = Needs assistance	
7. Standing unsupported feet together	Feet together without holding on	4 = Can place feet together and stand 1 min safely; 3 = Can place feet together independently and stand 1 min without support; 2 = Can place feet together but unable to hold 30 sec; 1 = Needs assistance to attain but able to stand 15 sec together; 0 = Needs assistance to attain/cannot hold for 15 sec	
8. Reaching forward with outstretched arm while standing	Lift arm(s) to 90°. Record distance of forward reach	4 = Can reach forward 10"; 3 = Can reach 5"; 2 = Can reach forward 2"; 1 = Reaches but needs support; 0 = Loses balance, needs assistance	
9. Pick up object from floor standing	Pick up object from floor	4 = Able to pick up safely; 3 = Able but needs support; 2 = Unable but reaches 1–2" from object and keep balance; 1 = Unable and needs support; 0 = Unable and needs assistance	
10. Turn to look over shoulders standing	Turn to look directly behind you over left shoulder. Repeat right	4 = Looks behind from both sides and weight shifts well; 3 = Looks back 1 side only; 2 = Turns sideways only but keeps balance; 1 = Needs support; 0 = Needs assistance	
11. Turn 360°	Turn completely in a full circle. Repeat other way		
12. Place foot on stool while standing unsupported	Place each foot alternately on step/stool. Continue until each foot touches step 4 times	4 = Stands independently and safely 8 steps in 20 sec; 3 = 8 steps in >20 sec; 2 = Completes 4 steps without and with support; 1 = Completes >2 steps, needs min assistance; 0 = Needs assistance to prevent fall	
13. Standing unsupported one foot in front	Place 1 foot directly in front of other. If unable, step far enough so heel is ahead of toe	4 = Can place foot tandem independently and hold 50 sec; 3 = Can place foot ahead independently 50 sec; 2 = Small step independently 30 sec; 1 = Assist to step but holds 15 sec; 0 = Loses balance	
14. Standing on one leg	Stand on one leg as long as able without holding on	4 = Lifts leg independently >10 sec; 3 = Lifts leg independently and hold 5–10 sec; 2 = Lifts leg independently and hold 3+ sec; 1 = Tries to lift, made to hold 3 sec but stands independently; 0 = Unable	

Equipment needed: Ruler, two standard chairs (one with arm rests, one without), footstool or step, stopwatch or wristwatch, 15 ft walkway.

Completion:
- Time: 15–20 minutes
- Scoring: A five-point scale, ranging from 0 to 4.

- "0" indicates the lowest level of function
- "4" the highest level of function.
- Total Score = 56 Interpretation:
- 41–56 = low fall risk
- 21–40 = medium fall risk
- 0–20 = high fall risk

PHYSIOTHERAPY ASSESSMENT FOR ATAXIA (CLINICAL TESTS FOR IDENTIFICATION OF SIGNS OF ATAXIA)

Symptom/sign	Test	How to	Observation/inference
Dysmetria	Finger–nose test (Fig. 7.3)	Ask the patient to touch with one index finger his tip of the nose to the tip of the other index finger	There will be overshooting or undershooting indicating dysmetria
Dyssynergia	Finger–nose—therapist finger	Ask the patient to touch with one index finger his tip of the nose to the tip of therapist finger and his other index finger	There will be no coordination and the patient misses the sequence on repetitions
Rebound phenomenon	Fist making	Ask the patient to make a tight fist and relax	There will be automatic reflex opening of the fist
Dysdiadokinesia	Supination-pronation test	Ask the patient to tap on his thigh by alternative supination and pronation	Patient cannot do the alternative movements
Nystagmus	Focus test	Ask the patient to focus an object (which is moved rhythmically by a therapist to and fro, and stopped suddenly) for a few seconds without blinking	There will be rhythmical, oscillatory movement of the eyeball
Ataxic gait (Fig. 7.4)	Walking on figure of eight	Draw a figure of eight and ask the patient to walk on it	Patient tends to fall due to lack of balance. The stepping patterns will be irregular and directionless.

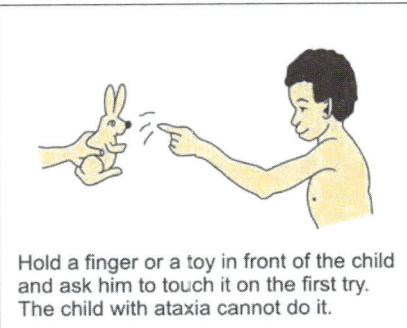

Hold a finger or a toy in front of the child and ask him to touch it on the first try. The child with ataxia cannot do it.

Fig. 7.3: Test for ataxia

To keep her balance, the child with ataxia walks bent forward with feet wide apart. She takes irregular steps, like a sailor on a rough sea or someone who is drunk.

Fig. 7.4: Ataxic gate pattern

- A change of 8 points is required to reveal a genuine change in function between 2 assessments.

The tests consist of 14 tasks:
1. Sitting to standing
2. Standing unsupported for 2 min without holding to support
3. Sitting unsupported with feet touching the floor for 2 min
4. Standing to sitting
5. Transfers
6. Standing with eyes closed
7. Standing with feet together
8. Reaching forward with outstretched arm
9. Retrieving object from floor
10. Turning to look behind

11. Turning 360°
12. Placing alternate foot on stool
13. Standing with one foot in front
14. Standing on one foot

SCALE FOR THE ASSESSMENT AND RATING OF ATAXIA (SARA)[8-10]

1. Gait

Patient is asked (1) to walk at a safe distance parallel to a wall including a half-turn (turn around to face the opposite direction of gait) and (2) to walk in tandem (heels to toes) without support.

  Normal, no difficulty in walking, turning and walking tandem (up to one misstep allowed)
  1—Slight difficulties, only visible when walking 10 consecutive steps in tandem.
  2—Clearly abnormal, tandem walking more than 10 steps not possible.
  3—Considerable staggering, difficulties in half-turn, but without support.
  4—Marked staggering, intermittent support of the wall required.
  5—Severe staggering, permanent support of one stick or light support by one arm required.
  6—Walking more than 10 m only with strong support (two special sticks or stroller or accompanying person).
  7—Walking less than 10 m only with strong support (two special sticks or stroller or accompanying person).
  8—Unable to walk, even supported.

2. Stance

Proband is asked to stand (1) in natural position, (2) with feet together in parallel (big toes touching each other) and (3) in tandem (both feet on one line, no space between heel and toe). Proband does not wear shoes, eyes are open. For each condition, three trials are allowed. Best trial is rated.

  0—Normal, able to stand in tandem for >10 s.
  1—Able to stand with feet together without sway, but not in tandem for >10 s.
  2—Able to stand with feet together for >10 s, but only with sway.
  3—Able to stand for >10 s without support in natural position, but not with feet together.
  4—Able to stand for >10 s in natural position only with intermittent support
  5—Able to stand >10 s in natural position only with constant support of one arm.
  6—Unable to stand for >10 s even with constant support of one arm.

3. Sitting

Proband is asked to sit on an examination bed without support of feet, eyes open and arms outstretched to the front.

  0—Normal, no difficulties sitting >10 s.
  1—Slight difficulties, intermittent sway.
  2—Constant sway, but able to sit >10 s without support.
  3—Able to sit for >10 s only with intermittent support.
  4—Unable to sit for >10 s without continuous support

4. Speech Disturbance

Speech is assessed during normal conversation.

  0—Normal.
  1—Suggestion of speech disturbance.
  2—Impaired speech, but easy to understand.
  3—Occasional words difficult to understand.
  4—Many words difficult to understand.
  5—Only single words understandable.
  6—Speech unintelligible/anarthria.

5. Finger Chase

Rated separately for each side: Proband sits comfortably. If necessary, support of feet and trunk is allowed. Examiner sits in front of proband and performs 5 consecutive sudden and fast pointing movements in unpredictable directions in a frontal plane, at about 50% of proband's reach. Movements have an amplitude of 30 cm and a frequency of 1 movement every 2 s. Proband is asked to follow the movements with his index finger, as fast and precisely as possible. Average performance of last 3 movements is rated.

0—No dysmetria.
1—Dysmetria, under/overshooting target <5 cm.
2—Dysmetria, under/ overshooting target < 15 cm.
3—Dysmetria, under/ overshooting target > 15 cm.
4—Unable to perform 5 pointing movements.

6. Nose–Finger Test

Rated separately for each side: Proband sits comfortably. If necessary, support of feet and trunk is allowed. Proband is asked to point repeatedly with his index finger from his nose to examiner's finger which is in front of the proband at about 90% of proband's reach. Movements are performed at moderate speed. Average performance of movements is rated according to the amplitude of the kinetic tremor.

0—No tremor.
1—Tremor with an amplitude <2 cm.
2—Tremor with an amplitude <5 cm.
3—Tremor with an amplitude >5 cm.
4—Unable to perform.
5—Pointing movements.

7. Fast Alternating Hand Movements

Rated separately for each side: Proband sits comfortably. If necessary, support of feet and trunk is allowed. Proband is asked to perform 10 cycles of repetitive alternation of pro- and supinations of the hand on his/her thigh as fast and as precise as possible. Movement is demonstrated by examiner at a speed of approximately 10 cycles within 7 s. Exact times for movement execution have to be taken.

0—Normal, no irregularities (performs <10 s).
1—Slightly irregular (performs <10 s).
2—Clearly irregular, single movements difficult to distinguish or relevant interruptions, but performs <10 s.
3—Very irregular, single movements difficult to distinguish or relevant interruptions, performs >10 s.
4—Unable to complete 10 cycles.

8. Heel-Shin Slide

Rated separately for each side: Proband lies on examination bed, without sight of his legs. Proband is asked to lift one leg, point with the heel to the opposite knee, slide down along the shin to the ankle, and lay the leg back on the examination bed. The task is performed 3 times. Slide-down movements should be performed within 1 s. If proband slides down without contact to shin in all three trials, rate 4.

0—Normal.
1—Slightly abnormal, contact to shin maintained.
2—Clearly abnormal, goes off shin up to 3 times during 3 cycles.
3—Severely abnormal, goes off shin 4 or more times during 3 cycles
4—Unable to perform the task

Tests for Coordination and Balance

The coordination is checked by equilibrium and non-equilibrium tests.

Non-Equilibrium Tests

These tests assess both static and mobile components of movement of body when it is not in upright position. These tests assess both fine and gross movements. The various tests are:

1. Finger to nose (patient)
2. Finger to therapist finger

3. Finger to finger (patient)
4. Alternate nostril to finger (patient)
5. Finger opposition (patient)
6. Tapping of foot and hand (patient)
7. Heel to shin test (patient)
8. Drawing a circle.

Equilibrium Tests

These tests assess both gross and fine movements. It also assesses both static and dynamic components of posture and balance when the body is in standing position. The tests include:

1. **Romberg's test:** Ask the patient to stand with eyes closed. If he cannot maintain standing balance with eyes closed, then the test is positive.
2. Standing with feet together
3. Standing with one foot directly in front of other.
4. Standing on one foot.
5. Walking by placing the heel of one foot directly in front of toe of other foot [tandem walking]
6. Walking along a straight line sideways, backwards
7. Walk on a circle on heels and later on toes.
8. Walk on a figure of eight.

Tests for Balance

Various tests of balance are as follows:

1. **Romberg test:** Here therapist instructs patient to stand straight with both feet together with eyes open and close. Therapist should stand adjacent to the patient so that the patient should not fall. This test is positive, if patient sways side-to-side or front and back with closed eyes.
2. **Single leg standing:** Here therapist instructs the individual to stand on one leg with eyes open and close. On eye closing, if patient sways in this position side-to-side or front and back, then the test is positive.
3. **Alternate single leg standing:** Here therapist instructs the patient to stand on single leg alternatively with eyes open and close. If patient falls down while eyes closed, then the test is positive.
4. **Wobble board:** Here therapist instructs the patient to stand on a wobble board and start moving the board on right and left side. If patient is able to maintain himself from falling down, the test is negative.
5. **Sharpened Romberg test:** In this test, the patient is made to stand with the feet in a tandem stance attitude, heel of one foot directly in front of the toes of other, arm folded across the chest and stand for one min without sway. If the patient sways, then the test is positive.
6. **Timed stance:** Timed stance test consists of making the patient stand in various foot positions in 8 different positions with eyes open and then eyes closed and the patient is asked to make balance for about 30 sec.
7. **Postural sway test:** It is a computerized test where a patient is asked to stand on a computer driven force plate for about 20–30 sec, sway with both eyes open and with eyes closed if often measured. The computer provides the graphical and numerical quantification of balance.
8. **Nudge test:** The patient is given perturbations in various positions and checked for balance.
9. **Functional reach test:** In this test, the patient is asked to reach an object from standing position as far as possible without losing the balance.
10. **Get up and go test:** This test detects the problems of balance in ADL activities. In this test, the patient is asked to sit in a chair with both feet touching the floor

and then asked to stand, walk for about 3 meters, turn around and return and sit. Performance is graded as 1. normal, 2. very slight abnormality, 3. mildly abnor-mal, 4. moderately abnormal, and 5. severely abnormal.

PHYSIOTHERAPY MANAGEMENT

Aims
1. To provide psychological support to the patients
2. To improve coordination and balance
3. To normalize tone in hypotonic ataxias
4. To improvise gait
5. To enhance the quality of living

Plans
The above goals are achieved by regaining coordination in the movement.

The coordination exercises are designed to improve neuromuscular coordination in order to attain smooth, accurate, purposeful movements.

The exercises are as follows:
1. Equilibrium exercises
2. Non-equilibrium exercises
3. Frenkel's exercises

Principles of Training Equilibrium and Non-Equilibrium Exercises
a. The patient must be fully relaxed.
b. He must be explained clearly the methods and instructions of the exercise.
c. The patient should start the exercise slowly and improve steadily and progressively.
d. Encourage the patient with verbal commands throughout the exercise.
e. Avoid over enthusiasm. Never allow the fatigue to set in.
f. Give as many number of breaks as patient wants.
g. Progress to more difficult exercise only after the previous one is mastered.

Equilibrium Exercises
i. Stand with both feet together with open and closed eyes
ii. Stand with heel of one feet exactly in front of the other
iii. Stand on one foot
iv. Walk on a straight line sideways and backwards
v. Walk along a circle
vi. Walk in a figure of eight
vii. Walk on the heels and toes alternatively

Non-Equilibrium Exercises
i. Finger to nose.
ii. Finger to finger.
iii. Finger to nose to therapist finger.
iv. Heel to shin.
v. Alternative pronation and supination.
vi. Tapping hand and foot.
vii. Drawing a circle.
viii. Fixation and position holding.
ix. Touch the therapist finger with toe.
x. Alternatively touch with heel of other foot.

Frenkel's Exercises
These are designed for sensory ataxia, there will be many activities involving lower and upper limbs. Progression is made by changing the speed, range or exercise.

Example:

Exercises for legs in lying:
- Position: Supine lying with head raised.
- Activity: Hip abduction and adduction on fully supported smooth surface.

Therapy for Balance and Proprioception
i. Therapy is used to improve balance and increase the independence of the patient using techniques focusing on balance, posture and increasing coordination.
ii. Therapeutic goals include improving balance and posture against outside stimuli, increasing joint stabilization,

developing independent, functional gait to promote independence.

iii. Training principles include progressing from simple to complex exercises, practicing exercises with eyes open and closed and providing support with home exercise and sports activities.

iv. Therapy involves plyometric exercises, balance board and mini-trampoline exercises, PNF.

v. Vibration and suit therapy are also used to improve proprioception, posture and movement.

vi. Yoga and other body-awareness exercises may also be included in the treatment plan to increase proprioception.

REFERENCES

1. Schmahmann JD. "Disorders of the cerebellum: ataxia, dysmetria of thought, and the cerebellar cognitive affective syndrome". J Neuropsychiatry Clin Neurosci 2004;16(3):36.
2. Fredericks CM. "Disorders of the Cerebellum and its Connections". In Saladin LK, Fredericks CM. Pathophysiology of the motor systems: Principles and clinical presentation, Philadelphia, FA Davis. 1996. ISBN 0-8036-0093-3.
3. Browne TR. "Clonazepam. A review of a new anticonvulsant drug". Arch Neurol 1976;33(5):326–32.
4. Gaudreault P, Guay J, Thivierge RL, Vardy. "Benzodiazepine poisoning. Clinical and pharmacological considerations and treatment". Drug Saf 1991;6(4):247–65.
5. Diez S. "Human health effects of methylmercury exposure". Rev. Environ. Contam. Toxicol. Reviews of Environmental Contamination and Toxicology 2009;198:111–32.
6. Walshe JM, Clarke CE, Nicholl DJ, (Eds.). "Wilson Disease". Birmingham Movement Disorders Coursebook.
7. Haldeman-Englert C. "Wilson's disease—PubMed Health". PubMed Health.
8. Schmitz-Hubsch T, Tezenas du Montcel S, Baliko L, Berciano J, Boesch S et al. Scale for the assessment and rating of ataxia: Development of a new clinical scale. Neurology, 2006;66:1717–20.
9. Schmitz-Hubsch T, Fimmers R, Rakowicz M, Rola R, Zdzienicka E, Fancellu R, et al. Responsiveness of different rating instruments in spinocerebellar ataxia patients. Neurology, 2010;74:678–84.
10. Wayer A, Abele M, Schmitz-Hubsch T, Schoch B, Frings M, et al. Reliability and validity of the scale for the assessment and rating of ataxia: A study in 64 ataxia patients. Mov C 2007; 22:1633–37.

8

Parkinsonism

LEARNING OBJECTIVES

At the end of this chapter, you will be able to:
1. List out various disorders of basal-ganglia
2. Describe the pathophysiology of Parkinson's disease
3. List out the various etiological factors of Parkinson's disease
4. Identify the various clinical manifestations of the Parkinson's diasese
5. Demonstrate the skills to assess the patients with Parkinson's disease
6. Design the treatment protocols for rehabilitating these patients.

INTRODUCTION

Parkinson's disease (also known as Parkinson disease or PD) is a degenerative disorder of the central nervous system that often impairs the sufferer's motor skills, speech, and other functions.

It is characterized by muscle rigidity, tremor, a slowing of physical movement (bradykinesia) and, in extreme cases, a loss of physical movement (akinesia).

The term *parkinsonism* is used for symptoms of tremor, stiffness, and slowing of movement caused by loss of dopamine.

Parkinson's disease was first described in England in 1817 by James Parkinson. The condition was popularly known as shaking palsy derived from Latin word paralysis agitans.

DEFINITION

It is a slowly progressing disease involving primarily a degeneration of cells in basal ganglia and substantia nigra leading to deficit of dopamine—a neurotransmitter which produces gradual weakness of the voluntary movement with muscular rigidity, tremors impairment in the balance and automatic reactions.

Under microscopic examination, the damaged neurons in the substantia nigra show a round inclusion called Lewy body in Parkinson's disease, hence sometimes it is also called Lewy body Prakinson's disease or simply Lewy body disease.

ETIOLOGY[1,2]

1. Idiopathic
2. Multi-infarct atherosclerosis

3. Viral infection (*encephalitis lethargica*)
4. Chemical toxicity: *Manganese, carbon-disulphide, carbon monoxide, cyanide*, etc.
5. Drug-induced parkinsonism: *Phenothiazines, butryophenones*, etc.
6. Secondary to *Alzheimer disease, Wilson's disease, Shy-Drager syndrome*, etc.
7. Metabolic causes: Abnormal calcium metabolism
8. Traumatic

Types of Basal Ganglia Disorders

Many brain disorders are associated with basal ganglia dysfunction. They include:
- Dystonia
- Huntington's disease
- Multiple system atrophy
- Parkinson's disease
- Progressive supranuclear palsy
- Wilson's disease

EPIDEMIOLOGY[4-6]

- PD is the second most common neurodegenerative disorder.
- Onset is usually in old age between 50 and 65 years.
- Males are more affected than females.
- More common in western countries less in Africans.

PATHOLOGY[7,8]

Due to degeneration of cells in basal ganglia and substantia nigra, the dopamine level which is a neurotransmitter responsible for coordination of normal movement is decreased.

Because of decrease in the dopamine, the dopamine receptors in the basal ganglia circuit called striatum are not adequately stimulated.

This causes decreased or absent inhibitory influence on the cholinergic pathway causing excessive excitation of the extrapyramidal system.

This impairment leads to rigidity, bradykinesia, tremors, and disorders of gait and balance.

CLINICAL MANIFESTATIONS (Figs 8.1 and 8.2)

Motor Symptoms

Tremor

Involuntary oscillations of body parts at frequency ranging from 3.5 to 7 Hz.[9]

- It is most commonly a *rest tremor*: The tremors become maximal when the limb is at rest and disappearing with voluntary movement and sleep.
- The tremors may be present when holding up the outstretched arms which are known as *postural or sustention tremors*.
- The tremors may be present when performing various movements which are known as *action tremors*.
- If the tremor is only felt by the patient and not seen by the examiner, it is called *internal tremor*.
- It is a pronation–supination tremor that is described as "*pill-rolling*".[9-11]
- Tremor affects to a greater extent the most distal part of the extremity and is typically unilateral at onset.

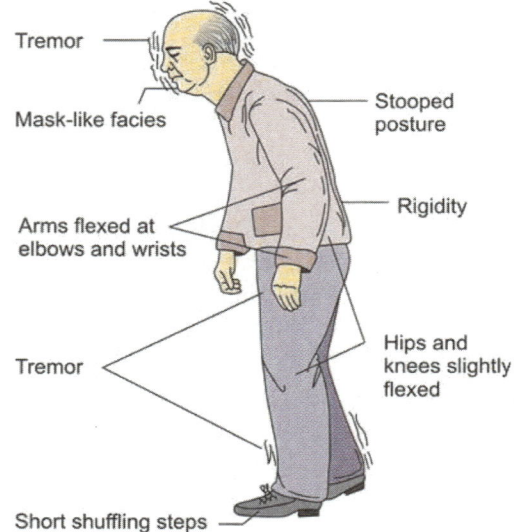

Fig. 8.1: Clinical manifestations of PD

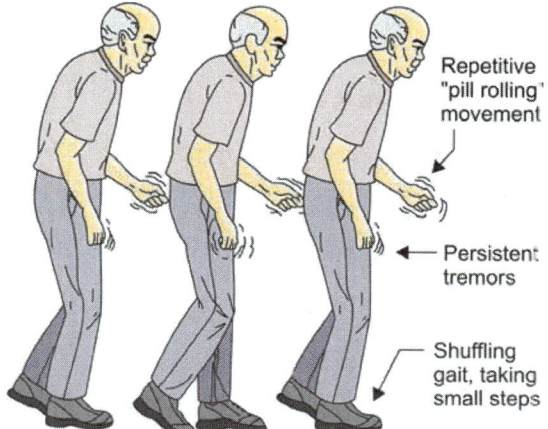

Fig. 8.2: Symptomatology of PD

Increased reaction time: The lapse in time between the desire and actual movement.

The patient usually complains difficulty in rolling in the bed and fine motor coordination like buttoning a shirt.

The bradykinesia may be due to inability or deficiency of the basal ganglia to integrate sensory information.

On EMG examination, there will be time lapse between the actual recruitment of motor units, firing by motor units and initiation of movement.

Akinesia is the inability to initiate a voluntary movement and is attributed to severe rigidity.

- Tremors usually begin in one hand and may later spread to other side. Feet, legs, jaw, head and neck can also show tremors.

Rigidity[12]
- It is an increased resistance when stretching a muscle passively throughout the range of motion.
- It is the contraction of agonist and antagonist muscles due to an increase in the supraspinal influences leading to increased tone in agonist and antagonist muscles. There is increased resistance throughout the range of motion and the limb is perceived as heavy or stiff by the patient.
- Rigidity can involve all parts of the body and be symmetrical or asymmetrical.
- The common forms of rigidity are:
  - *Lead pipe*: The rigidity where the resistance is constant and uniform.
  - *Cogwheel*: The rigidity where the resistance is intermittent but uniform.

Bradykinesia and Akinesia

Bradykinesia[12,13] is slowness in initiating the voluntary movement. There is decreased amplitude and intensity of contraction.

Postural Disturbances
- Failure of postural reflexes, along other disease-related factors such as orthostatic hypotension or cognitive and sensory changes, which lead to impaired balance and falls. It usually appears in the late stages of PD.
- There is failure to maintain static or dynamic postures.
- **Camptocormia:**[14,15] The patient develops a stooped, forward-flexed posture. In severe forms, the head and upper shoulders may be bent at a right angle relative to the trunk.
- **Scoliosis:** Gait and posture disturbances
- **Shuffling gait:**[15–20] Gait is characterized by short steps, with feet barely leaving the ground. Small obstacles tend to cause the patient to trip. There is decreased arm-swing.
- **Turning "en bloc":** Rather than the usual twisting of the neck and trunk and pivoting on the toes, PD patients keep their neck and trunk rigid, requiring multiple small steps to accomplish a turn.
- **Festination:**[15–20] Because of a combination of stooped posture, imbalance, and short steps, it results to a gait that gets progressively faster and faster, often ending in a fall.

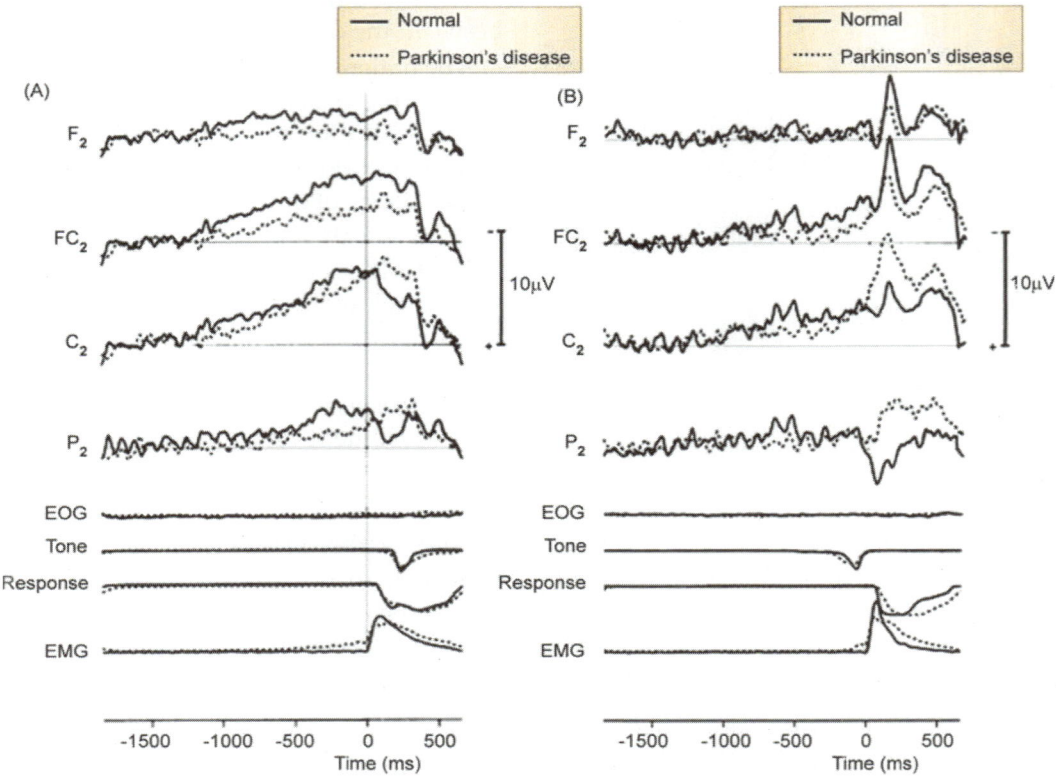

Fig. 8.3: Electrodiagnostic studies

- **Dystonia**: It is the abnormal, sustained, painful twisting muscle contractions, often affecting the foot and ankle (mainly toe flexion and foot inversion) which often interferes with gait.

Speech and swallowing disturbances

The characteristics of speech in Parkinson's disease are as follows:

- *Hypophonia* (soft speech), monotonic (speech quality tends to be soft, hoarse, and monotonous), festinating (excessively rapid, soft, poorly-intelligible speech).
- *Drooling*: There will be spilling of saliva from the mouth. The drooling is mainly due to decreased swallowing but not excessive production.
- *Dysphagia*: Difficulty in swallowing. Swallowing is an automatic and complex act involving the coordination of tongue and throat muscles. Because of the inability to coordinate, in PD there will be pooling of food in the throat.
- *Dysarthria*: Difficulty in articulation of speech.

Other Motor Symptoms

- Fatigue
- Hypomimia—a mask-like face
- Difficulty in rolling in bed or rising from a seated position.
- Micrographia—small, cramped handwriting.
- Impaired fine motor dexterity and motor coordination.
- Impaired gross motor coordination.
- Akathisia—an unpleasant desire to move.
- Re-emergence of primitive reflexes.

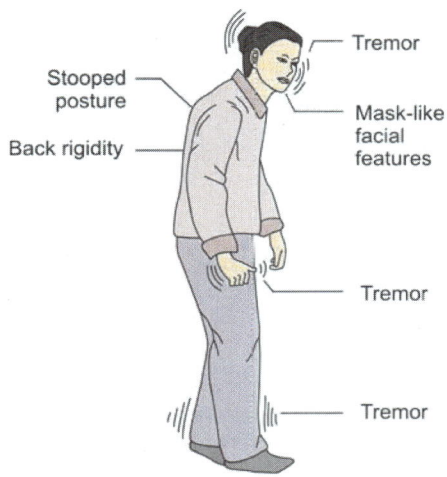

Fig. 8.4: Common clinical features of PD

Neuropsychiatric[9]

- *Dementia*: Starting with slowing of thought and progressing to difficulties with abstract thought, memory, and behavioral regulation.
- *Perceptual problems:*[9,21] There is perceptual deficits like difficulty in vertical perception, body scheme, body image disorder and topographical disorientation.

Autonomic

- Orthostatic hypotension
- Oily skin and dermatitis
- Urinary incontinence (typically in later disease progression) and nocturia (getting up in the night to pass urine)
- Sweating dysfunction

Secondary Complications

- **Seborrheic dermatitis:** It is a common disorder in PD patients due to excessive oily secretions particularly on the forehead and scalp.
- The skin becomes greasy, red itchy and flaky. There may be excessive dandruff in the scalp.
- **Ankle swelling:** There is excessive swelling in the feet due to reduced muscle movement of calf muscles and also due to anti-Parkinson's medications.

- **Visual problems:** The patients may complain of mild double vision.
- **Nystagmus:** Rhythmical oscillatory movements of the eyeball when fixing the gaze.
- There is decreased joint range of motion and flexibility leading to tightness and contracture.
- Muscle atrophy and weakness.
- Weight loss.
- Constipation due to decreased movements in the bowel and also attributed to anti-Parkinson's medications.
- Urinary urgency and frequency is observed due to normal reflex mechanisms controlling the bladder are impaired. There is slowness in voiding.
- Sexual desire may be reduced in Parkinson's disease. Inability to achieve and maintain erections are the common problems seen in men.
- Imbalance, lightheadedness, vertigo may be present in PD.
- Psychological depression and disturbances in sleep.
- There may be hallucinations and delusion. Paranoia and drug-induced psychosis also be presenting psychiatric symptoms in Parkinson's disease.
- Deconditioning of the cardiopulmonary system is common due to decreased activity and kyphotic posture that causes compression of the vital structures in thorax.
- Prolonged inactivity and poor diet intake causes osteoporosis.
- Increased risk of falls.

INVESTIGATIONS

1. Single photon emission computed tomography (SPECT)
2. Transcranial ultrasound
3. Routine blood investigations and clinical essays

MANAGEMENT

Medical Management

The following drugs are available for the Parkinsonism:
- Levodopa
- Dopamine agonists
- COMT (*catechol-O-methyltransferase*) inhibitors
- Amantadine
- Anticholinergics
- Monoamine oxidase inhibitors
- Selegiline

Surgical Management

Thalamotomy: It is the destruction of small group of cells in thalamus. This surgery is done to abolish tremors on the side of the body opposite to the surgery.

Pallidotomy: It is the surgical destruction of a group of cells in the internal globus pallidus. These procedures are most effective in relieving dyskinesias and tremors.

Deep brain stimulation (DBS) surgery: During deep brain stimulation surgery, electrodes are inserted into the targeted brain region using MRI and neuro-physiological mapping to ensure that they are implanted in the right place. A device called an impulse generator or IPG (similar to a pacemaker) is implanted under the collar bone to provide an electrical impulse to a part of the brain involved in motor function (Fig 8.5). Those who undergo the surgery are given a controller, which allows them to check the battery and to turn the device on or off. An IPG battery lasts for about three to five years and is relatively easy to replace under local anesthesia.

PHYSIOTHERAPY MANAGEMENT

For patients with PD, the main objective of the physiotherapy is to improve the quality of life by maintaining or increasing patient's functional independence, safety and well being.

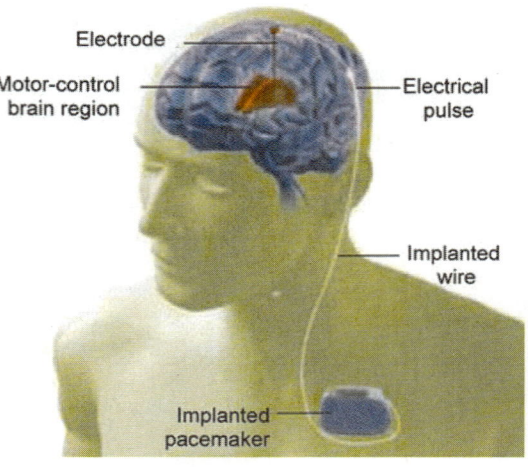

Fig. 8.5: IPG device

The physiotherapy management includes a thorough successive assessment and carrying out the treatment to achieve the goals.

PHYSIOTHERAPY ASSESSMENT

1. Obtain the demographic data of the patient by interviewing the patient, caretakers and thorough reviewing of the medical notes.
2. Take the history of the present illness, past illness irrelevant to the PD, family, personal, medical history.
3. Take a note of medications presently the patient is using.

Observations

1. Observe for the general built of the patient.
2. Observe the tremors and note the type—resting, postural, action or internal.
3. Observe for the tropical changes of the skin.
4. Observe for any disuse wasting of muscles.
5. Observe for facial expression, usually expression less or masked face is observed.
6. Observe for drooling of saliva.

7. Observe the posture of the patient in lying, supine and in standing.

Palpation

1. Check for tenderness and grade, if any.
2. Edema is usually found in feet due to reduced contractions of calf muscles and also due to anti-Parkinson drugs.

Examination

1. Examine the higher functions, memory, consciousness, intelligence behavior and orientation.
2. In patient's with PD, there may be memory loss, altered behavior (dementia, perceptual problems), disorientation and speech disorders.
3. The speech will be hypophonia (soft speech), monotonic (speech quality tends to be soft, hoarse, and monotonous), festinating (excessively rapid, soft, poorly-intelligible speech).
4. There will be dysarthria—difficulty in articulation of speech.
5. Muscle tone assessment: The tone is checked by repetitive passive movement. There will be rigidity. Examine the extent and distribution of rigidity and check whether it is cog wheel or lead pipe.
6. Examine the tightness of the muscle.
7. Examine the range of motion of all the joints by goniometry.
8. Examine the movement pattern to check bradykinesia or akinesia.
9. Testing for bradykinesia is done by timing the various activity undertaken by the patient, e.g. ask the patient to touch the opposite shoulder, trunk bending, hip adduction and flexion, etc.
10. Tapping the foot on the ground and then counting the number of tapping in a minute could also assess the bradykinesia.
11. Examine the superficial and deep reflexes. The reflexes are usually difficult to elicit. The plantar reflex would be downgrading.
12. Assess for the coordination and balance. Check the static and dynamic balance of the patient. Retropulsion test in which a sudden, unexpected, quick and firm jerk on the shoulder is given in a backward position.
13. Gait assessment: Observe for the festinating and shuffling gait. Examine for turning en bloc. Assess the various gait parameters like step length, stride length, step width, cadence, etc.
14. Timed up and go test is a reliable test in which gait and balance could be assessed.
15. Examine the cardiopulmonary assessment. Check for the breathing pattern, accessory muscle of respiration usuage, inspiratory–expiratory ratio, chest expansion symmetry and depth of excursion.
16. Endurance is checked by six-minute walk test.
17. Functional evaluation: ADL assessment is very important in PD. There are various scales available for functional evaluation.

Hoehn and Yahr Scale of Parkinsonism[22,23]

0	No clinical signs evident
I	Unilateral involvement
II	Bilateral involvement but no postural abnormalities
III	Bilateral involvement with mid-postural imbalance on examination or history of poor balance or falls; patient leads independent life
IV	Bilateral involvement with postural instability; patient required substantial help
V	Severe, fully developed disease; patient restricted to bed or wheelchair

Schwab and England Activities of Daily Living Scale[24]

100%	Completely independent. Able to do all chores w/o slowness, difficulty, or impairment. Essentially normal. Unaware of any difficulty.
90%	Completely independent. Able to do all chores with some degree of slowness, difficulty and impairment. May take twice as long. Beginning to be aware of difficulty.
80%	Completely independent in most chores. Takes twice as long. Conscious of difficulty and slowing.
70%	Not completely independent. More difficulty with some chores. X 3–4 as long in some. May spend a large part of the day with chores.
60%	Some dependency. Can do most chores, but exceedingly slowly and with much effort. Errors, some impossible.
50%	More dependent. Help with 1/2 of chores. Difficulty with everything.
40%	Very dependent. Can assist with all chores but a few alone.
30%	With effort, now and then does a few chores alone or begins alone. Much help needed.
20%	Nothing alone. Can do some slight help with some chores. Severe invalid.
10%	Totally dependent, helpless. Complete invalid.
0%	Vegetative functions such as swallowing, bladder and bowel function are not functioning. Bedridden.

Unified Parkinson's Disease Rating Scale (UPDRS)

I MENTATION, BEHAVIOR AND MOOD

1. **Intellectual impairment**
   0 = None
   1 = Mild, consistent forgetfulness
   2 = Moderate, difficulty with complex problems
   3 = Severe, disorientation for time and place
   4 = Severe, help with personal care. Cannot be left alone

2. **Thought disorder**
   0 = None
   1 = Vivid dreaming
   2 = "Benign" hallucination with insight retained
   3 = Hallucination or delusions without insight
   4 = Persistent hallucination, delusions, or florid psychosis

3. **Depression**
   0 = Not present
   1 = Periods of sadness or guilt >normal, never sustained for days/weeks
   2 = Sustained depression for >1 week
   3 = Vegetative symptoms (insomnia, anorexia, weight loss)
   4 = Vegetative symptoms and suicidal thoughts

4. **Motivation/initiative**
   0 = Normal
   1 = Less assertive than usual, more passive
   2 = Loss of initiative/disinterest in elective activities
   3 = Loss of initiative/disinterest in routine activities
   4 = Withdrawn, complete loss of motivation

II ACTIVITIES OF DAILY LIVING

5. **Speech**
   0 = Normal
   1 = Mildly affected
   2 = Moderately affected. Sometimes asked to repeat statements
   3 = Severely affected. Frequently asked to repeat statements
   4 = Unintelligible most of the time

6. **Salivation**
   0 = Normal
   1 = Slight saliva excess. Some nighttime drooling
   2 = Moderately excessive saliva; minimal drooling
   3 = Marked excess saliva with some drooling
   4 = Marked drooling, requires constant tissue/handkerchief

7. **Swallowing**
   0 = Normal
   1 = Rare choking
   2 = Occasional choking
   3 = Requires soft food
   4 = Requires nasogastric tube or gastrostomy feeding

Parkinsonism

8. **Handwriting**
   0 = Normal
   1 = Slightly slow or small
   2 = Moderately slow/small; all words are legible
   3 = Severely affected; not all words are legible
   4 = The majority of words are not legible

9. **Cutting food and handling utensils**
   0 = Normal
   1 = Somewhat slow and clumsy, but no help needed
   2 = Can cut most foods slowly; some help needed
   3 = Food must be cut by someone, but can still feed slowly
   4 = Needs to be fed

10. **Dressing**
    0 = Normal
    1 = Somewhat slow, but no help needed
    2 = Occasional assistance with buttons, arms in sleeves
    3 = Considerable help, can do some things alone
    4 = Helpless

11. **Hygiene**
    0 = Normal
    1 = Somewhat slow, no help needed
    2 = Help to shower/bathe
    3 = Assistance for washing hair, brushing teeth and hair
    4 = Foley catheter or pads

12. **Turning in bed and adjusting bed clothes**
    0 = Normal
    1 = Somewhat slow and clumsy, no help needed
    2 = Turns alone or adjusts sheets, but with difficulty
    3 = Can initiate, but not turn or adjust sheets alone
    4 = Helpless

13. **Falling**
    0 = None
    1 = Rare falling
    2 = Occasionally falls < 1 per day
    3 = Falls on average once per day
    4 = Falls > once per day

14. **Freezing when walking**
    0 = None
    1 = Rare freezing; may have start hesitation
    2 = Occasional freezing when walking
    3 = Frequent freezing. Occasional falls resulting
    4 = Frequent falls from freezing

15. **Walking**
    0 = Normal
    1 = Mild difficulty. May not swing arm or may drag leg
    2 = Moderate difficulty, but requires no assistance
    3 = Severe disturbance, requires assistance
    4 = Cannot walk, even with assistance

16. **Tremor (symptomatic complaint in any body part)**
    0 = Absent
    1 = Slight and infrequently present
    2 = Moderate; bothersome to patient
    3 = Severe; interferes with many activities
    4 = Marked; interferes with most activities

17. **Sensory complaints relating to parkinsonism**
    0 = None
    1 = Occasional numbness, tingling or aching
    2 = Frequent numbness, tingling or aching
    3 = Frequent painful sensations
    4 = Excruciating pain

III. MOTOR EXAMINATION

18. **Speech**
    0 = Normal
    1 = Slight loss of expression, diction or volume
    2 = Monotone, slurred but understandable
    3 = Marked impairment, difficult to understand
    4 = Unintelligible

19. **Facial expression**
    0 = Normal
    1 = Minimal hypomimia, could be 'poker face'
    2 = Definite diminution of expression
    3 = Moderate hypomimia; lips parted some of the time
    4 = Masked or fixed facies; lips parted ¼ inch or more

20. **Tremor at rest—right upper limb**
    0 = Absent
    1 = Slight, infrequently present
    2 = Mild amplitude and persistent or moderate and intermittent
    3 = Moderate amplitude, present most of the time
    4 = Marked amplitude, present most of the time

Tremor at rest—left upper limb
0 = Absent
1 = Slight, infrequently present
2 = Mild amplitude and persistent or moderate and intermittent
3 = Moderate amplitude, present most of the time
4 = Marked amplitude, present most of the time

Tremor at rest—right lower limb
0 = Absent
1 = Slight, infrequently present
2 = Mild amplitude and persistent or moderate and intermittent
3 = Moderate amplitude, present most of the time
4 = Marked amplitude, present most of the time

Tremor at rest—left lower limb
0 = Absent
1 = Slight, infrequently present
2 = Mild amplitude and persistent or moderate and intermittent
3 = Moderate amplitude, present most of the time
4 = Marked amplitude, present most of the time

21. **Action or posture tremor of hands—right hand**
    0 = Absent
    1 = Slight, present with action
    2 = Moderate in amplitude, present with action
    3 = Moderate in amplitude, with posture holding and action
    4 = Marked in amplitude; interferes with feeding

    Action or posture tremor of hands—left hand
    0 = Absent
    1 = Slight, present with action
    2 = Moderate in amplitude, present with action
    3 = Moderate in amplitude, with posture holding and action
    4 = Marked in amplitude; interferes with feeding

22. **Rigidity** (*judged on passive movement of major joints with patient relaxed in the sitting position*)
    Rigidity—neck
    0 = Absent
    1 = Slight, detectable only with mirror movements
    2 = Mild to moderate
    3 = Marked, but full range of movement easily achieved
    4 = Severe, range of movement achieved with difficulty

    Rigidity—right upper limb
    0 = Absent
    1 = Slight, detectable only with mirror movements
    2 = Mild to moderate
    3 = Marked, but full range of movement easily achieved
    4 = Severe, range of movement achieved with difficulty

    Rigidity—left upper limb
    0 = Absent
    1 = Slight, detectable only with mirror movements
    2 = Mild to moderate
    3 = Marked, but full range of movement easily achieved
    4 = Severe, range of movement achieved with difficulty

    Rigidity—right lower limb
    0 = Absent
    1 = Slight, detectable only with mirror movements
    2 = Mild to moderate
    3 = Marked, but full range of movement easily achieved
    4 = Severe, range of movement achieved with difficulty

    Rigidity—left lower limb
    0 = Absent
    1 = Slight, detectable only with mirror movements
    2 = Mild to moderate
    3 = Marked, but full range of movement easily achieved
    4 = Severe, range of movement achieved with difficulty

23. **Finger taps** (*patient taps thumb with index finger in rapid succession with widest amplitude possible*)
    Finger taps—right hand
    0 = Normal
    1 = Mild slowing and/or reduction in amplitude
    2 = Definite and early fatiguing; occasional arrests
    3 = Frequent hesitation in initiation or arrests in movement
    4 = Can barely perform the task

Finger taps—left hand
0 = Normal
1 = Mild slowing and/or reduction in amplitude
2 = Definite and early fatiguing; occasional arrests
3 = Frequent hesitation in initiation or arrests in movement
4 = Can barely perform the task

24. **Hand movements** (*patient opens and closes hands in rapid succession with widest amplitude possible*)
Hand movements—right hand
0 = Normal
1 = Mild slowing and/or reduction in amplitude
2 = Definite and early fatiguing; occasional arrests
3 = Frequent hesitation in initiation or arrests in movement
4 = Can barely perform the task
Hand movements—left hand
0 = Normal
1 = Mild slowing and/or reduction in amplitude
2 = Definite and early fatiguing; occasional arrests
3 = Frequent hesitation in initiation or arrests in movement
4 = Can barely perform the task

25. **Rapidly alternating hand movements** (*pronation–supination movements with as large an amplitude as possible*)
Rapidly alternating hand movements— right hand
0 = Normal
1 = Mild slowing and/or reduction in amplitude
2 = Definite and early fatiguing; occasional arrests
3 = Frequent hesitation in initiation or arrests in movement
4 = Can barely perform the task
Rapidly alternating hand movements— left hand
0 = Normal
1 = Mild slowing and/or reduction in amplitude
2 = Definite and early fatiguing; occasional arrests
3 = Frequent hesitation in initiation or arrests in movement
4 = Can barely perform the task

26. **Leg agility** (*rapid heel tapping. Amplitude ≥ 3 inches*)
Leg agility—right heel
0 = Normal
1 = Mild slowing and/or reduction in amplitude
2 = Definite and early fatiguing; occasional arrests
3 = Frequent hesitation in initiation or arrests in movement
4 = Can barely perform the task
Leg agility—left heel
0 = Normal
1 = Mild slowing and/or reduction in amplitude
2 = Definite and early fatiguing; occasional arrests
3 = Frequent hesitation in initiation or arrests in movement
4 = Can barely perform the task

27. **Arising from a chair** (*patient's arms across chest*)
0 = Normal
1 = Slow; or may need more than 1 attempt
2 = Pushes self up from arms of seat
3 = May fall back or try more than once to get up
4 = Unable to arise without help

28. **Posture**
0 = Normal erect
1 = Slightly stooped; could be normal for older person
2 = Moderately stooped; can be slightly leaning to 1 side
3 = Severely stooped with kyphosis; can be moderately leaning to one side
4 = Marked flexion with extreme abnormality of posture

29. **Gait**
0 = Normal
1 = Walks slowly, short steps but no festination
2 = Walks with difficulty but without assistance; festination, short steps or propulsion
3 = Severely disturbed gait; requires assistance
4 = Cannot walk even with assistance

30. **Postural stability** (*pull test, may have practice runs*)
0 = Normal
1 = Retropulsion, but recovers unaided
2 = Absence of posture response, would fall if not caught

3 = Very unstable, spontaneous loss of balance
4 = Unable to stand without assistance

31. **Body bradykinesia and hypokinesia** (*slowness, hesitancy, decreased arm swing, small amplitude and poverty of movement*)
    0 = None
    1 = Minimal slowness, deliberate character, possibly reduced amplitude
    2 = Mild slowness, poverty or small amplitude of movement
    3 = Moderate slowness, poverty or small amplitude of movement
    4 = Marked slowness, poverty or small amplitude of movement

IV. **COMPLICATIONS OF THERAPY (IN THE PAST WEEK)**

A. **Dyskinesias**

32. **Duration: What proportions of the waking day are dyskinesias present?**
    0 = None
    1 = 1–25% of the day
    2 = 26–50% of the day
    3 = 51–75% of the day
    4 = 76–100% of the day

33. **Disability: How disabling are the dyskinesias?**
    0 = Not disabling
    1 = Mildly disabling
    2 = Moderately disabling
    3 = Severely disabling
    4 = Completely disabled

34. **Painful dyskinesias: How painful are the dyskinesias?**
    0 = None
    1 = Slight
    2 = Moderate
    3 = Severe
    4 = Marked

35. **Presence of early morning dystonias**
    0 = No
    1 = Yes

B. **Clinical Fluctuations**

36. **Are any 'off' periods predictable as to timing after medication dosing?**
    0 = No
    1 = Yes

37. **Are any 'off' periods unpredictable as to timing after medication dosing?**
    0 = No
    1 = Yes

38. **Do any of the 'off' periods come on suddenly (seconds)?**
    0 = No
    1 = Yes

39. **What percentage of the waking day is the patient 'off' on average?**
    0 = None
    1 = 1–25% of the day
    2 = 26–50% of the day
    3 = 51–75% of the day
    4 = 76–100% of the day

C. **Other Complications**

40. **Does the patient have anorexia nauseas or vomiting?**
    0 = No
    1 = Yes

41. **Does the patient have any sleep disturbance?**
    0 = No
    1 = Yes

42. **Does the patient have symptomatic orthostasis?**
    0 = No
    1 = Yes

PHYSIOTHERAPY MANAGEMENT

Aims

1. To provide psychological support to the patient
2. To maintain or improve ROM at all the joints
3. Prevention of contractures and deformities
4. Prevent deconditioning of muscle and cardiovascular systems
5. To improve physical fitness of patients
6. Reduction of rigidity and tremors
7. Prevent the influence of psychosocial factors like depression over the patient
8. Gait training
9. To improve muscular and cardiovascular endurance

Plans

In order to achieve the above mentioned aims, the following methods are employed:

1. Develop rapport with the patient, explain the condition to the patient and the outcome of the treatment without either giving false hope or encouraging the negative thought.
2. Motivate the patient to actively participate in the treatment protocols.
3. To reduce rigidity:
    - Gentle rocking exercises and rotational exercises in slow and rhythmic patterns
    - PNF (rhythmic initiation) technique in which movement progresses from passive to active assisted and then to active movements.
    - Vestibular stimulations by means of exercises on Swiss ball.
    - At home, use of a rocking chair.
    - General relaxation techniques are taught and made to practice. Deep breathing exercises can be incorporated into rotational exercises to enhance relaxation. Meditation techniques—Jacobson's progressive relaxation techniques. Relaxation audiotapes—home exercise programmes.
4. To maintain and improve the joint ROM and flexibility of the muscles:
    - Gentle stretching of elbow flexors, hip, knee flexors and ankle plantar flexors.
    - Stretching can be combined with joint mobilization techniques to reduce tightness of the joint's capsule or of ligaments around a joint.
    - Autostretching or self-stretching. Maintain the stretch force at least 15–30 seconds.
    - Ideally the stretches are repeated at least 3–5 times.
    - Braces may be used for prolonged stretching of tight muscles.
    - Calisthenics exercises in supine, sitting and stand.
    - Standing erect with arms in elevation (over head) against a wall or corner of the room and the patient should try to stretch out his body.
    - Lie supine with pillow under the upper thorax.
    - **Passive positioning:** It is long duration technique to improve flexibility.
        a. To avoid phantom pillow posture, patient should be positioned in prone position.
        b. The patient with a developing lateral curvature can be positioned in side lying with a small pillow under the lateral trunk.
    - The overall focus is on improving mobility/controlled mobility function with specific emphasis on improving segmental mobility of the head, trunk and proximal segments (hips and shoulders). Relaxation exercises are important prerequisites to all mobility training.
        a. For thorax and neck extension: Prone on elbow, prone extension, standing wall push-ups or corner push-ups.
        b. For extremities: PNF techniques of rhythmic initiation.
        c. Bed mobility activities (rolling, supine to sit, etc.).
        d. Pelvic mobility (anterior, posterior and side-to-side tilt) exercises on Swiss ball.
        e. Weight shifting exercises and upper extremities reaching activities. Reaching should be practiced in all directions with emphasis on promoting rotational movement of trunk.
5. To improve balance, the following techniques are done:
    - The balancing training should always be beginning from lower COG to higher COG.
    - Training should begin with weight shifts in both sitting and standing in order to help the patient develop an appreciation of his limits of stability.

- By giving the slight push to patient. Patient tries to maintain the balance.
- Reaching activities.
- Activities on Swiss ball.
- Kitchen shink exercises: The patient can be instructed in standing heel rises toe offs, partial wall squats and chair rises, single limb stance with sidekicks or back kicks and marching in place, all while maintaining light touch down support of the hands.

6. To improve the motor control and motor endurance, the following techniques are used.
   - To teach the patient to do one movement at a time because he feels difficulty in carrying out simultaneous movements (dual task).
   - The combination of movements results in slowing of movements.
   - Teach the patient rolling, supine to sitting, sitting to standing, etc.
   - Avoid the movements where attention is divided; because learning activities is also become difficult. For example, a patient sitting on chair do not ask him to walk, first ask him to stand up from chair and then walk.

7. To improve cardiopulmonary endurance and to prevent deconditioning, the following techniques are used.
   - Diaphragmatic and segmental breathing exercises.
   - Air shifting techniques.
   - Deep breathing exercises to improve chest wall mobility and vital capacity
   - Chest mobility exercises.
   - PNF in respiration.
   - Incentive spirometry.
   - Balloon blowing.
   - Aerobic exercises, e.g. daily walking for short distance, ergometry, etc.

8. To improve the gait, the following techniques are incorporated:
   - The major goals are to lengthen stride, broaden base of support (BOS), improve stepping, improve heel–toe gait pattern, increase contralateral movement and arm swing and provide a program of regular walking.
   - Weight transfer; standing on single limb.
   - High stepping to strengthen the flexors.
   - Side stepping or crossed stepping with or without support.

9. To prevent the influence of the psychosocial factors on the patient and to improve the psychological well-being of the patient, the following methods are used.
   - Active participation of family members is required. Patient counseling to increase the confidence and independency.
   - Feeling of hopelessness, dependency and depression should be reduced.
   - Self-management skills should be promoted.
   - Participation of patient into various social activities should be encouraged. Coping skills (to compete with the variety of social and environmental factors) can be facilitated.

10. Home program and advices:
    - Teach the patient for relaxation, flexibility, strengthening, mobility and breathing exercises.
    - Avoid prolonged periods of inactivity.
    - The patient should be cautioned against over doing activity, which could result in excessive fatigue.
    - Early morning warm-up and calisthenic exercises.
    - Compensatory techniques or triggering maneuvers to overcome the crippling effects of bradykinesia and freezing.

REFERENCES

1. Kalia, LV; Lang, AE. "Parkinson's disease.". Lancet (London, England). 2015;386 (9996):896–912.
2. Barranco Quintana, JL; Allam, MF; Del Castillo, AS; Navajas, RF. "Parkinson's disease and tea: a quantitative review.". Journal of the American College of Nutrition, 2009.
3. Shulman JM, De Jager PL, Feany MB. "Parkinson's disease: genetics and pathogenesis.". Annual Review of Pathology 2011 Feb, 2010 Oct;26:193–222.
4. Samii A, Nutt JG, Ransom BR (29 May 2004). "Parkinson's disease". Lancet. 363 (9423): 1783–1193.
5. Yao, SC, Hart AD, Terzella MJ. "An evidence-based osteopathic approach to Parkinson disease". Osteopathic Family Physician. 2013;5(3):96–101.
6. De Lau LM, Breteler MM. "Epidemiology of Parkinson's disease". Lancet Neurol 2006;5(6):525–35.
7. Obeso JA, Rodríguez-Oroz MC, Benitez-Temino B, Blesa FJ, Guridi J, Marin C, Rodriguez M. "Functional organization of the basal ganglia: therapeutic implications for Parkinson's disease". Mov Disord 2008;23(Suppl 3):S548–59.
8. Obeso JA, Rodriguez-Oroz MC, Goetz CG, Marin C, Kordower JH, Rodriguez M, Hirsch EC, Farrer M, Schapira AH, Halliday G. "Missing pieces in the Parkinson's disease puzzle". Nat Med 210;16(6):653–61.
9. Jankovic J. "Parkinson's disease: clinical features and diagnosis". Journal of Neurology, Neurosurgery and Psychiatry 2008;79(4):368–376.
10. Samii A, Nutt JG, Ransom BR. "Parkinson's disease". Lancet 204;363(9423):1783–1193.
11. Cooper G, Eichhorn G, Rodnitzky RL (2008). "Parkinson's disease". In Conn PM. Neuroscience in medicine. Totowa, NJ: Humana Press. 2008;pp. 508–512.
12. Fung VS, Thompson PD. "Rigidity and spasticity". In Tolosa E, Jankovic. Parkinson's disease and movement disorders. Hagerstown, MD: Lippincott Williams & Wilkins. 2007;pp. 504–13. ISBN 0-7817-7881-6.
13. Armstrong RA. "Visual signs and symptoms of Parkinson's disease". Clin Exp Optom 2008;91(2): 129–38.
14. Shinjo Samuel Katsuyuki, Torres Silvia Carolina Ramos, Radu Ari Stiel. "Camptocormia: A Rare Axial Myopathy Disease". Clinics (Sao Paulo, Brazil) 2008-06-01;63(3):416–417.
15. Lenoir Thibaut, Guedj Nathalie, Boulu Philippe, Guigui Pierre, Benoist Michel. "Camptocormia: the bent spine syndrome, an update". European Spine Journal 2010-03-19;19(8):1229–1237
16. Morris M, Iansek R, Matyas T, Summers J. "Abnormalities in the stride length-cadence relation in Parkinsonian gait". Mov Disord 1998;13:61–69.
17. Koller WC, Glatt S, Vetere-Overfield B, Hassanein R. "Falls and Parkinson's disease". Clin Neuropharmacol 1989;12:98–105.
18. Hausdorff JM, Cudkowicz ME, Firtion R, Wei JY, Goldberger AL. "Gait variability and basal ganglia disorders: stride-to-stride variations of gait cycle timing in Parkinson's disease and Huntington's disease". Mov Disord 1998;13:428–437.
19. Stefan Kimmeskamp, Ewald M. Hennig "Heel to toe motion characteristics in Parkinson patients during free walking". Clinical Biomechanics, 2001;16(9).
20. Bloem BR, Hausdorff JM, Visser JE, Giladi N. "Falls and freezing of gait in Parkinson's disease: a review of two interconnected, episodic phenomena". Mov Disord 2004;19(8):871–84.
21. Caballol N, Martí MJ, Tolosa E (September 2007). "Cognitive dysfunction and dementia in Parkinson disease". Movement Disorders. 22 (Suppl 17): S358–S366
22. Hoehn M, Yahr M . "Parkinsonism: onset, progression and mortality.". Neurology 1967;17(5)427–42
23. Goetz CG, Poewe W, Rascol O, Sampaio C, Stebbins GT, Counsell C, Giladi N, Holloway RG, Moore CG, Wenning GK, Yahr MD, Seidl L. "Movement Disorder Society Task Force Report on the Hoehn and Yahr Staging Scale: Status and Recommendations. The Movement Disorder Society Task Force on Rating Scales for Parkinson's Disease.". Movement Disorders 204;19(9):1020–1028.
24. Dal Bello-Haas V, Klassen L, et al. "Psychometric Properties of Activity, Self-Efficacy, and Quality-of-Life Measures in Individuals with Parkinson Disease." Physiother Can 2011;63(1):47–57.

9

Spinal Cord Injuries

LEARNING OBJECTIVES

At the end of this chapter, you shall be able to:
1. List out various etiological factors of spinal cord injuries
2. List out the types of spinal cord injuries
3. Identify the various clinical manifestations of spinal cord injuries at various levels
4. Demonstrate the skills to assess the patients with spinal cord injuries
5. Design the treatment protocols for rehabilitating these patients.

INTRODUCTION

A spinal cord injury is damage to the spinal cord that causes changes in its function which may be permanent or temporary in the loss of muscle function, sensations, autonomic functions at various parts of the body below the level of lesion. The WHO recognizes it as a major musculoskeletal condition that presents a serious disease burden.

Injuries can occur at any level of spinal cord. In majority of cases, the damage is from trauma such as car accidents, gunshots, falls, sport injuries, assaults and stab injuries. Certain non-traumatic causes like infection, tumors can also lead to spinal dysfunction. The spinal cord injuries are well known for their catastrophic results. The advanced research into the treatments for spinal cord injuries includes stem cell implantation, engineered materials for tissue support and wearable robotic exoskeletons.

Definition

A spinal cord injury is damage or trauma to the spinal cord that results in a loss or impaired function of muscle, sensory loss or autonomic dysfunctions.

Etiology (Fig. 9.1)

a. Fractures or fractures with dislocation
b. Sports injuries
c. Industrial accidents
d. Stab injuries
e. Gunshot injuries
f. Surgical trauma, e.g. in corrective surgeries of scoliosis.

Epidemiology (Fig. 9.2)

- The prevalence of SCI in India is 1.85–2.19%.[1,2]
- SCI generally occurs in young adults between 20 and 40 years of age.

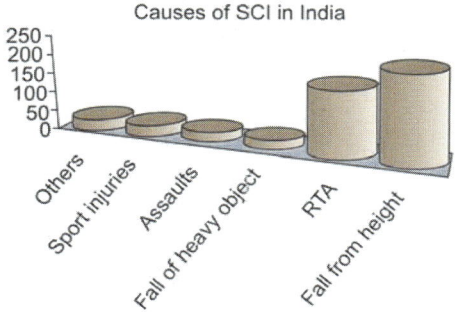

Fig. 9.1: Causes of SCI in India

Incidence of various level of spinal cord injuries in India

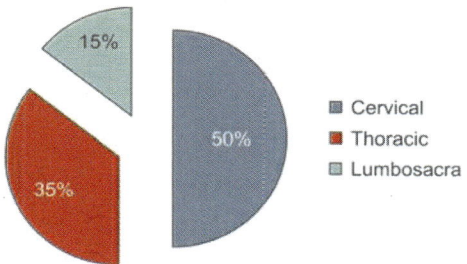

Fig. 9.2: Level of spinal cord injury

- Males are more affected than females.
- In India, fall from height is the most common cause of traumatic SCI

CLASSIFICATION OF TRAUMATIC SPINAL CORD INJURIES

The severity of spinal cord injury is based on the level of injuries and neurological category.

Neurological Level

The American Spinal Cord Injury Association (ASIA) defines neurological level as the lowest segment of the spinal cord with normal sensory and motor function on both sides of the body.

Neurological Category

Traumatic spinal cord injury is classified into five categories on the ASIA Impairment Scale.

Category	Description
A.	Indicates a "complete" spinal cord injury where no motor or sensory function is preserved in the sacral segments S4–S5
B.	Indicates an "incomplete" spinal cord injury where sensory but not motor function is preserved below the neurological level and includes the sacral segments S4–S5. This is typically a transient phase and if the person recovers any motor function below the neurological level, that person essentially becomes a motor incomplete, i.e. ASIA C or D
C.	Indicates an "incomplete" spinal cord injury where motor function is preserved below the neurological level and more than half of key muscles below the neurological level have a muscle grade of less than 3, which indicates active movement with full range of motion against gravity
D.	Indicates an "incomplete" spinal cord injury where motor function is preserved below the neurological level and at least half of the key muscles below the neurological level have a muscle grade of 3 or more
E.	Indicates "normal" where motor and sensory scores are normal. Note that it is possible to have spinal cord injury and neurological deficits with completely normal motor and sensory scores

CLINICAL CLASSIFICATION

- Anterior cord syndrome
- Posterior cord syndrome
- Brown-Sequard syndrome
- Central cord syndrome
- Conus medullaris syndrome
- Cauda equina syndrome

Anterior Cord Syndrome (Fig. 9.3)

- Anterior cord syndrome describes the damage to the spinothalamic tract and corticospinal tract.
- There is complete motor loss below the level of lesion due to involvement of corticospinal tract.

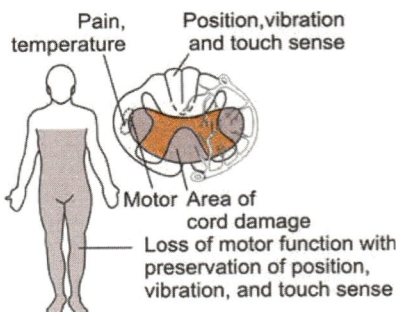

Fig. 9.3: Anterior cord syndrome

- There is loss of pain and temperature at and below the level of injury due to involvement of lateral spinothalamic tract.
- Preservation of the 2-point discrimination sense vibrations and proprioception senses due to intact posterior column.
- There is autonomic dysfunction leading to orthostatic hypotension.
- Bladder and bowel dysfunction and sexual dysfunction may arise depending on the level of lesion.

Posterior Cord Syndrome (Figs 9.4 and 9.5)

- This is a rare condition producing damage to the dorsal columns (sensations of light touch, proprioception and vibrations).
- There is preservation of motor function and pain and temperature pathways.

Brown-Sequard Syndrome (Fig. 9.6)

- It is a rare form of incomplete spinal cord injury which results after the

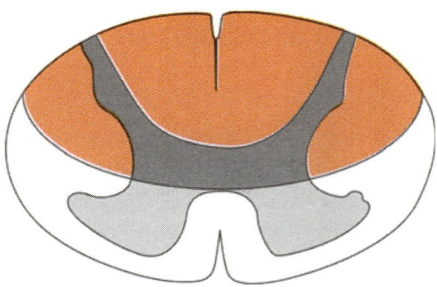

Fig. 9.4: Posterior cord syndrome

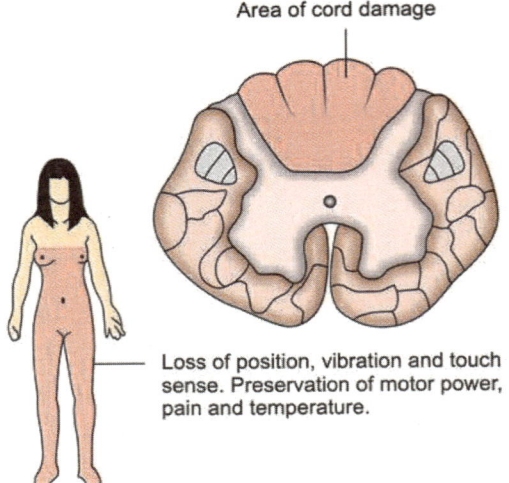

Fig. 9.5: Posterior cord syndrome

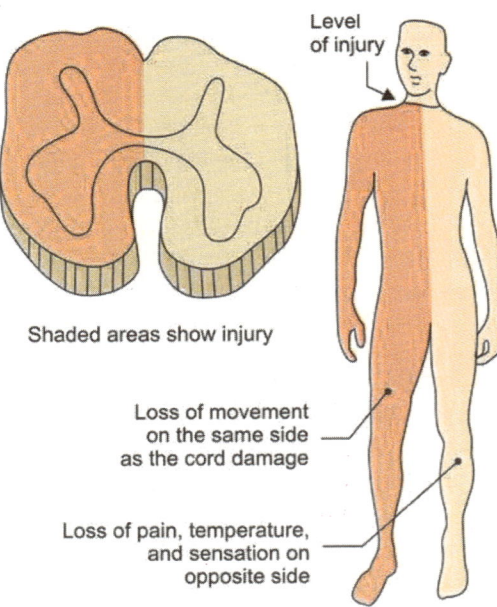

Fig. 9.6: Brown-Sequard syndrome

damage to one side of the spinal cord (hemisection). It accounts for up to 4% of all traumatic spinal cord injuries.
- It is characterized by a loss of sense of vibration, deep touch or pressure, joint position sense and motor paralysis below the level of spinal cord injury on the same side (ipsilateral).

- Loss of sense of light touch, pain and temperature on the opposite side (contralateral) of the body.

Central Cord Syndrome (Fig. 9.7)

It is the most common type of incomplete spinal cord injury. It is the resultant of the contusion of the central portion of the cervical spinal cord.

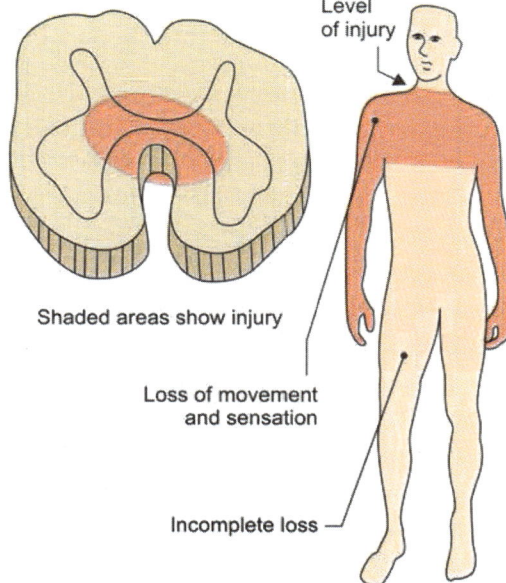

Fig. 9.7: Central cord syndrome

The patients present with upper limb weakness, urinary retention and sensory loss below the level of lesion.

The upper limbs are classically more affected than the lower limbs with motor dysfunction more than the sensory loss.

Conus Medullaris Syndrome

It is caused by the injury to the conus medullaris and lumbar nerve roots.

Injuries at the level of T12 to L2 vertebra are most likely to result in the conus medullaris syndrome.

Patients present with a combination of UMN and LMN palsies characters: Saddle anesthesia, urinary retention, loss of bowel reflex, lower limb motor weakness, paresthesia and numbness and chronic low back pain.

Cauda Equina Syndrome

It is caused by the injury in which there is damage to the cauda equina portion of the spinal cord.

Complete lesion of cauda equina is rare due to larger surface area and large number of nerve roots.

Characterized by muscles weakness/flaccidity in the lower limbs, wasting of muscles, and loss of reflexes.

In extramedullary lesions, there will be sensory loss and motor dysfunction more in the distal lower limbs than the proximal area.

In intramedullary lesions, there will be sensory loss and motor dysfunction more in the proximal than the distal lower limbs.

PHYSIOTHERAPY ASSESSMENT OF SPINAL CORD INJURIES

1. Take the demographic data of the patient. Obtain the data by interviewing the patient or caretaker or from the medical records.
2. Take a complete and detailed history which should include time of injury, cause of injury and details of first aid, if given any. The most important thing is ask the patient how he is carried to the hospital. If you are assessing the patient at his home or in your OPD, then take the history of medical/surgical intervention done after the injury.
3. Take a note of presenting problems conveyed as chief complaints of the patient.
4. **Observation:**
    a. Observe for the general built of the patient.
    b. Observe the posture and attitude of limbs.
    c. Observe for bedsores, wasting of muscles, swelling, deformities, scars and any other aids applied to the patient.
    d. Observe the tropical changes of the skin over the affected limb/limbs.

e. Observe the breathing pattern of the patient and signs of dyspnea.
f. Observe if the patient is having any urinary catheterization.

5. **Palpation:**
   a. Palpate the skin temperature and compare with unaffected areas.
   b. Palpate the edema and grade it.

6. **Examination:**
   a. Examine the vital signs and report—respiratory rate, temperature, pulse and blood pressure. Higher cervical (C3–C5), cord lesions affect the respiration because of paralysis of diaphragm.
   b. Examine the higher functions—memory, intelligence, level of consciousness, behavior, orientation and speech. (Most of the times spinal cord injuries are also associated with head injuries, assessment of higher functions would be required to rule out the head injuries.)
   c. Sensory examination would be very important to identify the level of lesion and type of lesion. A thorough examination of sensory system—superficial, deep and cortical senses should be evaluated.
   d. Examination of deep tendon reflexes is important to identify the level of lesion. Please note that the deep tendon reflexes are hypoactive at the level of lesion and hyperactive below the level of lesion and normal above the level of lesion. The superficial reflexes are hypoactive at and below the level of lesion and normal above the level of lesion. Please note that the abdominal reflex is difficult to observe in obese patients, patients with multiple pregnancy, scars over the abdominal region. So, abdominal reflex is not a reliable evaluation and the diagnosis should never be done alone on abdominal reflex in these cases. The plantar reflex is very important and reliable, in the spinal cord lesions as the reflex shows positive or extensor responses.
   PS: In spinal shock stage, all the reflexes are diminished.
   e. **Motor examination:** Muscle tone should be determined. Usually, at spinal shock stage, all the muscles at and below the level of lesions would be flaccid. Later the spasticity set in. At the level of lesion, there will be flaccidity and below the level of lesion, it is spasticity.
   f. **Muscle power:** The muscle power is examined by checking the MMT of muscles. Usually, group MMT according to the myotomes are examined. It is a definite sign to identify the level of lesion. It should be noted that the level of lesion is different in different definitions in concept of neurology and rehabilitation. According to neurology concept, the first spinal segmental level that shows the abnormal neurological loss is the level of lesion. Whereas the rehabilitation identifies as the lowest intact segment as the level of injury. For example, a person has loss of power (less than 3 on MMT grade is considered), the level of injury is C5 according to neurology and C4 according to the rehabilitation.
   g. **Voluntary control:** Grade the voluntary control, if the spasticity is set in. The grades of voluntary control are checked according to movement, e.g. shoulder flexion, shoulder extension, etc.

Grades	Description
0	No contraction
1	Initiation of contraction or flicker contraction
2	Half range of motion in pattern
3	Full range of motion in pattern
4	Initial half range in isolation and the later half in pattern.
5	Full range of motion in isolation but goes into pattern when resistance offered
6	Full range of motion in isolation against resistance.

h. Check the balance and coordination.
i. Assess the bladder and bowel dysfunction: This is the one of the most common clinical features of the spinal cord injuries. In spinal shock stage, there will be atonic bladder or flaccid bladder leading to overflow incontinence.
j. Automatic or reflex bladder is seen when there is damage to the spinal cord above the micturition center, i.e. S2–S4.
k. Autonomous bladder or areflex bladder is seen, if the lesion is at the level of S2–S4 spinal segments of the spinal cord leading to the overflow incontinence.
l. ADL assessment and level of ambulatory independence should be done.

PHYSIOTHERAPY MANAGEMENT

Short-term Goals

1. To provide psychological support to the patient.
2. To prevent abnormal movement of the spine.
3. To maintain normal muscle properties in the muscles affected.
4. To prevent secondary complications.
5. If high cervical lesions, to promote early independence from mechanical ventilation.

Long-term Goals

1. To continue psychological support and develop good rapport with the patient and caretakers.
2. To coordinate with the other members of rehabilitation and aid in early functional independence as a team work.
3. To improve and strengthen the muscles affected.
4. To maintain clear chest and prevent any chest complications.
5. To improve sensory system and prevent any damage to the skin.
6. To aid in mobility and ambulation for functional independence.
7. To provide social and economical rehabilitation.

Plans of Physiotherapy

1. The importance of effective communication with the patient is proven in physiotherapy care for better prognosis and improving the rapport with the patient. Have patience to listen to all the patient's complaints and problems. Try to answer and explain the doubts or queries of patient about his condition. Never give false promises or promise goals that cannot be achievable. Effective channels of communication and coordi-nation are required among the mem-bers of the multidisciplinary team and also with patient and guardians. In a research by Barnes et al. 2012, it was believed that facilitating a successful intervention might require a health care professional to enhance their communication skill.[3]
2. The spine should be immobilized by means of spinal corset. All precautions should be taken to avoid any mobility in the spine during transfers or physiotherapy maneuvers.
3. To maintain normal muscle properties and tone, the following techniques are used.

Hypotonicity

- Ice brisk stroking over the muscle
- Hacking over the muscle
- Weight bearing over a joint where the muscle is crossing
- Vibrations
- ENMS is found useful in developing the tone, but it is not advised in MND.
- Suspension therapy and aquatic exercises
- Functional re-education exercises
- Muscle energy techniques.
- EMG biofeedback techniques

Hypertonicity

- Cryotherapy in the form of prolonged icing—ice pack
- PNF techniques
- Passive stretching exercises
- Orthotic supports to maintain the length of the muscle

To prevent secondary complications: The possible complications due to prolonged bedrest would be:

1. Contractures and deformities
2. Pressure sores
3. Deep vein thrombosis
4. Psychological depression
5. Respiratory complications
6. Cardiovascular deconditioning
7. Disturbed bone metabolism

The contractures and deformities are prevented by proper positioning of the patient, passive movements, use of orthotic supports, passive mobilization techniques and stretching.

Pressure sores are deep wounds due to constant and continuous pressure on the skin overlying the bony prominences in the body. It is very common, painful complication of the patients who are on the prolonged bedrest. The pressure sores can aggravate the existing disability. It should be noted that the development of pressure sores is solely due to negligence of the caregivers and health providers but not accidental. Hence, the health care providers and caretakers should take all necessary precautions to prevent the pressure sores.

Precautions to prevent pressure sores:

1. Closely monitor all the areas of risk daily
2. Frequent change in the position of the patient
3. Small rubbing friction massage with a moisturizer for a small duration (15 sec) over the areas of risk shall improve the skin elasticity and pliability.
4. Use of water mattress or air bed
5. Proper bed making technique, avoid creases on the bed cover.

Even after taking necessary precautions, if the pressure sores appear to develop, then curative measures to be taken:

1. At redness stage: Stop the massage, start changing the positions frequently and take use of adaptive devices.
2. At blister stage, frequent dressing to be done. Hydrocolloid dressing would be helpful.

If an ulcer has already formed, debridement of dead tissues, dressing of the wound is done. UVR therapy, ultrasound have been found useful to promote healing.

Deep vein thrombosis is another major complication of immobility. It can be prevented by leg lifts, ankle pumping movements, toe curls, stretching and taking plenty of fluids would be helpful.

The respiratory complications could be avoided by teaching deep breathing exercises, postural drainage, suctioning, good coughing and huffing techniques. Chest muscle stretching exercises would be beneficial.

Cardiovascular deconditioning is prevented by regular endurance exercises.

If high cervical lesions, to promote early independence from mechanical ventilation: Respiratory muscle weakness in high cervical injury patients is common and can lead to respiratory dysfunction and respiratory complications. Diaphragmatic dysfunction is associated with dyspnea can decrease the lifespan. Ineffective cough can lead to respiratory infections like pneumonia (Senent et al., 2011). The condition can become worse, if there is shortening and stiffness of chest muscles and reduced lung compliance.

The main role of physiotherapists in respiratory management is to closely monitor for any symptoms of respiratory insufficiency and to prevent accumulation of secretions in the lungs.

Respiratory adjuncts such as PEP should not be indicated, instead manual assisted coughing should be used.

Regular postural drainage and deep breathing exercises are to be incorporated. Chest secretions may be mobilized by manual techniques like percussions, vibrations or mechanical assistive devices like mechanical exsufflator or a positive pressure device.

Stretching of pectoral, and sternocleidomastoid muscles could be helpful to improve chest mobility.

The early independence from mechanical ventilation can be achieved by neurophysiological stimulaton of respiration which is described below.

NEUROPHYSIOLOGICAL STIMULATION (NPF) OF RESPIRATION[4]

It is believed that cutaneous and proprioceptive stimulations reflexly increase the depth of breathing.

The perioral technique is thought to relate to the suckling reflex, and may facilitate slow as well as deep breathing. NPF also proven to improve coughing, swallowing and abdominal contractions.

Perioral technique: Moderate finger pressure is maintained inwards and downwards just above the lip as long as the patient is required to deep breathe. The effect may continue for some minutes afterwards.

Intercostal stretch: Apply pressure downwards towards the toes on the upper border of the rib at end, expiration, unilateral or bilateral but not on the floating ribs.

Co-contraction of abdominal muscles: Apply pressure laterally over the lower ribs and pelvis at right angles to patient, alternating right and left sides and maintain the pressure for up to 2 min or until desired effect is achieved.

Vertebral pressure: Finger pressure against thoracic vertebrae between T2 and T10.

Long-term Goals

1. The first and foremost aim for long-term management of the patient with spinal cord injuries is to develop good rapport and provide psychological support in order to accept the insult to the body.

   The patient must be motivated and supported to adjust and accept the life after spinal cord damage. The patient will frequently ask all the health care members when he will recover. The therapist is the member of health care team who spends more time and usually he is being questioned by patient about the recovery. The therapist should not give false promises; at the same time he should not allow the patient to develop negative thoughts in mind. The depression is the most common that the patients with spinal cord injuries, so motivation is very important in these patients and sometimes becomes very difficult to handle the situation. This is where the experience of the therapist combined with his/her wisdom is challenged. Make the patient first accept his situation and explain how easily he can adjust to the present situation. Show him different people who still lead a healthy productive life even after such a disaster. The support from the family members is equally important in rehabilitation of the patient. So along with patient his close family members should be counseled, explained about the importance of physiotherapy for recovery of the patient.

2. The strength of the muscles must be improved based on the recovery process. In case of paraplegics, it is must to increase the strength of the upper limbs in order to train him for ambulation by walking aids or parapodium.

In strengthening, there are various techniques that can be used. During the strengthening of the muscles in spinal cord injury patients, certain key points should be kept in mind before starting the exercise protocols:

a. Be careful with all the transfer activities on and off equipment and look for areas that could cause friction, pressure or shearing. Try to avoid all these in order to maintain a healthy and intact skin.
b. Do not be over enthusiastic and make the patients workout heavily which can lead to overuse of the muscles resulting in injuries. Always select exercises of low grade first and slowly progress in order to avoid over use.
c. Spasticity will be prevailing in chronic spinal cord injury patients which can interfere in the exercise. So keep an eye while using a machine to exercise which can cause over stretch.
d. Always empty the urine bag before starting the exercise because of dysreflexia.
e. Exercise sessions should be of shorter durations with rest intervals. Usually each bout can be for 5–10 min with a rest interval of 5 min. The intensity of the exercises should start with low intensity and then progress to moderate.
f. Along with strengthening exercises, flexibility exercises should also be incorporated daily.
g. Be realistic to your goals, do not be over enthusiastic and make sure that the patient works out regularly at least 5 days a week.
h. Build up the power slowly and gradually but constantly.
i. Make the patient feel interested in the exercise program. It is the skill of the therapist to design the exercise to be intresting to the patient. A soft music can be played while exercising to develop interest in the patient.
j. Always appreciate the patient when he achieves a goal in order to encourage him.

The protocol for strengthening program should have the following components:

a. Cardiovascular conditioning by aerobic exercises like wheelchair pushing, seated aerobics, arm ergometer, aquatic therapy, rowing, cycling, circuit training.
b. Functional electrical stimulation (FES[6]) delivers electrical current to the muscles via electrodes placed on the muscles of the limbs which drives the paralyzed muscles to perform movements that allow the patient to cycle or walk (Fig. 9.8).
c. Muscle strengthening program should include:
   - Free weights
   - Elastics/bands
   - Wall weights
   - Circuit training
   - Exercise machines
   - Springs and pulleys
   - Weight cuffs, etc.

Always start with the low weight and make the patient practice first in gravity

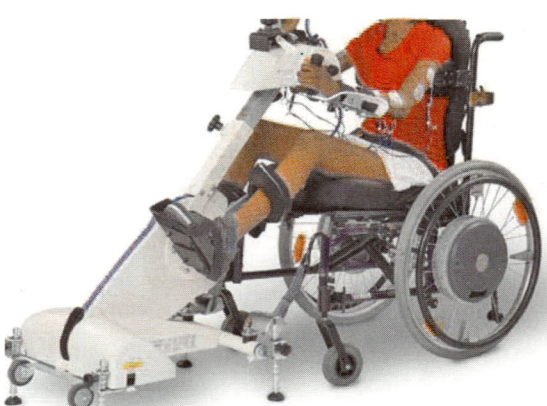

Fig. 9.8: FES exerciser

eliminated or gravity assisted and then progress to against gravity positions.

Always remember to maintain good quality of movment.

Avoid any painful positions.

Prevent over use.

Flexibility/stretching exercises allow the muscles to maintain good posture, balance and aids in performing ADL.

Gentle sustained stretching should be given regularly to the key muscles. Prolonged icing for muscles which are in spasticity and then making the muscle stretch could be useful.

CRYOKINETICS

It is a rehabilitative technique involving ice application followed by progressive active exercises. Cryokinetics allows the patient which is pain free. Cold should be applied for a maximum of 20 min which should be sufficient to produce the numbed response. The process can be repeated for 5 min to renumb the area, if necessary. The exercises performed during cryokinetics are active. Cryokinetics are proven to re-establish neuromuscular function.

Make the patient exercise on active–passive exerciser bike. The advantage of this bike is direct accessibility with the wheelchair. The machine has a sensor which can identify the active efforts of the patient and readjusts the movement by active–passive version, that means the range until the patient is able to put active effort, the resistance will be provided and where the patient cannot put the effort, the passive mode is activated (Fig. 9.9).

To Improve Sensory System and Prevent any Damage to the Skin

Sensory stimulation can be used to acquire afferent pathways with the aim of providing information that can subsequently be used to perform a motor activity with the direct

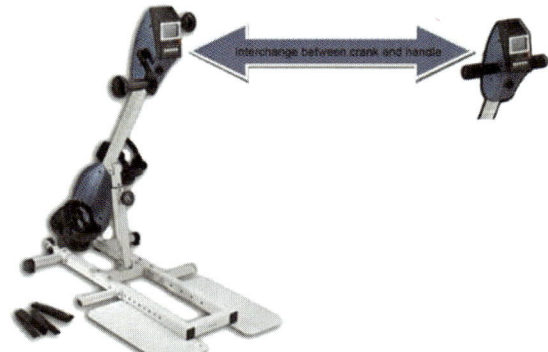

Fig. 9.9: Interchangeable crank exerciser

effect being improved motor and sensory functions (Issurin, 2005 and Peurala et al. 2002). Sensory re-education and integration techniques are employed by the therapists in an attempt to retain the sensory pathways or stimulate the unused pathways (Table 9.1). The therapist should also employ adaptive techniques to cope up with the sensory loss as a compensation. Sensory re-education techniques can include touching different textured objects, massage, vibrations, pressure, determining joint positions, identifying different temperatures and electrical stimulations. The transcutaneous electrical stimulation would be useful in the rehabilitation of patients with sensory loss (Ito et al. 2015).

Other Techniques

1. Feel an object and find the same matching object from a sand tray.
2. Ask the patient to close the eyes. Place two objects of different weights and ask the patient to recognize the difference.
3. Write with the finger an alphabet or a number over the dermatome of the

---

Aquatic Exercises for Spinal Cord Injury Patients

1. Aqua yoga pilates
2. Circuit training
3. Water walking inside the pool
4. Therapeutic aquatic body work

Table 9.1: Sensory re-education exercises

Training for differentiation of different textures	Use cotton, sandpaper, satin, velcrow rubber, velvet, wool. Train 2–4 times for 10 min in a quiet environment daily
Steriognosis	Hide objects with different shapes like marbles, currency coins, erasers in sand. Ask the patient to close eyes and try to find the objects with the hand feel and recognize. *Note:* The objects should be familiar to the patient. Frequency: 2 times for 10 min daily
Pressure	Apply pressure with fingers of different intensity over an area once with patient's eyes opened and patient's eyes closed. Ask the patient to feel the pressure and identify Frequency: 2 times for 10 min daily
Vibrations	Ask the patient to close the eyes and apply vibrations over a dermatome of the patient. The vibrations first should be of a powerful vibrator which is used for massage, later by mobile phone vibrations, tuning fork vibrations and vibrations as the patient progresses. Ask the patient to feel the vibrations once with eyes opened and then with eyes closed. Frequency: 2–3 times for 5 min daily
Temperature	Take two bowls of water of hot and cold temperatures and ask the patient to dip the hands or a test tubes can be used instead. Ask the patient to feel and recognize. Frequency: 2–3 times for 10 min daily
Joint position sense	Position the joint in normal and affected at the same time, ask the patient first to feel the position and then repeat only on the affected limb. Ask the patient to analyze and feel Frequency: 2–3 times for 10 min daily
Joint kinesthetic sense	Move the limbs once with eyes opened and then eyes closed. Ask the patient to feel and recognize Frequency: 2–3 times for 5 min daily

patient and ask the patient to feel and see and perceive. Now ask the patient to close the eyes and recognize the same.

To Aid in Mobility and Ambulation for Functional Independence

Ambulation is a commonly expressed goal of most people with paraplegia. Patients with injuries at T2 and above typically cannot achieve ambulation. Whereas patients with T3 to T11 affection are able to use orthotic supports for physiologic standing and therapeutic ambulation. The goals for individuals with a T12 to L2 injury is to achieve ambulation in household, while patients with injuries at L3 and lower are most likely to achieve ambulation in the community. The muscle activity in the back and lower extremities helps the therapists predict the orthotic support needed for ambulation. For example, patients with pelvic control and intact quadratus lumborum and abdominal muscles may walk using knee-ankle-foot orthoses and crutches. Patients with control of muscles across the hip joint like iliopsoas, hip adductors, sartorius and gluteus maximus can ambulate with KAFO and crutches. Patients with intact tibialis anterior and posterior can walk full-time with AFO (ankle-foot orthoses).

During the pre-gait training program, certain muscles should be strengthened like core trunk and abdominal muscles, shoulder depressors, scapular stabilizers, triceps and wrist extensors. During gait training, control of pelvis is a critical factor for a successful ambulation. Gait training includes standing balance training in parallel bars and practice gait patterns in parallel bars once inside the bars and later outside. Assistive devices depending upon the independence like canes, crutches or walkers may be used for ambulatory training.

Ambulation with ORLAU Parawalker

The ORLAU parawalker is a type of reciprocal walking orthosis. Reciprocal walking orthosis helps to take steps alternatively with one foot and then the other as in normal walking. This orthosis is primarily for paraplegics who have no voluntary control of the muscles in the lower body. This orthosis is worn outside clothing and fits closely from just under the axilla and down to the feet. It has locking knee joints that keep the legs straight during standing and walking special joints at the hips allow fore and after movements to permit a stepping pattern. The knee and hip joints have quick-release mechanisms to allow the patient to sit on a chair or couch.

Robot assisted walking therapy (Fig. 9.10) is a form of physical therapy that uses a robotic device to help a person whose ability to walk has been impaired by neurological or orthopedic conditions. Commercially, it is called LOKOMAT®. In a randomized control trail by Esclarin–Ruz et al. 2014 observed that robotic walking therapy yielded better results in the 6 min walk test and improved endurance in patients with UMN and LMN lesions.

To Improve the Functional Independence

Functional re-education exercises, training for the transfers, are very important (refer to transverse myelitis for details).

Bladder Management

A bladder management is very important in rehabilitation to prevent infections, avoid bladder accidents and gaining functional independence.

Depending upon the level of injury, the bladder management can be planned:
  a. Foley's catheter is a tube that is inserted through the urethra into the bladder, where a balloon on the end holds in position.
  b. **Condom catheters:** It is a cone made of latex rubber or silicone that covers the penis and attaches to a tube for drainage into the urinary bag.
  c. **Intermittent catheterization technique:** This is the technique of removing urine or emptying the bladder several times by inserting a small rubber or plastic tube (Fig. 9.11).

Stimulated voiding: Voiding can be encouraged by anal or rectal stretch and crede and tapping technique.

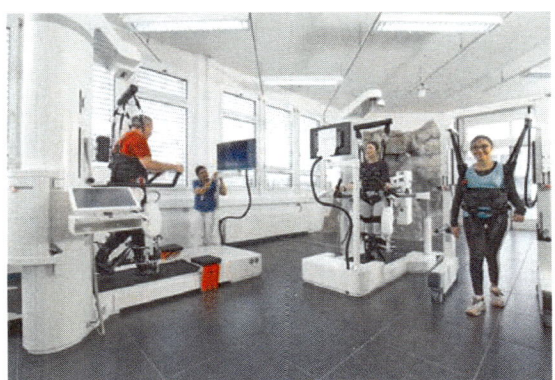

Fig. 9.10: Robotic assisted walking therapy

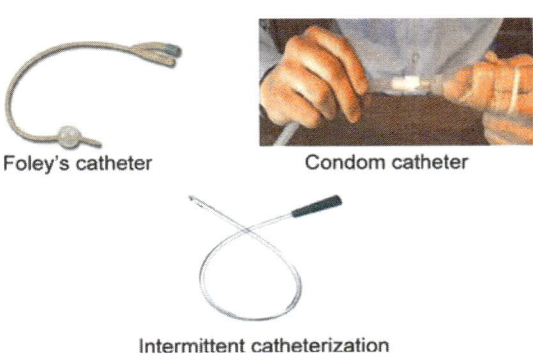

Fig. 9.11: Intermittent catheterization

Bowel Management

Bowel programs involve training bowels to have a bowel movement and involves the following:

1. **Manual removal:** Physical removal of the stool from the rectum.
2. **Digital stimulation:** Circular motion with the index finger in the rectum which causes the anal sphincter to relax.
3. **Suppository:** These stimulate the nerve endings in the rectum causing a contraction of the bowel.
4. **Mini-enemata:** It softens, lubricates and draws the water into the stools to stimulate evacuation.

Sexuality and Reproductive Health

A spinal cord injury can affect the sexuality of the person both physically and psychologically. Erectile dysfunction can be corrected by penile injection therapy, medications, vacuum pump, etc. Alternative sexual positions can be taught to the patient.

REFERENCES

1. Mukherjee AK. Spine Injury and Disability Care. Vikas Publishing House Pvt. Ltd., New Delhi, 1999.
2. Singh R, Sharma SC, Mittal R, Sharma A. Traumatic Spinal Cord Injuries in Haryana: An Epidemiological Study. Indian Journal of Community Medicine 2003;28(4):10–12.
3. Barnes et al. Enhancing patient-professional communication about end-of-life issues in life-limiting conditions: a critical review of the literature. Pain Symptom Manage 2012;44(6):866–79.
4. Chang AI, Paratz J, Rollston J. Ventilatory effects of neurophysiological facilitation and passive movement in patients with neurological injury. A Physiother 2002;48(4):305–10.
5. Nagai MK, Marquez–Chin C, Popovic MR. "Why is functional electrical stimulation therapy capable of restoring motor function following severe injury to the central nervous system"? Translational Neuroscience, Mark Tuszynski, Ed. Springer Science and Business Media LLC 2016;pp. 479–498.
6. Bailey SN, Hardin EC, Kobetic R, Boggs LM, Pinault G, Triolo RJ. Neurotherapeutic and neuroprosthetic effects of implanted functional electrical stimulation for ambulation after incomplete spinal cord injury. Journal of rehabilitation research and development 2010;47:7–16.

10

Motor Neuron Disease

LEARNING OBJECTIVES

At the end of this chapter, you will be able to:
1. Define motor neuron disease
2. Describe the pathophysiology of MND
3. List out the various etiological factors of MND
4. Identify the various clinical manifestations of the MND
5. Demonstrate the skills to assess the patients with MND
6. Design the treatment protocols for rehabilitating these patients.

INTRODUCTION

Motor neuron disease which is also known as *amyotrophic lateral sclerosis* is characterized by a selective degeneration of the motor neuron involving both the corticospinal pathways, neurons that originate from the motor nuclei of the brainstem and anterior horn cells of the spinal cord. It is not a painful condition as no sensory neurons are directly involved. The motor neuron disease is crippling and when the disease progresses the patient will not be able to move, communicate, swallow and finds difficult to breath in. In up to 15% of cases, MND is also associated with dementia[1,2] which usually affect the personality and behavior of the patient. ALS has an incidence of 0.6 to 2.6 per 100,000 of the population, and a mean onset of 47–52 years in familial cases, and 58–63 years in sporadic cases, but can occur at any age. Male sex, increasing age and hereditary disposition are the main risk factors (Anderson et al. 2007; Phukan and Hardiman, 2009).[3,4]

Definition

It is a neurodegenerative group of disorders characterized by selective degeneration of motor neurons of corticospinal pathways, motor nuclei of brainstem and anterior horn cells of the spinal cord.

Disorders

1. Progressive muscular atrophy
2. Progressive bulbar palsy
3. Amyotrophic lateral sclerosis
4. Primary lateral sclerosis

AETIOLOGY[7,8]

MND can affect adults at any stage, but is most commonly appear in adults of age

50–60. Males are more affected than females. In India, *Madras MotorNeuron Disease*[9,14] is a rare disorder that occurs at young age and takes a relatively benign course. MND has predominant geographic distribution in south India.

The majority of cases of MND are considered to be caused by several factors that contribute solely or combined to this disease. They include genetic, environmental and lifestyle influences.

Genetics: For approximately 5–10% of people suffering with MND, the genes are thought to play a role in the disease. The defective genes may increase a person's susceptibility to developing disease or slowing down the progression of the disease. Researches have shown that genes called SOD1, TDP-43, FUS, VCP, C9ORF72, SQSTM1, Profilin 1, MATR3 and TUBA4A have been identified as a cause of this disease. (MND Association, Northampton, NN1 2PR).

Diet: Studies on MND have shown that people who ate a healthier diet, with fruits and vegetables were less likely to develop the disease. People whose diet is not balanced with vitamins and minerals have chances of developing the disease.

Professional sportsmen: Research in 2008 identified that professional Italian football players develop MND more often than the general population. The reason beyond is yet to be identified.

Environmental factors and lifestyle: Several researchers have identified possible links with prior exposure to mechanical and or electrical trauma, military services, smoking, agricultural, high levels of physical activity, exposure to chemicals and heavy metals with development of MND. But it is not known how environmental and lifestyle factors could increase the risk of MND. One possible explanation is that these factors can cause weakening of nerve cells leading to degeneration.

Cellular Mechanisms

1. Cellular scaffolding called cytoskeleton maintains the shape and structure of motor neurons. If the structure is not held, it can lead to degeneration.
2. Defective RNA can lead to adverse faulty production of proteins leading to MND.
3. Disruption of chemical communication networks due to high concentrations of glutamate.

Other causes: Viral infections can also cause MND.

PATHOLOGY

The gross changes in spinal cord on naked eye inspection are slight, but on section of gray matter of anterior horns appear smaller than normal and wasting occurs in the anterior roots.

Microscopically, there is severe degeneration of motor neurons of anterior horn most commonly in the cervical enlargement, but can be widespread often. The ganglion cells number reduced due to chromatolysis. There is degeneration of white horn which is usually marked in anterior and lateral columns. There is degeneration in the corticospinal pathways both direct and crossed which is not equal and symmetrical at all levels. The motor nuclei of medulla, midbrain (3rd and 4th cranial nuclei) also show degeneration.

CLINICAL MANIFESTATIONS[11-13]

Onset is usually insidious, rarely subacute. The clinical features may depend upon the type of lesions (UMN and LMN) and their distribution. The detailed clinical picture of various clinical catagories of MND is described below.

Progressive Muscular Dystrophy

a. The disease starts with initial signs and symptoms of lower motor neuron lesion resulting in progressive muscular atrophy.

b. The early symptoms may be weakness of hand muscles, stiffness or clumsiness of movement of fingers and wasting.
c. Fasciculation or twitching may be present.
d. Cramps in the lower limbs may be early sign.
e. When it progresses to the bulbar nuclei, the first symptom observed is dysarthria (difficulty in articulation of speech), dysphagia (difficulty in swallowing), difficulty in chewing of food materials.
f. The progression of wasting of muscles first with the thenar muscles of the hand (sometimes symmetrical) and then progresses to forearm where the flexors are more affected than the extensors and less often the shoulder girdle and arm muscles are involved.
g. Later stages, it may progress to lower limb muscles, but initial involvement of lower limb muscles is rare. The anterior tibial group and peroneal muscles are first to be affected resulting in bilateral foot drop.
h. Finally there will be weakness of trunk muscles and respiratory muscle involvement may characterize the terminal stage.
i. The fasciculation can be the important sign which may be seen in a group of muscles or widespread to many groups indicating the progress of degeneration of motor nuclei.

In early stages, the fasciculation can be evoked by sharply tapping the muscles.

Progressive Bulbar Palsy

a. The bulbar palsy is the consequence of the progression of degeneration of the motor nuclei.
b. The foremost sign can be seen in the tongue. The tongue wastes, shrunken and wrinkled and shows fasciculation (Figs 10.1 and 10.2). Protrusion of tongue becomes weak and later lost.
c. The orbicularis oris can also be affected at the same time with tongue muscles, other facial muscles are involved later but may be with less severity.
d. The palate, intrinsic muscles of the pharynx and larynx may be involved at later stages resulting in dysarthria, dysphagia and food tends to regurgitate through the nose.

Progressive bulbar palsy must be distinguished from pseudobulbar palsy. The later usually occurs from diffused cerebrovascular disease affecting the hemispheres, and brainstem. In pseudobulbar palsy, there is emotional lability which is not seen in bulbar palsy.

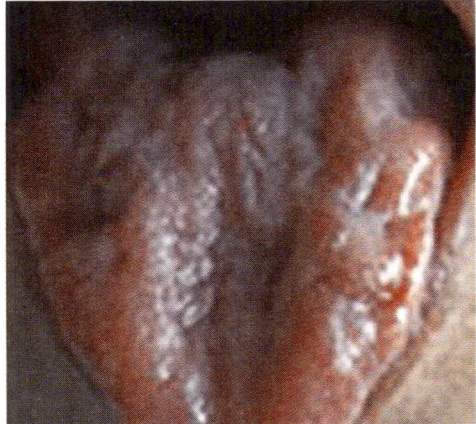

Fig. 10.1: Tongue wasting seen in MND

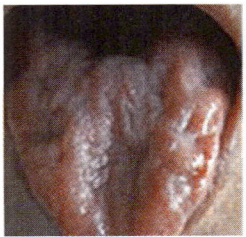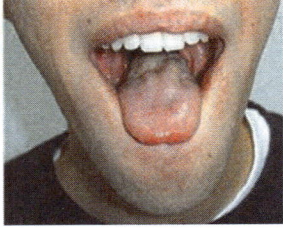

Fig. 10.2: Tongue in progressive bulbar palsy and amyotrophic lateral sclerosis

Amyotrophic Lateral Sclerosis

a. The involvement of UMN of lateral column results in amyotrophic lateral sclerosis.
b. There will be degeneration of the LMN in the spinal cord and brainstem.
c. There is lateral column degeneration or sclerosis.

> There will be a combination of UMN and LMN signs in ALS. The combination of exaggerated reflexes with muscular atrophy is not seen in any other neurological disorders other than ALS.

d. Also known as *Lou Gehrig's disease and Charcot disease*.
e. The clinical manifestations may be a combination of UMN and LMN lesions.
f. Spasticity and weakness of muscles in combination could be seen in the ALS.
g. Persistent weakness or spasticity or weakness in an arm or leg.
h. The reflexes depend upon the predominance of upper and lower motor degeneration.
i. Corticospinal tract degeneration results in exaggerated tendon reflexes.
j. It is not uncommon in ALS, to observe wasting of muscles along with the exaggerated deep tendon reflexes.
k. The muscles innervated from medulla may have effects of UMN or LMN degeneration or a combination of both.
l. A lesion of both corticospinal tracts above the medulla causes weakness of the bulbar muscles and leads to dysarthria and dysphagia. But the muscles are not wasted as in bulbar palsy, but are spastic.
m. The tongue appears small but is not wasted, wrinkled and shows the evidence of fasciculation (Fig. 10.2).
n. The jaw-jerk, and gag reflex are exaggerated, sneezing and coughing may be excited reflexly with abnormal readiness.
o. The sphincter muscles in early stage are not affected, but some cases reported difficulty in micturition.
p. The subcutaneous fat tends to disappear due to muscle waste and ocular sympathetic palsy is seen.

> Some cases of ALS have been reported with alterations in thinking or may have reported abnormal behavior pattern which is referred as *frontotemporal dementia*. But memory loss is not a classical feature of ALS.

Primary Lateral Sclerosis

Rare form of MND.
- It is a progressive, degenerative disease involving the upper motor neurons.
- It is characterized by spasticity.
- The progression of disease is development of spasticity in lower extremities followed by trunk followed by upper extremities and bulbar muscles.
- The deep tendon reflexes are exaggerated. No fasciculation or wasting is seen.

DIAGNOSIS OF MOTOR NEURON DISEASE

The motor neuron disease has to be distinguished and differentiated chiefly from other neurological disorders leading to wasting of muscles.

Syringomyelia: The syringomyelia also shows wasting of muscles, but **fasciculation** is rarely seen. Also there will be sensory loss in syringomyelia which is not seen in any MND.

Intramedullary tumor: It results in sensory loss.

Cervical spondylosis: Only upper limb muscle wasting is present, whereas lower limb involvement is also present in the MND. The radiological changes in cervical spondylosis, i.e. development of osteophytes, could be useful for differentiating it from MND. But sometimes a patient can exhibit both cervical spondylosis and MND.

Cervical rib: It is an occasional cause of muscle wasting but fasciculation is absent and pain is a striking feature in cervical rib, whereas MND is a painless disease.

Other hereditary myopathies: Often the age of incidence of hereditary myopathies is much earlier than that of MND except Madras Motor Neuron disease.

Myasthenia gravis: The bulbar palsy signs and symptoms are present but does not usually lead to muscular wasting.

Table 10.1: The revised El Escorial research diagnostic criteria for ALS with the Awaji electrodiagnostic algorithm (Anderson et al. 2012)[3,4]

Clinically definite ALS	Clinically definite ALS UMN and LMN clinical signs or electrophysiological evidence in three regions
Clinically definite ALS—laboratory supported	UMN and LMN clinical signs or electrophysiological evidence in one region and the patient is a carrier of a pathogenic SOD1-gene mutation
Clinically probable ALS	UMN and LMN clinical signs or electrophysiological evidence by LMN and UMN signs in two regions with some UMN signs rostral* to the LMN signs
Clinically possible ALS	UMN and LMN clinical signs or electrophysiological evidence in one region only, or UMN and LMN clinical signs in two regions with no UMN signs rostral to LMN signs. Neuroimaging and laboratory studies have excluded other diagnoses

* Rostral situated to the near front end of the body

Investigations

The following investigations may be required for differential diagnosis:
1. Blood investigations to check the thyroid levels and creatinine kinase.
2. Magnetic resonance imaging (MRI)
3. EMG and NCV
4. Lumbar puncture
5. Muscle biopsy

Prognosis

Though the motor neuron disease is invariably progressive, the rate of progress is different from case-to-case. Most patients survive for 2–3 years after onset and some survived extremely longer but to a maximum of 10 years as per the records till now.

MANAGEMENT

As the cause of the disease is not clearly understood, the treatment is limited to dealing with the symptoms alone.

Medical Management

Riluzole[15,16] is the only medication approved for treatment of motor neuron disease in advanced countries like USA, Europe and Australia.

Antioxidants[17] have been considered and used for many years as potential treatment of MND, such as vitamin C and vitamin E at very high doses (>1000 mg/daily). But a study by Orrell et al. have identified nine randomized controlled trails that these doses can cause increased risk of hemorrhagic stroke and premature death.

Ciliary neurotrophic growth factor is a neurotrophic factor which was considered to promote motor neuron survival and believed to slow the progression of the disease. Bongioanni et al identified two randomized controlled trails demonstrated that no significant efficacy is observed at lower doses and higher doses adverse effects may be present.

Quinine sulphate may alleviate symptoms of muscle spasms and cramps.

Baclofen, tizanadine may relieve spasticity.

Amitriptyline, fluvoxamine, dextromethorphan-quinidine may alleviate emotional lability.

Precautions

The patients diagnosed with MND should consider some factors as precautionary measures:

1. Avoid any activities that cause fatigue
2. Avoid exposure to cold
3. Avoid air travel
4. During dysphagia, solid foods are avoided.

PHYSIOTHERAPY MANAGEMENT

In MND, the physiotherapy is mainly aimed to improve the mobility of the body and maintain the physical activity. The physiotherapy shall be helpful to prevent joint stiffness, respiratory and other complications. Although some authors believe physiotherapy has only a psychological value, physiotherapy in proper protocols have been proved to improve the quality of living, improved self-esteem and prolonged life.

The physiotherapy management includes a thorough assessment, setting goals, planning the treatment and carrying out.

Principles of Physiotherapy Assessment of MND

1. Introduce yourself to the patient in the first meeting. Make the patient feel comfortable and adhere to privacy principles of the hospital.
2. The subjective assessment should start with the primary personal details of the patient like age, sex, occupation, socio-economic status, address, etc.
3. Start the initial conversation by asking the patient what he already knows or suspects about his condition.
4. Take time to go through the medical and nursing notes in detail and summarize.
5. Question the patient about the symptoms, functional limitations, expectations of the patient with physiotherapy.
6. The objective assessment should be done to assess musculoskeletal, respiratory and neurological status.
7. Assess the patient's functional abilities of balance, mobility, exercise tolerance and activities of daily living.

Short-term Goals

1. To develop good rapport with the patient and provide psychological support to the patient.
2. To maintain normal tone in the muscle
3. To promote mobility and enhance motor skills
4. To maintain clear chest and prevent respiratory complications

Long-term Goals

1. To improve the confidence levels in the patient
2. To prevent complications due to immobility or bedrest
3. To maintain normal joint ROM and prevent contractures in the muscles
4. To improve the aerobic capacity of the patient
5. To modify the lifestyle and promote functional capabilities in the patient in activities of daily living
6. To prevent early fatigue in the muscles and improve muscular and cardiovascular endurance levels

Summary of Objective Assessment

Objective assessment	Outcome measures
Assessment of higher function	Memory, intelligence, consciousness, behavior, orientation and speech
Sensory examination	Superficial, deep and cortical senses are assessed
Reflex examination	Superficial, deep tendon reflexes are assessed
Motor examination	Identify the tone and report grade on modified Ashworth scale. Identify the muscles wasted and power in case of spastic muscles, identify the grade of voluntary control
Coordination	Identify the level of coordination by equilibrium and non-equilibrium tests
Pain	Visual analog scale
Fatigue	Fatigue severity scale
Joint mobility	Joint range of motion assessed by the goniometry
Body posture	Body chart
Balance	Berg balance scale
Mobility	Modified reivermead mobility index, 10 m walk test, timed up and go test
Exercise tolerance	6 m walk test, modified BORG scale for dyspnea
ADL	Barthel index
Sputum retention	Auscultation, pulse oximetry
Effectiveness of cough	Peak cough flow
Nocturnal desaturation	Modified BORG scale of dyspnea, spirometry, pulse oximetry

Recommendations for Physiotherapists

1. Patients with MND should be seen as a priority
2. Good communication and coordination within the multidisciplinary team, patient and family
3. Always set goals depending upon the level of activity/participation of the patient
4. Always be ready to attend the MND case whenever necessary
5. Passive movements should be applied to all joints that patient cannot move
6. Stretching and ROM exercises
7. Electrical stimulation of muscles is better avoided in MND
8. Active/passive trainer or unloaded cycling is ideal for home use
9. Regular review of the patient and suggesting the aids and appliances depending upon the functional level may be helpful
10. Postural control and training for good alignment of body segments
11. Closely monitor the patient for signs and symptoms of respiratory insufficiency
12. Pain can be managed as similar to any pain of musculoskeletal origin

Plans of Treatment

To Develop Good Rapport with the Patient and Provide Psychological Support to the Patient

The importance of effective communication with the patient is proven in physiotherapy care for better prognosis and improving the rapport with the patient. Have patience to listen to all the patient's complaints and problems. Try to answer and explain the doubts or queries of patient about his condition. Never give false promises or promise goals that cannot be achievable. Effective channels of communication and coordination are required among the members of the multidisciplinary team and also with patient and guardians. In a research by Barnes et al, 2012, it was believed that facilitating a successful intervention might require a health care professional to enhance their communication skills.

To Maintain Normal Tone in the Muscle

The goal in the early stages of MND is to maintain and optimize mobility and function. To maintain good mobility, the tone in the muscles must be maintained at normal levels. Muscle weakness results in the limitations in function and mobility in patients with MND. Disuse atrophy and deconditioning resulting in wasting of muscles is common in MND (McDonald, 2002). The contractures may develop due to imbalance in the opposing muscle groups either because of wasting or spasticity in a muscle group.

The tone can be maintained by the following techniques.

Hypotonicity:
- Ice brisk stroking over the muscle
- Hacking over the muscle
- Weight bearing over a joint where the muscle is crossing
- Vibrations
- ENMS is found useful in developing the tone, but it is not advised in MND.
- Suspension therapy and aquatic exercises
- Functional re-education exercises
- Muscle energy techniques.
- EMG biofeedback techniques

Hypertonicity:
- Cryotherapy in the form of prolonged icing: Ice pack
- PNF techniques
- Passive stretching exercises
- Orthotic supports to maintain the length of the muscle

To Promote Mobility and Enhance Motor Skills

Muscle weakness results in decreased mobility in patients with MND and thus lead to limitations in ADL. The development of contractures due to imbalance in the muscle groups can also lead to difficulty in mobility. The mobility is maintained by relaxed passive movements at regular intervals to maintain normal muscle properties and range of motion.

The use of standing equipment such as tilt table to support a stretch and weight bearing over the muscle groups would be helpful.

Individualized stretching and strengthening exercises at early stage would be effective in improving the function of the patients.

Supported treadmill training in early stages will be useful in early stages of MND.

Robot assisted walking therapy is a form of physical therapy that uses a robotic device to help a person whose ability to walk has been impaired by neurological or orthopedic conditions. Commercially, it is called LOKOMAT®. In a randomized controlled trail, by Esclarin–Ruz et al. 2014, observed that robotic walking therapy yielded better results in the 6 min walk test and improved endurance in patients with UMN and LMN lesions.

Many patients may be requiring assistive devices to improve and promote mobility, e.g. ankle foot orthosis, foot-up splint, cervical collar (soft). In later stages, a rigid cervical collar (headmaster or Hessinger collar) may be needed.

Rolling exercises, pelvic rocking, bridging exercises must be trained and be practiced by the patient to maintain mobility.

To Maintain Clear Chest and Prevent Respiratory Complications

Respiratory muscle weakness in MND patients is common and can lead to respiratory dysfunction and respiratory complications. Diaphragmatic dysfunction associated with **dyspnea** can decrease the lifespan. Ineffective cough can lead to respiratory infections like pneumonia (Senent et al, 2011). The condition can become worse, if there is shortening and stiffness of chest muscles and reduced lung compliance.

The main role of physiotherapists in respiratory management is to closely monitor for any symptoms of respiratory insufficiency and to prevent accumulation of secretions in the lungs.

Respiratory adjuncts such as PEP should not be indicated, instead manual assisted coughing should be used.

Regular **postural drainage** and deep breathing exercises are to be incorporated. Chest secretions may be mobilized by manual techniques like percussions, vibrations or mechanical assistive devices like mechanical exsufflator or a positive pressure device.

Stretching of pectoral, and sternocleido-mastoid muscles could be helpful to improve chest mobility.

Neurophysiological stimulation (NPF) of respiration: It is believed that cutaneous and proprioceptive stimulations reflexly increase the depth of breathing.

The perioral technique is thought to relate to the suckling reflex, and may facilitate slow as well as deep breathing. NPF also proven to improve coughing, swallowing and abdominal contractions.

Perioral technique: Moderate finger pressure is maintained inwards and downwards just above the lip as long as the patient is required to deep breathe. The effect may continue for some minutes afterwards.

Intercostal stretch: Apply pressure downwards towards the toes on the upper border of the rib at end expiration, unilateral or bilateral but not on the floating ribs.

Cocontraction of abdominal muscles: Apply pressure laterally over the lower ribs and pelvis at right angles to patient, alternating right and left sides and maintain the pressure for up to 2 min or until desired effect is achieved.

Vertebral pressure: Finger pressure against thoracic vertebrae between T2 and T10.

To Prevent Complications due to Immobility or Bedrest

Even though early mobilization is done, the course of MND is such that at later stages, the patient mobility level may decrease. The physiotherapist should take all the possible precautions to prevent such complications arising from prolonged bedrest. The possible complications due to prolonged bedrest would be:
1. Contractures and deformities
2. Pressure sores
3. Deep vein thrombosis
4. Psychological depression
5. Respiratory complications
6. Cardiovascular deconditioning
7. Disturbed bone metabolism

The contractures and deformities are prevented by proper positioning of the patient, passive movements, use of orthotic supports, passive mobilization techniques and stretching.

Pressure sores are deep wounds due to constant and continuous pressure on the skin overlying the bony prominences in the body. It is very common, painful complication of the patients who are on the prolonged bed rest. The pressure sores can aggravate the existing disability. It should be noted that the development of pressure sores is solely due to negligence of the caregivers and health providers but not accidental. Hence, the health care providers and caretakers should take all necessary precautions to prevent the pressure sores.

Precautions to prevent pressure sores:
1. Closely monitor all the areas of risk daily.
2. Frequent change in the position of the patient.
3. Small rubbing friction massage with a moisturizer for a small duration

(15 sec) over the areas of risk shall improve the skin elasticity and pliability.
4. Use of water mattress or air bed.
5. Proper bed making technique, avoid creases on the bed cover.

Even after taking necessary precautions, if the pressure sores appear to develop then curative measures to be taken:
1. At redness stage: Stop the massage, start changing the positions frequently and take use of adaptive devices.
2. At blister stage, frequent dressing to be done. Hydrocolloid dressing would be helpful.

If an ulcer has already formed, debridement of dead tissues, dressing of the wound is done. UVR therapy, and ultrasound have been found useful to promote healing.

Deep vein thrombosis is another major complication of immobility. It can be prevented by leg lifts, ankle pumping movements, toe curls, stretching and taking plenty of fluids would be helpful.

The respiratory complications could be avoided by teaching deep breathing exercises, postural drainage, suctioning, good coughing and huffing techniques. Chest muscle stretching exercises would be beneficial.

Cardiovascular deconditioning is prevented by regular exercises as mentioned before.

To Improve the Aerobic Capacity of the Patient

Respiratory muscle training with spirometry, assisted treadmill walking, frequent stretching exercise, active and passive mobility exercises, deep breathing exercises would be beneficial to improve the aerobic capacity of the patients.

To modify the lifestyle and promote functional capabilities in the patient in activities of daily living.

In training, the patients with ADL activities and enabling the patient with MND to be active, the major hindrance would be fatigue which is a commonly experienced debilitating symptoms. The level of fatigue experienced is significantly greater than healthy individuals (Ramirez et al., 2008[18]). Excessive activity, stress, depression, pain, poor cardiovascular endurance, some medications, and disturbed sleep cycles would be contributing factors to fatigue (Mitsumoto et al, 2000[19]). The physiotherapist should inquire the levels of fatigue and factors contributing by closely monitoring and interviewing the patient. Teaching energy conservation techniques may encourage the patients to properly pace their activities and reduce levels of fatigue. Active use of assisted devices would be helpful for improving the functional capabilities of patient in performing the ADL.

REFERENCES

1. Gajdusek DC. Foci of motor neuron disease in high incidence in isolated populations of East Asia and the Western Pacific. Adv Neurol 1982;36:363–393.
2. Buckley J, Warlow C, Smith P, Hilton-Jones D, Irvine S, Tew JR. Motor neuron disease in England and Wales, 1959–1979. J Neurol Neurosurg Psychiatry 1983;46:197–205.
3. Anderson PM, Borasio GD, Dengler R, Hardiman O, Kollewe K, Leigh PN, Silani V, Tomik B. EALSC Working Group. Good practice in the management of amyotrophic lateral sclerosis: clinical guidelines. An evidence based review with good practice points. EALSC Working Group. Ayotrophic Lateral Sclerosis 2007;8(4):195–213.
4. Anderson PM, Abrahams S, Borasio GD, De Carvaldo M, Chio A, Van damme P, Hardiman O, Kollewe K, Morrison K, Petri S, Pradat PF, Silani V, Tomik B, Wasner M, Weber M. EFNS guidelines on the Clinical Management of Amyotrophic Lateral Sclerosis (MALS): Revised report of an EFNS task force. European Journal of Neurology 2012;19:360–375.

5. Brownell B, Oppenheimer DR, Hughes JT. The central nervous system in motor neurone disease. J Neurol Neurosurg Psychiatry 1970;33:338–357.
6. Rowland LP. Motor neuron diseases: the clinical syndromes. In: DW Mulder (Ed.) The Diagnosis and Treatment of Amyotrophic Lateral Sclerosis. Houghton Mifflin Professional Publishers, Boston 1980;7–27.
7. Mantovani S, Garbelli S, Pasini A, Alimonti D, Perotti C, Melazzini M, et al. Immune system alterations in sporadic amyotrophic lateral sclerosis patients suggest an ongoing neuroinflammatory process. J Neuroimmunol 2009;210(1–2):73–9.
8. Orrell RW. Motor neuron disease: systematic reviews of treatment for ALS and SMA. Br Med Bull. 2010;93:145–59.
9. Meenakshisundaram E, Jagannathan K, Ramamurthy B. Clinical pattern of motor neuron disease seen in younger age groups in Madras. Neurol India 1970;18(Suppl 3):109–112.
10. Millecamps S, Boillée S, Le Ber I, Seilhean D, Teyssou E, Giraudeau M. Phenotype difference between ALS patients with expanded repeats in C9ORF72 and patients with mutations in other ALS-related genes. J Med Genet 2012;49(4):258–63.
11. de Carvalho M, Matias T, Coelho F, Evangelista T, Pinto A, Luís ML. Motor neuron disease presenting with respiratory failure. J Neurol Sci 1996;139:117–22.
12. Wijesekera LC, Leigh PN. Amyotrophic lateral sclerosis. Orphanet J Rare Dis 2009;4:3.
13. Josephs KA, Dickson DW. Frontotemporal lobar degeneration with upper motor neuron disease/primary lateral sclerosis. Neurology 2007;69(18):1800–1.
14. M Gourie-Devi and TG Suresh, Madras pattern of motor neuron disease in South India. J Neurol Neurosurg Psychiatry 1988;51(6):773–777.
15. Lacomblez L, Bensimon G, Leigh PN, Guillet P, Meininger V. Dose-ranging study of riluzole in amyotrophic lateral sclerosis. Amyotrophic Lateral Sclerosis/Riluzole Study Group II. Lancet 1996;347:1425–31.
16. Miller RG, Mitchell JD, Lyon M, Moore DH. Riluzole for amyotrophic lateral sclerosis (ALS)/motor neuron disease (MND). Cochrane Database Syst Rev 2007(1):CD001447.
17. Orrell RW, Lane RJ, Ross M. Antioxidant treatment for amyotrophic lateral sclerosis/motor neuron disease. Cochrane Database Syst Rev 2007(1):CD002829.
18. Ramirez C, Pimentel Piemonte ME, Callegaro D, Da Silva HCA. Fatigue in amyotrophic lateral sclerosis: frequency and associated factors. Amyotrophic Lateral Sclerosis 2008;9(2):75-80.
19. Mitsumoto H, Del Bene M. Improving the quality of life for people with ALS: the challenge ahead. Amyotrophic Lateral Sclerosis and Other Motor Neuron Disorders 2000;1(5):329–336.

11

Multiple Sclerosis and Demyelinating Diseases

LEARNING OBJECTIVES

At the end of this chapter, you will be able to:
1. Define multiple sclerosis (MS)
2. Describe the pathophysiology of multiple sclerosis
3. List out the various etiological factors of multiple sclerosis
4. Identify the various clinical manifestations of the multiple sclerosis
5. Demonstrate the skills to assess the patients with multiple sclerosis
6. Design the treatment protocols for rehabilitating these patients.

INTRODUCTION

Multiple sclerosis is one of the commonest nervous diseases. It is not a contagious disease and does not spread from one person to the other. Usually, this disease does not shorten the life expectancy. This disorder frequently has the tendency of remissions and relapses and hence the course of the disease is considered to be prolonged or chronic. Multiple sclerosis does not necessarily produce serious disability. The course of MS varies widely from person-to-person. Some people only experience mild symptoms over their lifetime while others will have relapses followed by incomplete remission. Most of the patients suffering with MS experience slowly progressive worsening of disability over many months or years.

Definition

It is the disorder of nervous system characterized by the widespread occurrence of patches of **demyelination** followed by **gliosis** in the white matter of the nervous system.

Incidence

In western countries, it is about 1 in 1000 population. It is common in people originating from western Europe and rare in Asians.[1,2] The disease attacks the young adults and is more common in women. Usually, the disease is affected at the age of 20–40 rather than the later decades. Occasionally, it is also seen in children at ages of 12 and 15. Very rarely it is seen to be prevalent at late ages of 50–60. The prevalence in India is found to be 1.33/1,00,000 in Mumbai and 3.2 in south India (Singhal et al). In the Parsi community in India, the prevalence is found 21/1,00,000 (Bharucha et al[3]).

> The risk of developing multiple sclerosis in identical twins is 31%, non-identical twins is 5%, first degree siblings is 3 or 4%.

ETIOLOGY[4-6]

The causes of MS are still not known. But researchers believe that both genetic and environmental factors and their interactions can lead to the disease. Some studies report that this could be autoimmune in which the body produces misdirected immune system which attacks on its own tissue. A reaction to virus hidden in the central nervous system is also suspected but there are no evidences. However, in persons with MS, the immune system appears to be normal in all other aspects.

MS is not directly considered hereditary disorder, but there is increased risk in close family members which is not attributed to single gene but is related to several genes whose function is not well understood.

It is found that the multiple sclerosis is common in temperate climate, i.e. at higher altitudes particulary in the northern hemisphere and is rare in tropical countries. However, there are some areas of low prevalence in northern climates, e.g. Iceland and high prevalence in the mediterranean areas, e.g. Sardinia and Sicily.

There are some precipitating factors which were found to precede the onset of illness and hence considered as precipitating factors. These include influenza and infections of upper respiratory tract, pregnancy, some fevers, extraction of teeth, electric shock, trauma, etc. However, the mode of operation or how they are connected to MS is still unknown.

Some biochemical factors are also considered as a cause for MS though the relation is still not clearly understood. People who have low intake of dietary polysaturated fats are found to be attacked with the disease. Other unexplained biochemical defects in multiple sclerosis patients include increased platelet adhesiveness.

PATHOLOGY[8,9]

1. It is an inflammatory, demyelinating disease of the CNS.
2. In MS, there will be patches in the white matter of brain and spinal cord.
3. There is myelin loss, destruction of oligodendrocytes and reactive astrogliosis, often with relative sparing of the axon cylinder. In some cases, the axons are aggressively destroyed.
4. In acute stage, the myelin sheaths degenerate and the perivascular spaces contain fagranule cells and lymphocytes.
5. As neural inflammation resolves in MS, the damaged myelin is removed and a glial overgrowth leads to a sclerotic plaque.
6. For naked eye, this will be slightly sunken, grayish and more transculent than normal nervous tissue.
7. The location of lesions in the CNS usually dictates the clinical deficit.
8. The recovery is believed due to neural plasticity rather than remyelination that occurs in MS.
9. MS is also characterized by periventricular infiltration of lymphocytes and macrophages.
10. Infiltration of inflammatory cells occurs in the parenchyma of the brain, brainstem, optic nerves and spinal cord.

Common Types of Multiple Sclerosis

A. Relapsing-remitting multiple sclerosis (RRMS)
B. Secondary-progressive multiple sclerosis (SPMS)
C. Primary-progressive multiple sclerosis (PPMS)
D. Benign multiple sclerosis (BMS)

E. Progressive-relapsing multiple sclerosis (PRMS)
F. Malignant or fulminant multiple sclerosis

Relapsing-remitting MS

In this, there will be relapses with a flare up of old symptoms or in some patients, new symptoms are developed and followed by a remission with resolution or reduction of symptoms. Some stability of symptoms is observed in between the attacks.

Secondary-progressive MS

In this type, after an initial course of relapsing/remitting MS, there will be development of slowly progressive disability after a duration of long years. There may be relapses in between the episodes.

Primary-progressive MS

There is insidious onset and there will be slow progressing of symptoms leading to worsening of disability without distinct attacks.

Benign MS

There will be a few attacks and there will be a little or no disability after a duration of 20 years.

Progressive-relapsing MS

There will be progression of symptoms from the onset and is combined with occasional acute symptoms flare ups.

Malignant or Fulminant MS

It is a rapidly developing and progressive MS.

CLINICAL MANIFESTATIONS

The clinical features of MS differ from person-to-person and the part of CNS involved. Some of the most common clinical features are

A. **Vision:** It is due to optic nerve, cervical cord, brainstem and cerebellar peduncle involvement. The first feature shall be optic neuritis of varying severity. There may be pain and visual disturbances in affected eye. On progression, the disease results in dysmetria, nystagmus, internuclear ophthalmoplegia and diplopia. The patient usually experiences a visual image with a dark, blank area in the middle.

B. **Sensory defects:** Sensory abnormalities that are experienced in the patients with MS are numbness, pins and needles, feeling of tightness around a limb. Most of the times, the patient complains of feeling of water running down the limb. As the disease progresses, the joint position sense may be affected.

C. **Motor dysfunction and incoordination:** The motor symptoms would be more disabling than the sensory symptoms in patients with MS. The neurological conditions where the white matter is affected often result in spasticity. The spasticity is most common in lower limbs and may or may not accompany weakness. The other motor problems would be:
   i. Loss of balance
   ii. Intentional tremors
   iii. Ataxic gait
   iv. Vertigo
   v. Clumsiness of movement in limbs
   vi. Incoordination
   vii. Tightness, stiffness or pull of muscles
   viii. Involuntary muscle spasms which are usually observed in night.

Charcot's Triad

Dr Jean Charcot described the characteristics of multiple sclerosis which is described as *Charcot's triad*
   a. Paralysis
   b. Intentional tremors
   c. Scanning speech
   d. Nystagmus

D. **Fatigue:** It is the one of the most common symptom of MS. The fatigue in MS is often described as a feeling of extreme mental or physical exhaustion. It affects a patient ability in ADL and social and personel life. The physical fatigue leads to limbs feel heavy and hard to use, whereas cognitive fatigue could include difficulty following a conversation or thinking of words or numbers.

E. **Dysphagia:** The difficulty in swallowing (dysphagia) may be experienced by the patients at later stages.

F. **Bladder and bowel disturbances:** The bladder disturbances in multiple sclerosis are very common and can be either detrusor hyperreflexia or detrusosphincter dyssynergia.

The bladder and bowel is affected due to damge of myelin in the CNS resulting in interruption of messages between the brain, spinal cord that controls the bladder (sacral segments). The following common bladder problems could be observed:
   i. Urgency: Feeling of having to empty the bladder immediately.
   ii. Frequency: It is an increase in the number of times the micturition occurs.
   iii. Nocturia: It is the increased frequency of micturition in the night.
   iv. Incontinence: It is the inability to hold urine in the bladder.
   v. Hesitancy: It is the difficulty in beginning to urinate.

The bowel problem in MS patients is constipation.

G. **Emotional and cognitive problems:** The emotional and cognitive problems may be the hidden symptoms in many patients with MS. These changes can affect the way people feel about themselves and alter their cognitive functions. The emotional changes would range from anxiety and stress to depression and attention deficits. There could also be mood changes from happy to sad to angry.

The cognitive functions that would be affected are:
- Memory
- Attention and concentration
- Word-finding difficulties
- Difficulty in speed of information processing
- Abstract reasoning and problem solving
- Visual spatial disabilities
- Difficulty in executive functions.

Pain

Pain is the common symptoms that is encountered by two-thirds of the people suffering with MS. The patients usually suffer steady and achy types of pain and may be due to fatigued and stretched muscles. Some patients may also experience more stabbing type of pain due to defective nerve signals due to MS lesions in the brain and spinal cord. The most common pain syndromes experienced by people with MS include:
- Headache
- Continuous burning pain in the extremities
- Back pain
- Painful tonic muscle spasms
- Pain in the eye.

Sexual Dysfunction

The sexuality of patients with MS is a direct result of neurologic changes that affect the sexual response. This may include a decrease or loss of sex drive, decreased or unpleasant genital sensations and diminished capacity of orgasm. Women may experience decreased vaginal lubrication, loss of vaginal muscle tone and diminished clitoral engorgement. At later stages, tertiary sexual dysfunction can result due to disability-related psychological issues.

INVESTIGATIONS

CSF Analysis

CSF analysis after lumbar puncture reveals excess of lymphocytes and increased protein content, abnormally high IgG proportions.

MRI Scan

The MRI scan shows irregular areas of increased signal density probably due to presence of inflammation, edema or gliosis.

Evoked Potentials

Visual evoked potentials would be helpful in optic neuritis with a clearcut delay. Auditory evoked potentials would be beneficial for diagnosing brainstem MS. Sensory evoked potentials recorded from surface electrodes over the cervical cord could also show objective evidence of the presence and site of lesion in multiple sclerosis.

Medical Management

Presently, there is no absolute cure or prophylaxis management for multiple sclerosis. The treatment usually focuses on speeding recovery from attacks and slowing the progress of the disease.

Treatment for MS Attacks[10]

1. **Corticosteroids:** Oral prednisolone and IV methylprednisolone are used to reduce nerve inflammation.
2. **Plasmapheresis:** Plasma exchange may be used, if the symptoms are new, severe and are not responding to steroids.
3. Disease-modifying therapies would be helpful in relapsing-remitting MS.
4. Beta-interferons can reduce the frequency and severity of relapses.
5. Glatiramer acetate can help to block the immune system's attack on myelin.
6. Dimethyl fumarate and fingolimod can reduce the relapse rate.
7. **Mitoxantrone:** Usually not advised because of its adverse effects, but used to treat severe, advanced MS.
8. Other drugs for symptomatic relief like muscle relaxants, medications to reduce fatigue.

Differential Diagnosis of MS

Compression of the spinal cord: Myelography would be helpful in diagnosing MS

Cervical spondylosis: Characteristic X-ray changes like development of osteophytes and subchondral sclerosis

Hereditary ataxia: These are often familial and symptoms tend to begin at an earlier age and run a slow progression.

Friedreich's ataxia: Loss of ankle jerks, and later of the knee jerks, scoliosis which are not seen in MS.

Vitamin B_{12} deficiency: Begins at a late age than MS, paresthesia exist and remains persistent.

Hysteria: Giddiness, paresthesia and paresis rarely present in MS.

PHYSIOTHERAPY MANAGEMENT

Physiotherapy plays an important role in rehabilitation of multiple sclerosis patients, since the majority of patients fall in the age groups of 20 and 50. The physiotherapist should have information, experience relating to the variability of symptoms between individuals and to the unpredictable and fluctuating nature of the progressive disease.

There is no fixed protocols for treatment of people with MS or any time limits set up, in other words, there are no tailor made treatments for MS patients. The physiotherapist should employ numerous problem solving skills, interventions and resources. The coordination and communication with other health care providers extremely important.

PHYSIOTHERAPY ASSESSMENT

At the initial session, taking a thorough history is critical. The history should include diagnosis, date and nature of initial symptoms, other health conditions, medications, prior level of activity and intereference of disease with quality of life.

Apart from the routine neurological examination, some standardized assessment tools recommended for assessment of patients with MS are:
- MS functional composite which includes the 25 feet walk
- Expanded disability status scale
- MS fatigue impact scale
- Disease steps
- MS walking scale -12
- Berg balance scale
- Tinetti gait and balance assessment
- Activities specific balance confidence
- Timed up and go
- Dynamic gait index
- Six spot step test
- Functional independence measure
- 6-minute walk test
- Fatigue severity scale

Posture, balance and trunk control: The balance impairments are common in MS and the patients would develop high risk to falls. Hence, the balance, trunk control and transfers to and from bed, chair, toilet and floor should be evaluated.

Ambulation: Ambulation is very important and should be assessed whether the ambulation is independent or dependent should be assessed. The vision, senses and vestibular equilibrium should be assessed for ambulation.

Range of motion: Both passive and active range of motions should be assessed by goniometer.

Motor functions: Assessment should focus on gross strength with emphasis on function. The power and voluntary control should be assessed. The tone of the muscles should also be assessed by modified Ashworth scale.

Respiratory functions: Recognize the respiratory problems in the patients with MS.

Other considerations: Psychological assessment, social support, and patient's safety profile should be assessed.

Physiotherapy management: Same as multiple sclerosis.

REFERENCES

1. Goldberg L, Edwards N, Fincher C, Doan Q, Al-Sabbagh A, Meletiche D. Comparing the cost effective of disease modifying drugs for the firstline treatment of relapsing-remitting multiple sclerosis. J Manag Care Pharm 2009;15:543–555.
2. Weiner HL. A shift from adaptive to innate immunity: a potential mechanism of disease progression in multiple sclerosis. J Neurol 2008;255:3–11.
3. Rohit Bhatia, Prerna Bali and Rima M Chaudhari. Epidemiology and genetic aspects of multiple sclerosis in India. Ann Indian Acad Neurol 2015;18(Suppl I):S6–S10.
4. Duddy M, Niino M, Adatia F, Herbert S, Freedman M, Atkins H, Kim H, Bar-Or A. Distinct effector cytokine profiles of memory and naive human cell subsets and implication in multiple sclerosis. J Immunol 2007;178:6092–6099.
5. Magliozii R, Howell O, Vora A, Sefafini B, Nicholas R, Puopolo M, Reynolds R, Aloisi F. Meningeal B-cell follicles in secondary progressive multiple sclerosis associate with early onset of disease and severe cortical pathology. Brain 2007;130:1089–1104.
6. Hawker K. B cells as a target of immune modulation. Ann. Indian Acad Neurol 2009;12:221–225.
7. Barcellos LF, Oksenberg JR, Begovich AB, Martin ER, Schmidt S, Vittinghoff E, Goodin DS, Pelletier D, Lincoln RR, Bucher P, Swerdin A, Pericak-Vance MA, Haines JL, Hauser SL, HLA-DR2

dose effect on susceptibility to multiple sclerosis and influence on disease course. Am J Hum Genet 2003;72:710–16.
8. Codarri L, Fontana A, Becher B. Cytokine networks in multiple sclerosis: lost in translation. Curr Opin Neurol 2010;23:205–11.
9. Gandhi R, Laroni A, Weiner HL. Role of the innate immune system in the pathogenesis of multiple sclerosis. J Neuroimmunol 2009;221:7–14.
10. Ingrid Loma and Rock Heyman. Multiple Sclerosis: Pathogenesis and Treatment, Curr Neuropharmacol 2011;9(3):409–16.
11. Freeman JA, Langdon DW, Hobart JC, Thompsson AJ. The impact of inpatient rehabilitation on progressive multiple sclerosis. Ann Neurol 1997;42:236–44.
12. Freeman JA, Langdon DW, Hobart JC, Thompsson AJ. Inpatient rehabilitation in multiple sclerosis. Do the benefits carry over into the community? Neurology 1999;52:50–56.
13. Kraft GH. Rehabilitation still the only way to improve function in multiple sclerosis. Lancet 1999;354: 2016.
14. Di Fabio RP, Choi T, Soderberg J, Hansen CR. Health-related quality of life for patients with progressive multiple sclerosis. Influence of rehabilitation. Phys Ther 1997;77: 704–16.
15. Petajan JH, Gappmaier E, White AT, Spencer MK, Mino L, Hicks RW. Impact of aerobic training on fitness and quality of life in multiple sclerosis. Ann Neurol 1996;39:432–41.
16. Svensson B, Gerdle B, Elert J. Endurance training in patients with multiple sclerosis. Five case studies. Phys Ther 1994;74:1017–26.
17. Solari A, Fillippini G, Gasco P, Colla L, Salmaggi A, La Mantia L et al. Physical rehabilitation has a positive effect on disability in multiple sclerosis patients. Neurology 1999;52:57–62.

12

Transverse Myelitis

LEARNING OBJECTIVES

At the end of this chapter, you will be able to:
1. Define transverse myelitis
2. Describe the pathophysiology of transverse myelitis
3. List out the various etiological factors of transverse myelitis
4. Identify the various clinical manifestations of the transverse myelitis
5. Demonstrate the skills to assess the patients with transverse myelitis
6. Design the treatment protocols for rehabilitating these patients.

INTRODUCTION

Transverse myelitis is a rare inflammatory disease causing injury to spinal cord resulting in varying degrees of weakness, sensory disturbances and autonomic dysfunction. The first case of transverse myelitis was described in 1882. There were many cases reported in England around 1922 and 1923 after vaccination to smallpox and rabies vaccines as complications to vaccine. In later

> 'Suchett-Kaye' first coined the term "transverse myelitis" in 1948.

years, with eventual development of medicine, transverse myelitis is believed to occur post-infection and agents including measles, rubella and mycoplasma. In 1948, An English neurologist described it as a case of rapidly developing progressive paraparesis occurring as a post-infectious complication of pneumonia.

Definition

It is an idiopathic inflammation of the spinal cord, usually involves both the gray and white matter extending transversely over the spinal cord. The lesion may be limited to a few segments longitudinally. When it ascends upwards it is called ascending myelitis. Transverse myelitis usually occurs in thoracic region and can result in paraparesis, paraplegia and sensory impairment in the segments involved.

Epidemiology

It is estimated that there will be 1–8 new cases per million per year or approximately 1400 new cases each year. Although all age groups are affected, but is peak observed between the ages of 10–19 years and 30–39 years. There is no gender or familial association with transverse myelitis.[1] In 75% of the cases, it is monophasic, but a few cases have shown recurrence of symptoms.

Etiology[2-8]

Transverse myelitis is due to large variety of causes, some have not yet been identified.
1. Bacterial infections
2. Post-TB infection
3. Post-meningovascular syphilis
4. Predisposing demyelinating disorder of spinal cord.
5. Post-acute disseminated encephalomyelitis
6. Viral infection with known neurotrophic propensitis.
7. Rarely to post-vaccination as complication
8. Vascular etiology
9. Autoimmune disorders

Neuromyelitis optica (Devic's disease) is a condition that causes inflammation and loss of myelin around the spinal cord and the nerve in your eye that transmits information to the brain.

Pathology

On naked eye examination, the spinal cord at the site of involvement shows edema and hyperemia and in severe cases there will be softening. Microscopically, the leptomeninges are congested and infiltrated with inflammatory cells. The substance of the cord exhibits congestion or thrombosis of the vessels, with perivascular inflammatory infilteration and edema.

CLINICAL MANIFESTATIONS[9,10]

The disease is acute or subacute. The symptoms usually develop over the course of hours or days and may progress after weeks.

Systemic Symptoms

1. Pain
2. Hyperalgia

Complications
- Hypertension
- Urinary tract infections
- Urolithiasis
- Pneumonia
- Deformities
- Myocarditis
- Paralytic ileus

---

3. Fever
4. Seizures
5. Ophthalomoplegia

Neurological Symptoms

The disturbances in neurological function involve sensory, motor and autonomic dysfunction and purely depend upon the level of lesion.

Motor Dysfunction

Cervical segments: If upper cervical cord is involved, all four limbs show flaccid paresis which may be complete or incomplete. There is risk of respiratory paralysis due to involvement of C3–C5 segments which innervate the diaphragm which is the chief muscle of respiration.

Lesions at lower cervical region (C5–T1) will result in the combination of upper and lower motor signs in the upper limbs and UMN lesion signs in the lower limbs.

Thoracic region: A lesion of the thoracic spinal cord (T1–T12) can lead to UMN signs in lower limbs leading to spastic diplegia (lower limbs are more affected than the upper limbs). There is also involvement of autonomic nerves at this level resulting in impairment of sphincter control, often resulting in complete paralysis of the bladder and rectum. A high blood pressure due to ANS involvement is observed.

Lumbar cord level (L1–S5): A lesion in the lower part of the spinal cord often produces a combination of UMN and LMN lesion type signs in the lower limbs.

Sensory Symptoms

The sensory symptoms depend upon the disruption in the sensory tract pathways but is often observed that there is loss of pain and light touch below the level of lesion.

The sensory impairments may be numbness, tingling, coldness and feeling of tight wrapping the skin over the chest, abdomen or legs. Diminished sensations of vibrations, and joint position sense.

Reflexes:
- The reflexes are usually diminished at first below the level of lesion and later exaggerated.
- The Babinski reflex is positive (extensor response).

Diagnosis

- The diagnosis of transverse myelitis is done by taking medical history and performing a thorough neurological assessment.
- Diagnostic imaging of the brain and spinal cord using MRI (magnetic resonance imaging) (Fig. 12.1) could be useful in ruling out structural lesions, e.g. tumors, herniated disks, slipped disks, stenosis, abscesses, abnormal collection of blood vessels.
- The multiple sclerosis could be ruled out by MRI of brain.
- If a patient is contraindicated for MRI, e.g. patients with implanted pacemaker, CT (computed tomography) scan of the spine, myelography, myelograms could be done.
- Blood investigations may be performed to rule out various SLE, HIV, vitamin B_{12} deficiency. A blood test for NMO–IgG is also necessary.
- CSF (cerebrospinal fluid) analysis for protein and WBC could be of diagnostic value of transverse myelitis. In transverse myelitis, the proteins and WBC count increases.
- A spinal tap may be useful to exclude infections and identify certain diseases as multiple sclerosis.
- EMG (electromyography) can document anterior horn cell involvement.

Medical Management

1. Intravenous steroid treatment in acute TM. Corticosteroids have multiple mechanisms of action including anti-inflammatory, immunosuppressive and antiproliferative actions.
   Methylprednisolone (1000 mg) or dexamethasone (200 mg) for 3–5 days
2. PLEX (plasma exchange) is often initiated in patients having moderate to severe TM.[11]
3. Immunomodulatory therapy,[11] e.g. azathioprine (150–200 mg/d), methotrexate (15–20 mg/wk).

Prognosis

Some patients with transverse myelitis may experience recovery in neurological dysfunction regardless of therapy instituted. Recovery should begin within 6 months and show restoration of neurologic function within 8 weeks (JHTMC case series).

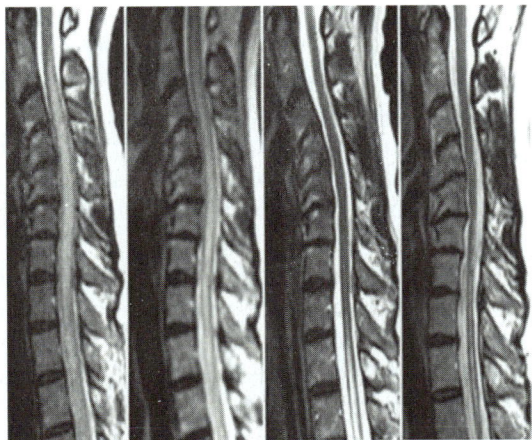

Fig. 12.1: MRI scan showing transverse myelitis

PHYSIOTHERAPY MANAGEMENT

Rehabilitation is very important component of management in patients where the neurological function is compromised. A comprehensive multidisciplinary rehabilitation program designed with suitable protocols is needed for rehabilitation of these patients. The main role of physiotherapy would be strengthening of muscles, improving muscular and cardiovascular endurance, balance training, improving coordination, maintaining normal joint range of motion, functional independence and prevention of secondary complications (Fig. 12.2).

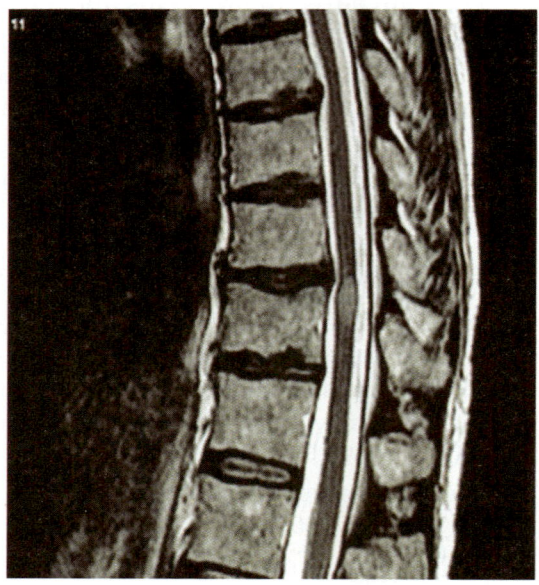

Fig. 12.2: Radiological changes[5]

Apart from a physiotherapist, occupational therapist, orthotist, psychologist also play a role in rehabilitation of patients with transverse myelitis. The physiotherapy management includes a thorough assessment, setting goals, planning the treatment and carrying out.

Principles of Assessment of Patient with Transverse Myelitis

Assessment is a key factor for planning and designing the protocol in the therapeutic program in rehabilitation of the patient with transverse myelitis. It may not be possible for complete assessment of patient in single session. So a periodic assessment sessions may be required to assess the patient. The periodic assessment sessions will also help in assessment in prognosis.

1. Prior to the assessment of the patient. The physiotherapist should take the complete personal information and detailed history including present, past, family, personal and medications presently used.
2. Identify after inquiry with the patient the chief complaints or the present functional problems the patient is facing.
3. Kindly note that the therapist assessment skills should concentrate not only on what patient cannot do but also what the patient is able to do.
4. Observe the patient for abnormal attitude of limbs, bedsores, wasting of muscles, swelling deformities, scars any other aids applied to the patient.
5. Palpate the patient to check the tenderness over the muscles, edema, and vital signs.
6. Check for the higher functions and cranial nerves to rule out whether any higher centers are involved.
7. Sensory examination of superficial, deep and cortical senses for all dermatomes and mark on the body chart the dermatomes involved in sensory loss.
8. Reflex examination is important to identify the level of lesion.
9. Motor examination of tone, power and voluntary control are assessed. Grade the muscle tone by modified Ashworth scale.
10. Check for the balance and coordination, equilibrium reactions. Balance is checked by Berg balance scale.
11. Bladder and bowel assessment is also important.
12. Assess the gait and grade the level of independence in ambulation.

GAITRite portable walkway system:[12] It is a commercial, economical device used internationally to obtain quantitative gait parameters. The system can assess the subject's gait parameters like step, cycle, swing and stance time, velocity, cadence, step and stride length, H–H dynamic base, single and double support, toe in and out degrees, step/extremity ratio and functional ambulation performance score.

Lower limb spasticity measurement system: The LLSMS is a modern technique of assessment of quality of muscles around the ankle.

13. Assess the ADL activities by a functional assessment scale, e.g. Barthel index.
14. Additional assessment may be required for respiratory status, if the higher cervical segments are involved.

Treatment Goals

Short-term Goals

1. To provide psychological support to the patient.
2. To maintain clear airways.
3. To prevent damage to skin and joints
4. To care for bladder and bowel function

Long-term Goals

1. To continue to give support and try to help acceptance of the situation of the patient and the relatives
2. To normalize the tone and maintain normal muscle properties
3. To maintain normal joint range of motion
4. To improve the sensory integration
5. To prevent secondary complications
6. To promote and train for transfer activities and ambulation.
7. To reduce pain
8. To improve functional independence of the patient.

Principles of Treatment for Physiotherapist

1. The emphasis should be on activity to gain as much independence as possible.
2. The activities should be achievable and challenging and should be progressive.
3. Adjustments to sitting and standing positions must be acquired, balance and posture training are given.
4. Be responsible for care of the patient and also train the patient for self-care of skin, joints, bladder and bowel health.
5. Teach the transfer activites without damage to the skin and joints.
6. Ambulation in a self-propelled or mechanized wheelchair is acquired, in some cases ambulation in orthosis is encouraged.
7. Resettlement at home and if possible in a suitable occupation should be aimed for.
8. Follow up after care will be needed when a patient is discharged from the hospital. Regular reviews are necessary and home visit may be planned by the therapist.

Plans of the Treatment

To Provide Psychological Support to the Patient

The importance of effective communication with the patient is proven in physiotherapy care for better prognosis and improving the rapport with the patient. Have patience to listen to all the patient's complaints and problems. Try to answer and explain the doubts or queries of patient about his condition. Never give false promises or promise goals that cannot be achievable. Effective channel of communication and co-ordination is required among the members of the multidisciplinary team and also with patient and guardians. In a research by Barnes et al, 2012, it was believed that facilitating a successful intervention might require a health care professional to enhance their communication skill.

To Maintain Clear Airways

Pulmonary compromise is a leading cause of morbidity in patients with spinal cord lesions like transverse myelitis. Respiratory complications are very common, if the higher cervical segments are involved. Accumulation of secretions leading to respiratory infections are also common problem encountered by patients with transverse myelitis at lower cervical segments involvement due to immobility and decreased thoracic mobilitiy or prolonged bedrest. It is also evident in patients with pre-existing chronic respiratory disorders, lowered resistance to infection. The main aim of chest care in these patients is to prevent internal secretions causing obstruction, improve ventilation to the lungs and improve vital capacity.

Deep breathing exercises, frequent change of positions would assist in clearing bronchial secretions and decrease paradoxical movement of the rib cage (Doughlas et al. 1977). Nebulizations, postural drainage with manual techniques like clapping, vibrations to remove the accumulated secretions would be useful. **Spirometric** and chest mobility exercises are to be encouraged to improve ventilation capabilities. Coughing and huffing techniques are taught to the patient and he/she is encouraged for effective coughing.

A manual abdominal thrust to assist the patient with coughing has been shown to be effective in increasing the expiratory force to the patient and can generate aid in clearing secretions (Kirby et al, 1966). **Glossopharyngeal breathing** would be helpful to increase vital capacity and enhance the effectiveness of a patient's cough, provide the patient with ability to sigh and improve lung and chest wall compliance. Diaphragmatic breathing execises would improve strength and endurance, decrease the work of breathing and increases vital capacity of the patient. Manual stretching of chest wall and use of air shift maneuvers assist in maintaining the mobility and compliance of chest wall. A use of abdominal corset is found useful in sitting position to improve the patient's tidal volume and decrease shortness of breath (Alvarez et al. 1981).

To Normalize the Tone and Maintain Normal Muscle Properties

Hypotonicity:
- Ice brisk stroking over the muscle
- Hacking over the muscle
- Weight-bearing over a joint where the muscle is crossing
- Vibrations
- ENMS is found useful in developing the tone.
- Suspension therapy and aquatic exercises
- Functional re-education exercises
- Muscle energy techniques.
- EMG biofeedback techniques

Hypertonicity:
- Cryotherapy in the form of prolonged icing—ice pack
- PNF techniques
- Passive stretching exercises
- Othotic supports to maintain the length of the muscle

To Prevent Damage to Skin and Joints

The care of skin and joints is taught by giving instructions of careful handling. All insensitive parts should be protected by vigilant observation, avoidance of extreme temperatures, careful positioning and inspection. The patient and caregiver are instructed to inspect regularly the areas prone to skin damage and report to the nursing staff immediately, if any sign of skin damage is suspected. There should be a firm base mattress to the bed and should be devoid of creases in the bed sheet. If the patient is

Fig. 12.3: Wheelchair seat

wheelchair bound and spends most of the time in sitting in wheelchair, the skin over the ischial tuberosities must be protected by using a wooden based, sheep-skin covered firm sorbo cushion or cushions with a convex base which compensate for the sag in the wheelchair seat (Fig. 12.3).

Regular lifts performed by the patient himself shall relieve pressure and allow return of circulation to the skin. The lifts are maintained for 15–30 sec every 15 min interval. Teach the patient prone lying to prevent the sacral pressure.

Training in the care of the hands including gloves and special mittens are worn during the wheelchair propelling.

To Care of Bladder and Bowel Function

The bladder and bowel care is essentially a nursing and medical concern, but the physiotherapist needs to be aware of the management required to avoid retention of urine or bladder infections. The therapist should also employ skills during handling the patient, transfer activities, positioning and other activities and also during application of orthotic supports.

To Maintain Normal Range of Motion

Achieving adequate joint ROM is necessary for functional independence and ambulatory training. The joint range of motion is maintained and improved by stretching exercises. Use of appropriate orthotic supports would be useful to maintain the joint range. Strengthening exercises of weak muscles would be necessary to prevent the contractures and deformities in order to maintain full range of motion. It is mostly observed that shoulder elevation, scapular depression, scapular retraction, elbow extension, supination, hip extension, ankle dorsiflexion and great toe extension are the common movements that are impaired and result in contractures and deformities in spinal cord lesions like transverse myelitis. In some cases, tightness of muscles may be preserved for functional independence like finger flexor in order to produce strong tenodesis grasp in individuals where C6 spinal segments are involved. Tightness in back extensors would be helpful for upright back sitting.

In rare cases, overstretching is encouraged like hip external rotation in order to help the ADL activities like putting on shoes and socks. It should be kept in mind that aggressive ROM exercises are contraindicated, if the patient has unstable fractures, active heterotropic ossification and deep vein thrombosis.

In case of spastic muscles, daily stretching and terminal sustained stretch is to be considered for rehabilitation for limitation of joint range of motion.

Strengthening program: To target the muscle strengthening and facilitation exercises, all spared or intact muscles are given progressive resistance exercises, active-assisted ROM exercises. It is observed that with strengthening program, adaptive muscle changes are seen which include increased cross-sectional area of type II fibers, increased metabolic capacity due to increased concentrations of adenosines 5-triphosphate and creatinine phosphate and increased levels of myofibrilar proteins.

The strengthening program is designed with static or dynamic training through isometric, isotonic and isokinetic techniques depending upon the degree of neurological deficit.

To Improve Sensory Integration and Sensory Re-education Training

Sensory stimulation can be used to acquire afferent pathways with the aim of providing information that can subsequently be used to perform a motor activity with the direct effect being improved motor and sensory functions (Issurin, 2005 and Peurala et al. 2002). Sensory re-education and integration techniques are employed by the therapists in an attempt to retain the sensory pathways or stimulate the unused pathways. The therapist should also employ adaptive techniques to cope up with the sensory loss as a compensation. Sensory re-education techniques can include touching different textured objects, massage, vibrations, pressure, determining joint positions, identifying different temperatures and electrical stimulations. The transcutaneous electrical stimulation would be useful in the rehabilitation of patients with sensory loss (Ito et al. 2015).

Other techniques:

1. Feel an object and find the same matching object from a sand tray.
2. Ask the patient to close the eyes. Place two objects of different weights and ask the patient to recognize the difference.
3. Write with the finger an alphabet or a number over the dermatome of the patient and ask the patient to feel and

Sensory Re-education Exercises

Training for differentiation of different textures	Use cotton, sandpaper, satin, velcrow rubber, velvet, wool. Train 2–4 times for 10 min in a quiet environment daily
Steriognosis	Hide objects with different shapes like marbles, currency coins, erasers in sand. Ask the patient to close eyes and try to find the objects with the hand feel and recognize. Note: The objects should be familiar to the patient Frequency: 2 times for 10 min daily
Pressure	Apply pressure with fingers of different intensity over an area once with patient's eyes opened and patient's eyes closed. Ask the patient to feel the pressure and identify Frequency: 2 times for 10 min daily
Vibrations	Ask the patient to close the eyes and apply vibrations over a dermatome of the patient. The vibrations first should be of a powerful vibrator which is used for massage, later by mobile phone vibrations, tuning fork vibrations as the patient progresses. Ask the patient to feel the vibrations once with eyes opened and then with eyes closed Frequency: 2–3 times for 5 min daily
Temperature	Take two bowls of water of hot and cold temperatures and ask the patient to dip the hands or a test tubes can be used instead. Ask the patient to feel and recognize Frequency: 2–3 times for 10 min daily
Joint position sense	Position the joint in normal and affected at the same time, ask the patient first to feel the position and then repeat only on the affected limb. Ask the patient to analyze and feel Frequency: 2–3 times for 10 min daily
Joint kinesthetic sense	Move the limbs once with eyes opened and then eyes closed. Ask the patient to feel and recognize Frequency: 2–3 times for 5 min daily

Principles of Training of Transfer Technique for Physiotherapists

Thoroughly assess the patient's ability and preparations for resettlement for transfer activities.

Functional charts useful to record these activities which could be useful to check the prognosis and also helps to stimulate the patient's interest.

Correct positioning of the chair, ensuring that the brakes are fully on and lifting the legs with the hands on plinth, placement of legs on foot-rest would be the principles of safe training.

During the transfer techniques, care must be taken not to knock the leg or drag them along a hard surface.

Check that the tyres of wheelchair are not worn out and the floor should not be slippery.

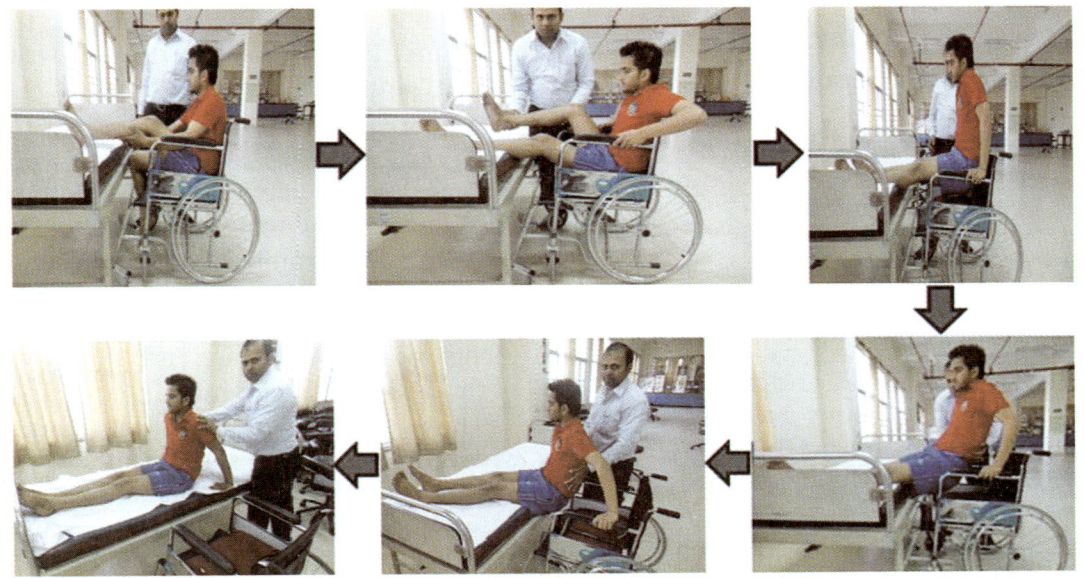

Fig. 12.4: Transfer technique from wheelchair to bed

see and perceive. Now ask the patient to close the eyes and recognize the same.

Compensatory Techniques:
- Using the vision to observe motion and location of body parts
- Using your unaffected side to check the temperatures of the objects
- Using unaffected side to handle sharp objects
- Change of position to prevent pressure on affected side
- Cushioning to avoid abnormal pressures

Training for Transfer Activities

These are taught from and to the wheelchair and the bed, plinth, bath/lavatory seat and motorized vehicle. Transfer from a mat to wheelchair and vice versa are also taught for various treatment sessions. Sitting up from a lying position is a prerequisite for independent dressing and transfers. A thorough evaluation should be conducted to determine the most appropriate transfer technique for any individual with paralysis. Dependent transfers include sliding transfers and dependent standing pivot technique or use of Hoyer lifts. Transfers that require some active patient participation include the two men lift, sliding board transfer or assisted pivot transfer. Always remember the aim of assisted transfer is to gradually reduce the assistance required until the patient can perform the transfer independently. Floor to chair transfer training is very important for anyone who falls out of wheelchair or otherwise ends upon the floor and needs to get back to the chair.

To Prevent Secondary Complications

The possible complications due to prolonged bedrest would be:
1. Contractures and deformities
2. Pressure sores
3. Deep vein thrombosis
4. Psychological depression
5. Respiratory complications
6. Cardiovascular deconditioning
7. Disturbed bone metabolism

The contractures and deformities are prevented by proper positioning of the patient, passive movements, use of orthotic supports, passive mobilization techniques and stretching.

Pressure sores are deep wounds due to constant and continuous pressure on the skin overlying the bony prominences in the body. It is a very common, and painful complication of patients who are on the prolonged bedrest. The pressure sores can aggravate the existing disability. It should be noted that the development of pressure sores is solely due to negligence of the caregivers and health providers but not accidental. Hence, the health care providers and caretakers should take all necessary precautions to prevent the pressure sores.

Precautions to prevent pressure sores:
1. Closely monitor all the areas of risk daily
2. Frequent change in the position of the patient
3. Small rubbing friction massage with a moisturizer for a small duration (15 sec) over the areas of risk shall improve the skin elasticity and pliability
4. Use of water mattress or air bed
5. Proper bed making technique, avoid creases on the bed cover.

Even after taking necessary precautions, if the pressure sores appear to develop, then curative measures to be taken:
1. At redness stage: Stop the massage, start changing the positions frequently and take use of adaptive devices.
2. At blister stage, frequent dressing to be done. Hydrocolloid dressing would be helpful.

If an ulcer has already formed, debridement of dead tissues, dressing of the wound is done. UVR therapy and ultrasound have been found useful to promote healing.

Deep vein thrombosis is another major complication of immobility. It can be prevented by leg lifts, ankle pumping movements, toe curls, stretching and taking plenty of fluids would be helpful.

The respiratory complications could be avoided by teaching deep breathing exercises, postural drainage, suctioning, good coughing and huffing techniques. Chest muscle stretching exercises would be beneficial.

Cardiovascular deconditioning is prevented by regular endurance exercises.

To Improve Functional Independence

To improve functional independence, mat activities are taught depending upon the level of independence (Fig. 12.5). Training activities done on adjustable therapy mat and are composed of sequenced activities that progress from the easiest to the most difficult. The usual progression is from bed mobility to rolling, prone lying, long sitting, short sitting, kneeling, half kneeling and standing. Muscles needed for individuals with lower limb paralysis to be able to move or position

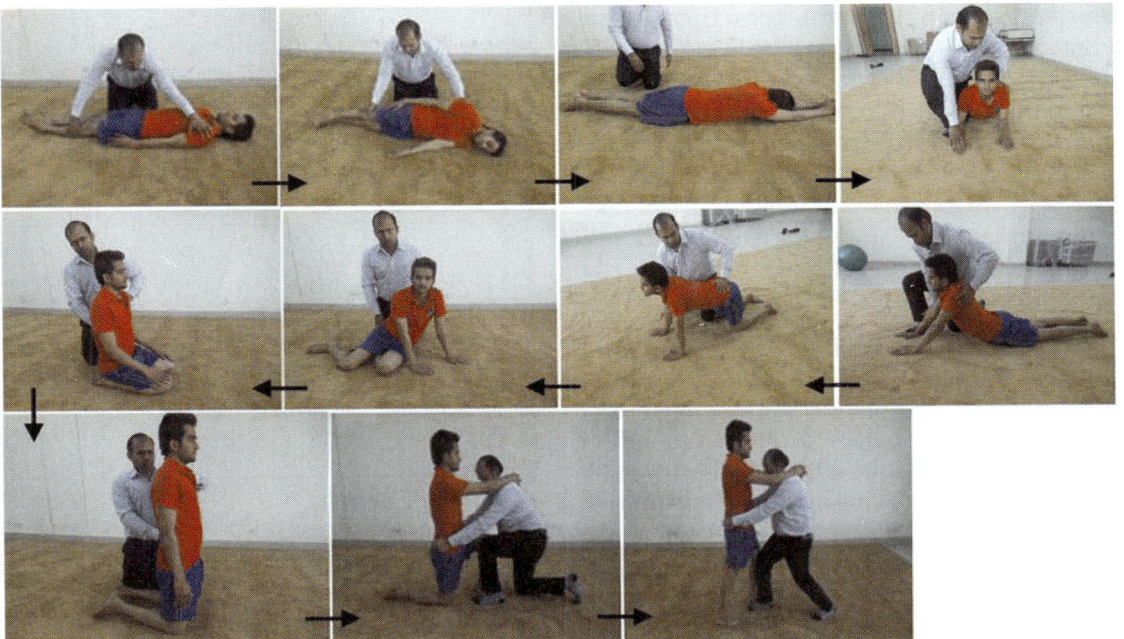

Fig. 12.5: Mat activities

their legs in bed are wrist extensors, biceps, anterior deltoid, middle deltoid, and shoulder girdle stabilizers. Individuals with tetrapelgia are taught to use their arms, head and neck for momentum to roll in bed with keeping the elbow straight while the shoulder is flexing across the body.

To Improve and Train Ambulatory Activities

Ambulation is a commonly expressed goal of most people with transverse myelitis. Patients with injuries at T2 and above typically cannot achieve ambulation, whereas patients with T3 to T11 affection are able to use orthotic supports for physiologic standing and therapeutic ambulation. The goals for individuals with a T12 to L2 injury is to achieve ambulation in household, while patients with injuries at L3 and lower are most likely to achieve ambulation in the community. The muscle activity in the back and lower extremities helps the therapists predict the orthotic support needed for ambulation. For example, patients with pelvic control and intact quadratus lumborum and abdominal muscles may walk using knee-ankle-foot orthoses and crutches. Patients with control of muscles across the hip joint like iliopsoas, hip adductors, sartorius and gluteus maximus can ambulate with **KAFO** and crutches. Patients with intact tibialis anterior and posterior can walk full-time with **AFO** (ankle-foot orthoses).

During the pre-gait training program, certain muscles should be strengthened like core trunk and abdominal muscles, shoulder

---

Features of ORLAU

1. A rigid body brace which helps to maintain the relative abduction of the legs during the swing phase of gait cycle
2. A hip joint with a limited flexion-extension range and friction free operation
3. Stabilization of knees and ankles
4. A shoe plate incorporating rocker sole
5. Simple fastening arrangements to ease putting on and taking off the orthosis.

depressors, scapular stabilizers, triceps and wrist extensors. During gait training, control of pelvis is a critical factor for a successful ambulation. Gait training includes standing balance training in parallel bars and practice gait patterns in parallel bars once inside the bars and later outside. Assistive devices depending upon the independence like canes, crutches or walkers may be used for ambulatory training.

Ambulation with ORLAU parawalker: The ORLAU parawalker (Fig. 12.6) is a type of reciprocal walking orthosis. Reciprocal walking orthosis helps to take steps alternatively with one foot and then the other as in normal walking. This orthosis is primarily for paraplegics who have no voluntary control of the muscles in the lower body. This orthosis is worn outside clothing and fits closely from just under the axilla and down to the feet. It has locking knee joints that keep the legs straight during standing and walking. Special joints at the hip allow fore and after movements to permit a stepping pattern. The knee and hip joints have quick release mechanisms to allow the patient to sit on a chair or couch.

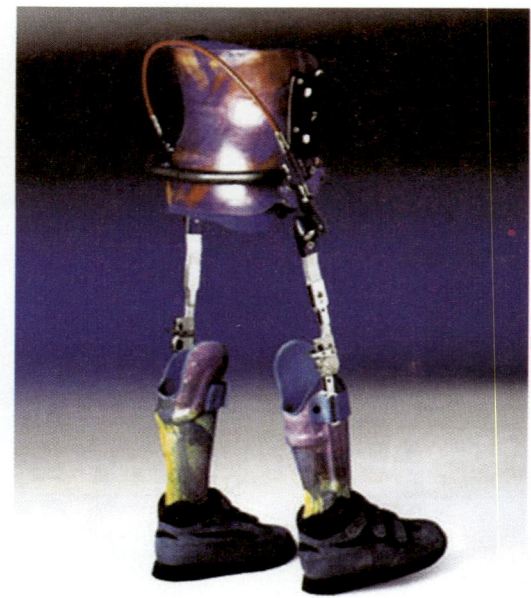

Fig. 12.6: ORLAU parawalker

REFERENCES

1. Bhat A, Naguwa S, Cheema G, Gershwin ME. The epidemiology of transverse myelitis, Autoimmun Rev 2010;9(5):A395–99.
2. Pandit L. "Transverse myelitis spectrum disorders". Neurol India 2009;57(2):126–33.
3. Greenberg BM, Frohman EM. "Immune-mediated myelopathies". Continuum (Minneapolis, Minn) 2015;21(1):121–32.
4. Blanc F, Froelich S, Vuillemert F, et al. "(Acute myelitis and Lyme disease)". Rev. Neurol. (Paris) (in French) 2007;163(11):103:47.
5. Atlas, Scott W. Magnetic Resonance Imaging of the Brain and Spine. Lippincott Williams & Wilkins.
6. "Transverse Myelitis Fact Sheet: National Institute of Neurological Disorders and Stroke (NINDS)"(2009-01-01) www.ninds.nih.gov. Retrieved 2015–08.
7. Agmon-Levin N, Kivity S, Szyper-Kravitz M, Shoenfeld Y. "Transverse myelitis and vaccines: a multi-analysis". Lupus 2009;(13):1198–1204.
8. Schlossberg, David. Infections of the Nervous System. Springer Science & Business Media 2012-12-06.
9. Dale RC, Vincent A. Inflammatory and Autoimmune Disorders of the Nervous System in Children. John Wiley & Sons 2010;pp 96–106.
10. Alexander MA, Matthews DJ, Murphy KP. Pediatric Rehabilitation, Fifth Edition: Principles and Practice. Demos Medical Publishing 2015;pp 523–24.
11. Alireza Mlinagar J, Steven Alexander. Inflammatory Disorders of the Nervous System: Pathogenesis, Immunology and Clinical Management, Springer Science & Business Media 2007.
12. Bilney B, Morris M, Webster K. Concurrent related validity of the GAITRite walkway system for quantification of the spatial and temporal parameters of gait, Gait Posture 2003;17(1):68–74.

13

Syringomyelia

LEARNING OBJECTIVES

At the end of this chapter, the student will be able to:
1. Define syringomyelia and discuss briefly the etiopathogenesis
2. Describe the clinical features of syringomyelia
3. Assess the patient with syringomyelia
4. Plan and carryout a detailed physiotherapy management

INTRODUCTION

Syringomyelia is the development of a fluid cavity or syrinx within the spinal cord. Hydromyelia is a dilatation of the central canal by cerebrospinal fluid (CSF) and may be included within the definition of syringomyelia.

Definition
Syringomyelia is a chronic progressive degenerative disorder in which there is formation of a cyst or cavity in the spinal cord. This cyst or cavity is known as syrinx (Fig. 13.1).

Etiology
The syringomyelia origin is considered in two ways: A. Congenital, B. Acquired.

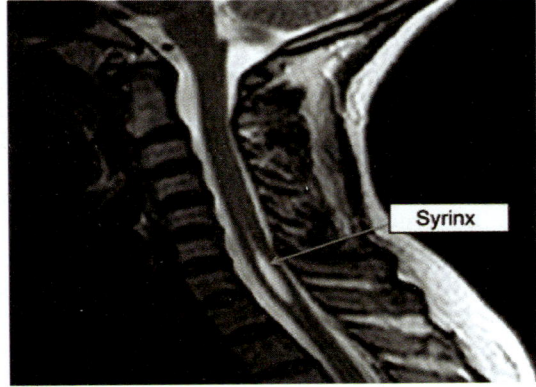

Fig. 13.1: Syrinx

Congenital
Arnold-Chiari malformation
Familial (rare)
Hydrocephalus
Spina bifida

Acquired
Trauma
Meningitis
Hemorrhage
Tumor
Scoliosis

TYPES OF SYRINGOMYELIA

Syringomyelia with fourth ventricle communication: About 10% of syringomyelia cases are of this type. There is a communication between the formed syrinx and fourth ventricle. In some cases, a blockage of CSF circulation occurs.

Syringomyelia due to blockage of CSF circulation[1]**:** Here there is formation of syrinx due to the blockage of CSF flow. There is no communication with the fourth ventricle. It is common in 50% of syringomyelia cases. Arnold-Chiari malformations, basal arachnoiditis, meningeal carcinomatosis, and arachnoid cysts are the causes of this type of synringomyelia.

Syringomyelia due to spinal cord injury: Less than 10% of cases belong to this category. Mechanisms of injury include:
1. Spinal trauma
2. Radiation necrosis
3. Hemorrhage
4. Infections
5. Cavitation following ischemic injuries or degenerative diseases.

Syringomyelia and spinal dysraphism: Spinal dysraphism may cause syringomyelia through a variety of mechanisms.

Syringomyelia due to intramedullary tumors: Fluid accumulation is usually caused by secretions from the neoplastic cells or hemorrhage. Ependymoma and hemangioblastoma are the common tumors associated with syringomyelia.

Idiopathic syringomyelia:[2] The cause is unknown and the mechanism is not clear for formation of syrinx.

PATHOPHYSIOLOGY

Although many mechanisms for syrinx formation have been postulated, still there is a debate on exact cause of formation of syrinx. The most accepted theories are given below.

Gardner's Hydrodynamic Theory

1. Due to the blockage of foramen of Magendie, there is block of flow of CSF from the fourth ventricle to subarachnoid space.
2. This results in "water-hammer" like transmission of pulsatile CSF via a communication between the fourth ventricle and the central canal of spinal cord which results in formation of syrinx.

William's Theory[3]

According to this theory, the syrinx development particularly in patient with Arnold–Chiari malformation follows a difference between intracranial pressure and spinal pressure caused by a valve-like action at the foramen magnum. This results in formation of syrinx.

Oldfield's Theory[4]

According to this theory[4], there are oscillatory movements caused by cerebellar tonsils which create a piston effect in the spinal subarachnoid space that acts on the surface of the spinal cord and forces CSF through the perivascular and interstitial spaces into the syrinx raising the intrameduallry pressure.

Intramedullary Pulse Pressure Theory[5]

According to this theory, the syringomyelia is caused by increased pulse pressure in the spinal cord and the syrinx consists of extracellular fluid. A new principle is introduced implying that the distending force in the production of syringomyelia is a relative increase in pulse pressure in the spinal cord compared to that in the nearby subarachnoid space. The formation of a syrinx then occurs by the accumulation of extracellular fluid in the distended end.

From overall postulates given above, the comprehensive summary of mechanism of formation of syrinx:

- CSF usually flows in a pulsatile manner in the subarachnoid space and envelops the brain and spinal cord as helps in protection and nutrition to brain and spinal cord.
- Excess CSF (normal volume = 500 mL) in the central canal of spinal cord is called **hydromyelia**.
- When this excess fluid dissects outside into white matter, it forms a syrinx or cyst known as syringomyelia.

Prevalence

The estimated prevalence of this disease is about 8.4 cases per 100,000 people and occurs more frequently in men than in women.

The disease usually appears in the third or fourth decade of life, with a mean age of onset of 30 years. Rarely, synrigomyelia may develop in childhood or late adulthood.

Clinical Features

- The clinical features depend on the extent of spinal cord injury.
- Patients may experience severe pain/abnormal sensations/complete loss of sensations.
- If autonomic system is disturbed, it results in the abnormal body temperatures, bowel and bladder dysfunction, erectile dysfunction, etc.
- Muscle weakness or plegia depending upon the level of lesion.
- Ataxia may be seen in some patients.
- If syrinx is formed in the brainstem, it is called **syringobulbia**.
- It results in the vocal cord paralysis, cranial nerve involvement usually facial nerve palsy is seen, ipsilateral tongue muscle wasting.
- Classically, the syringomyelia spares the dorsal column of spinal cord, so patients shall have intact pressure, vibration, touch proprioception sense.
- Sometimes, due to loss of pain sensory fibers of the joints, it results in **neuropathic arthropathy** also called **Charcot joint**.[7,8]
- Common joints affected are shoulder joints.

Prognosis[6]

Prognosis depends on the underlying cause, the magnitude of neurological dysfunction, and the location and extension of the syrinx. Patients presenting with moderate or severe neurological deficits face much worse than those with mild deficits. Some studies suggest that 20% of patients died at an average age of 47 years. Mortality rates are likely lower in today's patients because of surgical interventions and better treatment option.

Investigations

- MRI scan
- Additional tests usually include EMG (electromyography) to measure the extent of muscle weakness
- Lumbar puncture for CSF analysis
- Myelogram

Physiotherapy Assessment of Syringomyelia Patients

Physiotherapy assessment is same in line with that of multiple sclerosis and transverse myelitis.

PHYSIOTHERAPY MANAGEMENT

- The physiotherapy management includes complete assessment of the patient and planning the treatment.
- The common problems associated with syringomyelia include:
  1. Anesthetic hands (loss of sensation)
  2. Muscle weakness
  3. Abnormal sensations (paresthesia)
  4. Charcot joint
  5. Ataxia
  6. Decrease joint ROM
  7. Breathing problems

Aims of Physiotherapy

- To provide psychological support to the patient
- To take care of anesthetic hand
- To give sensory re-education
- To reduce pain
- To maintain normal joint range of motion
- To improve the muscle power
- To retrain for functional independence for ADL
- To develop the coordination of movements
- To improve ventilatory effort and maintain clear chest

Plans of Physiotherapy

- Counseling to gain maximum cooperation from the patient and to prevent him from going to depression, psychological counseling is very important.
- Without encouraging the negative thoughts into the patient's mind, the therapist should explain the patient his condition and importance of physiotherapy.

Care of Anesthetic Hand

- The involved part should be inspected regularly for wounds or skin color changes.
- In case of any wound, immediate antiseptic precautions should be used.
- The patient should be asked to avoid extreme temperatures.
- Protective gloves and soft shoes must be worn to prevent injury by sharp objects.
- The skin must be kept moist but not wet.

Sensory Integration

- For sensory re-education, the methods of stimulation such as contrast baths, exposure of the parts to soft and hard surfaces, whirlpool baths and different textures of material—rough hessian, cotton wool, fur and toweling and asking the patient to note the difference and identify the material without using the eyes.
- Shape and size differentiation sometimes be difficult and patients can be encouraged to handle objects of identical shape but a different size such as a large beach ball and a small tennis ball.

Charcot Joint Care

- Full range of motion in pain-free range is encouraged.
- Pain relieving techniques like IFT, moist heat (with care), ultrasound, etc. are used.
- Joint mobilization techniques and frequent active exercises to prevent joint adhesions and maintain normal range.
- Improving co-ordination.

Principles of Training Equilibrium and Non-Equilibrium Exercises

a. The patient must be fully relaxed.
b. He must be explained clearly the methods and instructions of the exercise.
c. The patient should start the exercise slowly and improve steadily and progressively.
d. Encourage the patient with verbal commands throughout the exercise.
e. Avoid over enthusiasm. Never allow the fatigue to set in.
f. Give as many number of breaks as patient wants.
g. Progress to more difficult exercise only after the previous one is mastered.

Equilibrium Exercises

i. Stand with both feet together with open and closed eyes.

ii. Stand with heel of one feet exactly in front of the other.
iii. Stand on one foot.
iv. Walk on a straight line sideways and backwards.
v. Walk along a circle.
vi. Walk in a figure of eight.
vii. Walk on the heels and toes alternatively.

Non-Equilibrium Exercises

i. Finger to nose
ii. Finger to finger
iii. Finger to nose to therapist finger
iv. Heel to shin
v. Alternative pronation and supination
vi. Tapping hand and foot
vii. Drawing a circle
viii. Fixation and position holding
ix. Touch the therapist finger with toe
x. Alternatively touch with heel of other foot

Frenkel's Exercises

These are designed for sensor ataxia, there will be many activities involving lower and upper limbs. Progression is made by changing the speed, range or exercise.

Example: Exercises for legs in lying:

- *Position*: Supine lying with head raised
- *Activity*: Hip abduction and adduction on fully supported smooth surface

Therapy for Balance and Proprioception

i. Therapy is used to improve balance and increase the independence of the patient using techniques focusing on balance, posture and increasing co-ordination.
ii. Therapeutic goals include improving balance and posture against outside stimuli, increasing joint stabilization, developing independent, functional gait to promote independence.
iii. Training principles include progressing from simple to complex exercises, practicing exercises with eyes open and closed and providing support with home exercise and sports activities.
iv. Therapy involves plyometric exercises, balance board and mini-trampoline exercises, PNF.
v. Vibration and suit therapy are also used to improve proprioception, posture and movement.
vi. Yoga and other body awareness exercises may also be included in the treatment plan to increase proprioception.
vii. Strengthening program.
viii. To target the muscle strengthening and facilitation exercises, all spared or intact muscles are given progressive resistance exercises, active-assisted ROM exercises. It is observed that with strengthening program, adaptive muscle changes are seen which include increased cross-sectional area of type II fibers, increased metabolic capacity due to increased concentrations of adenosine 5-triphosphate and creatinine phosphate and increased levels of myofibrilar proteins.
ix. The strengthening program is designed with static or dynamic training through isometric, isotonic and isokinetic techniques depending upon the degree of neurological deficit.

For other plans of management, please refer to transverse myelitis.

REFERENCES

1. Nakanishi K, Uchiyama T, Nakano N et al. Spinal syringomyelia following subarchanoid hemorrhage. J Clin Neurosci 2012;19(4):594–97.
2. Kim J, Kim CH, Jahng TA, Chung CK. Clinical course of incidental syringomyelia without predisposing pathologies. J Clin Neurosci 2012.
3. Gardner WJ. Hydrodynamic mechanism of syringomyelia: its relationship to myelocele. J. Neurol. Neurosurg Psychiatry 1965;28:247–59.
4. Williams B. Progress in syringomyelia. Neurol Res 1986;8(3):130–45.
5. Oldfield EH, Muraszko K, Shawker TH, Patronas NJ. Pathophysiology of syringomyelia associated with Chiari I malformation of the cerebellar tonsils. Implications for diagnosis and treatment. J. Neurosci 1994;80(1):3–15.
6. Sixt C, Riether F, Will BE, Tatagiba MS, Roser F. Evaluation of quality of life parameters in patients who have syringomyelia J Clin Neurosci 2009.
7. Nacir B, Arslan Cebeci S, Centinkaya E, Karagoz A, Erdem HR. Neuropathic arthropathy progressing with multiple joint involvement in the upper extremity due to syringomyelia and type I: Arnold-Chiari malformation. Rheuymatol Int 2009.
8. Ono A, Suetsuma F, Ueyana K, Yokoyana T, Aburakawa S, Numasawa T. Surgical outcomes in adult patients with syringomyelia associated with Chiari malformation type I: the relationship between socio and neurological findings. J Neurosurg Spine 2007;6(3):216–21.
9. Akyuz G, Kenis O. Physical Therapy Modalities and Rehabilitation Techniques in the Treatment of Neuropathic Pain Int J Phys Med Rehabil 2013;1:124. doi:10.4172/jpmr.1000124.
10. Rebecca Smith, et al. Established Methods of Physiotherapeutic Management for Long-term Neurological Conditions Applicable to 'Orphan' Conditions such as Syringomyelia? Physiotherapy Research International 2014;21(1):pp. i-i, 1–64.

14

Myasthenia Gravis

LEARNING OBJECTIVES

At the end of this chapter, you will be able to:
1. Define and classify myasthenia gravis.
2. Identify the clinical manifestations of myasthenia gravis
3. Identify the various investigations done for diagnosing myasthenia gravis
4. Medical management of myasthenia gravis
5. Detailed physiotherapy assessment and management of myasthenia gravis

INTRODUCTION

Myasthenia gravis is a chronic autoimmune neuromuscular disease. The name was originated from Greek and Latin which literally means grave muscle weakness. It is caused by a breakdown in the normal communication between nerves and muscles that control and perform voluntary movements (Fig. 14.1).

Definition

It is an autoimmune disorder caused by circulating antibodies that block the acetylcholine receptors at the postsynaptic neuromuscular

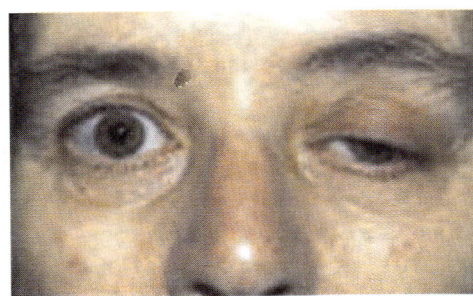

Fig. 14.1: Myasthenia gravis

junction resulting in fluctuating muscle weakness and fatigability.[1,2]

Epidemiology[2-4,8]

- Prevalence of 1 in 7500.
- Affects all age groups with peak incidence between 20 and 40 years of age.
- Women are affected more frequently than men with an approximately ratio of women : men = 3 : 2.
- Familial occurrence is rare.

Etiology[5]

- Antibodies[5] against tyrosine kinase—a muscle-specific receptor.
- Thymus gland pathology like thymomas (tumors of thymus gland).
- Genetic factors.

Pathology[6,7]

- Myasthenia gravis is caused by impaired transmission of nerve impulse to muscle.
- The antibodies block, alter or destroy the receptors for acetylcholine at the neuromuscular junction.
- The amount of acetylcholine release per impulse normally declines on repeated activity which is known as presynaptic rundown.
- This prevents the muscle contraction.

Classification of Myasthenia Gravis

The most widely accepted classification of myasthenia gravis is the Myasthenia Gravis Foundation of America Clinical Classification.[9]

- **Class I:** Any eye muscle weakness, possible ptosis, no other evidence of muscle weakness elsewhere
- **Class II:** Eye muscle weakness of any severity, mild weakness of other muscles
- **Class IIa:** Predominantly limb or axial muscles
- **Class IIb:** Predominantly bulbar and/or respiratory muscles
- **Class III:** Eye muscle weakness of any severity, moderate weakness of other muscles
- **Class IIIa:** Predominantly limb or axial muscles
- **Class IIIb:** Predominantly bulbar and/or respiratory muscles
- **Class IV:** Eye muscle weakness of any severity, severe weakness of other muscles
- **Class IVa:** Predominantly limb or axial muscles
- **Class IVb:** Predominantly bulbar and/or respiratory muscles (can also include feeding tube without intubation)
- **Class V:** Intubation needed to maintain airways

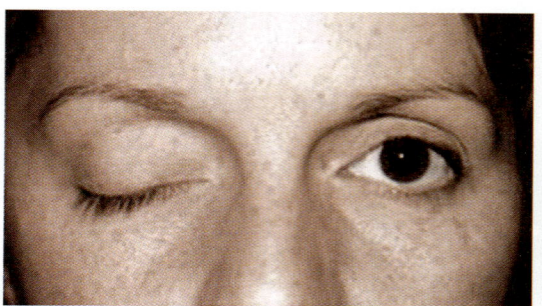

Fig. 14.2: Ptosis

Clinical Features

The initial symptoms of myasthenia gravis are:

- Difficulty in speaking [dysarthria]
- Difficulty in swallowing [dysphagia]
- Drooping of eyelids [ptosis (Fig. 14.2)]
- Nasal sound speech
- Fatigability
- Development of progressive weakness in the muscles of arms, trunk, limbs within a year of onset.
- The muscles will become progressively weaker while activity and recover at rest.
- Muscles responsible for breathing are also affected.
- Unstable or waddling gait
- Expression less face

Examination of Case

Muscle fatigability can be tested by many ways:

- Looking upward and sideward for 30 seconds—ptosis and diplopia.
- Looking at the feet while lying on the back for 60 seconds
- Keeping the arms stretched forward for 60 seconds
- 10 deep knee bends
- Walking 30 steps on both the toes and the heels
- 5 sit-ups
- Lying down and sitting up completely

- **Peek sign:** After complete initial apposition of the lid margins, they quickly (within 30 seconds) start to separate and the sclera starts to show drying response.

Complications
- **Myasthenic crisis:** It is a medical emergency that involves the respiratory muscle paralysis.
- Other complications may include choking, pneumonia, food aspiration.

Investigations
- **Blood tests:** Serum antibodies sensitivity to acetylcholine
- EMG studies
- Chest X-ray
- Pulmonary function tests

Management
Medical:
- Administration of cholinesterase inhibitors, e.g. neostigmine
- Immunosuppressive drugs, e.g. prednisolone
- Thymectomy

PHYSIOTHERAPY MANAGEMENT

Aims
- To provide psychological support to the patient
- To prevent respiratory complications
- To improve the vital capacity of the patient
- To avoid undue fatigue in the muscles
- To train for oromotor control
- To advise for assistive devices
- To maintain normal muscle properties
- To improve functional capacity of the patient.

Plans
- To gain maximum cooperation from the patient and prevent him from going into depression, psychological counseling is very important.
- Patient must be encouraged to move his limbs actively with or without assistance. If necessary, passive movements are given by the therapist.
- Passive stretching plays a very important role to maintain normal length of soft tissues. Certain corrective splints are given to maintain the stretch.
- Chest physiotherapy in the form of deep breathing exercises like diaphragmatic, VMT, *pranayama* are made to practice by the patient to prevent respiratory complications.
- Postural drainage, coughing, huffing are done to remove the secretions and maintain good bronchial hygiene.
- All precautionary measures should be taken to prevent pressure sores:
    - Frequent change in position.
    - Good skin hygiene is maintained.
    - Prevent creases over the bed linen.
    - Waterbed or air bed is advised.
- If pressure sore has already occurred, then regular dressing of wound:
    - Cryotherapy
    - IR radiations
    - UVR therapy
- Care of the bladder is very important. The patient should be taught self-clean intermittent catheterization.
- To maintain functional independence the following exercises are employed:
    - Mat exercises
    - Strengthening exercises
    - Weight bearing
    - Gait training
    - Transfer techniques.

REFERENCES

1. Kandel E, Schwartz J, Jessel T, Siegelbaum S, Hudspeth A. Principles of Neural Science (5 ed.), 2012;318–19.
2. Vrinten C, van der Zwang AM, Weinreich SS, Scholten RJ, Vershuuren JJ. "Ephedrine for myasthenia gravis, neonatal myasthenia and the congenital myasthenic syndromes". The Cochrane database of systematic reviews 2014;(12):CD010028.
3. Spillane J, Higham E, Kullmann DM. "Myasthenia gravis". The BMJ 2012;345:e8497.
4. Cea Gabriel, Benatar Michael, Verdugo Renato J, Salinas Rodrigo A. "Thymectomy for non-thymomatous myasthenia gravis". Cochrane Database of Systematic Reviews. John Wiley & Sons Ltd. 2013.
5. Information, National Center for Biotechnology; Pike, US National Library of Medicine 8600 Rockville; MD, Bethesda; USA, 20894. "Myasthenia Gravis—National Library of Medicine". PubMed Health.
6. Evoli A, Tonali PA, Padua L. Clinical correlates with anti-Musk antibodies in generalized aeronegative myasthenia gravis. Brain 2003;26(Pt 10):2304–11 (Medicine).
7. Sanders DB, Howard JF, Massey JM. Seronegative myasthenia gravis. Ann Neurol 1987; 22:126.
8. Bershad EM, Feen ES, Suarez JI. Myasthenia gravis crisis. South Med J 2008;101(1):63–69.
9. Jaretzki A 3rd, Barohn RJ, Ernstoff RM, et al. Myasthenia gravis: recommendations for clinical research standards. Task Force of the Medical Scientific Advisory Board of the Myasthenia Gravis Foundation of America. Neurology 2000;55(1):16–23.

15

Peripheral Nerve Injuries

LEARNING OBJECTIVES

At the end of this chapter, you will be able to:
1. Describe various etiological factors and types of PNI
2. Identify various clinical manifestations of the patients with PNI
3. Identify the various investigations done in patients with PNI
4. Demonstrate the skills to assess the patients with PNI
5. Design treatment protocols for rehabilitating the patients with PNI

INTRODUCTION

The peripheral nervous system is a network of spinal and cranial nerves that connect the brain, spinal cord to the entire human body.

ANATOMY OF PERIPHERAL NERVE[1-4] (Fig. 15.1)

1. A spinal nerve is formed by union of ventral nerve root and dorsal nerve root.
2. After emerging from the intervertebral foramen, it divides into a dorsal ramus and a ventral ramus.
3. Because of their position, ventral ramus is usually affected by injuries.
4. Each peripheral nerve is enveloped by myelin sheath making them myelinated nerves.
5. Peripheral nerve consists of number of nerve bundles or fasiculi, each bundle consists of several nerve fibers.
6. Each large nerve is surrounded by epineurium.
7. Each fasciculus is surrounded by perineurium.
8. Each nerve fiber is covered by endoneurium.
9. The myelin sheath is deficient at certain points, such points are called nodes of Ranvier.

CLASSIFICATION OF PERIPHERAL NERVE INJURIES[5,6]

The peripheral nerve injuries are classified according to Seddon or Sutherland classification:

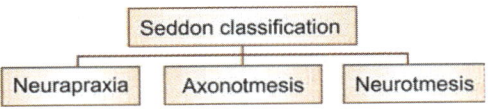

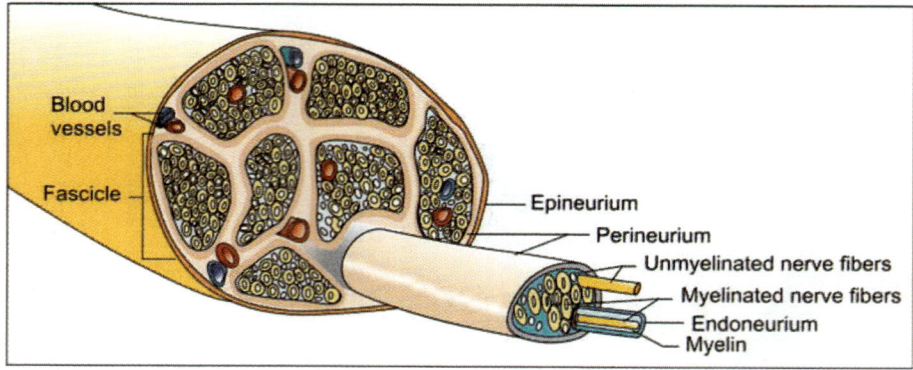

Fig. 15.1: Structure of peripheral nerve

Neurapraxia

a. There is block in the conduction of impulses down the nerve fiber.
b. There will be recovery without Wallerian degeneration.
c. The reason is a biochemical lesion which have been due to concussion or shock like injury to the nerve fiber.
d. The neurapraxia is usually brought by compression or stretch.
e. Saturday night palsy, peroneal palsy are some examples of neurapraxia.
f. There is no degeneration of the axon and hence the recovery shall be good.

Axonotmesis (Fig. 15.2)

a. There is loss of continuity of the axon with its myelin sheath but there is intact connective tissue framework.

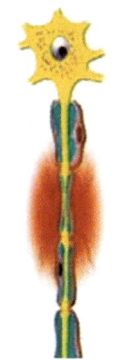

Fig. 15.2: Axonotmesis

b. There is Wallerian degeneration (Fig. 15.3)

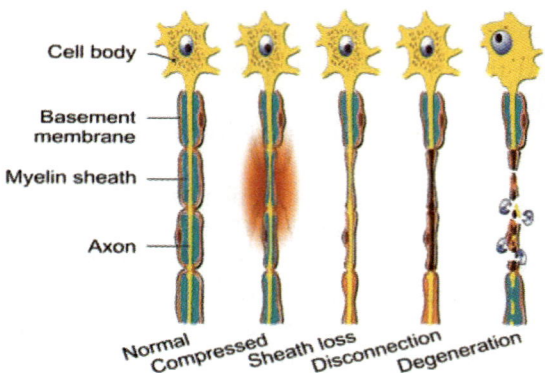

Fig. 15.3: Wallerian degeneration

c. Loss of motor, sensory and in some cases autonomic dysfunction can occur.
d. Recovery following axonotmesis depends upon the rate and extent of regeneration.
e. If the lesion is proximal, then the regeneration may be at a rate of 2 to 3 mm per day but if the lesion is distal, then it takes place at a rate of 1 to 1.5 mm per day.

Neurotmesis

a. There is a loss of continuity in the axon as well as investing connective tissue framework caused by nerve contusion, severe stretch or laceration.
b. Not all the neurotmeses involve complete transaction of the nerve trunk

but rather internal disruption of the nerve leading to perineurium and endoneurium disruption may also occur.

c. Spontaneous reversal of the changes in the electrodiagnosis and nerve conduction velocity are unlikely to occur because regenerating axons become mixed in a swirl of regenerating fibroblasts and collagen producing a disorganized repair site or neuroma.
d. In most of the cases, the axons may reach the distal stump but they often fail to find their earliest pathways and may even behave as free nerve endings causing abnormal sensory manifestation in the patient.
e. The neurons may fail to regain sufficient axonal diameter and myelination because of endoneurial proliferation and contraction of distal nerve sheath.

Sutherland's Classification

- **Grade I:** It is the first degree of injury which corresponds to neurapraxia.
- **Grade II:** It is the second grade of injury and involves loss of axon continuity with preservation of endoneurium and fascicular structure.
- **Grade III:** It is a third degree injury and is a mixed axonotmetic–neurotmetic type of injury wherein both axons and endoneurium are damaged with perineurium and fascicular structures intact.
- **Grade IV:** It is fourth grade of injury and involves loss of axons, endoneurium, perineurium with absence fascicular structure with intact epineurium.
- **Grade V:** It is a complete transection of the nerve trunk and so resembles a neurotmesis.

PRINCIPLES OF ASSESSMENT OF A PERIPHERAL NERVE INJURY[7]

History

After taking the demographic information, chief complaint, history of the present illness, past history of neurological dysfunction, personal history and medical history should be taken.

Take a detailed history of onset of presenting symptoms and their severity and episodes.

The common symptoms would be:
a. Numbness/tingling or prickling sensation
b. Causalgia
c. Anesthesized skin
d. Inability to perform activities
e. Weakness
f. Lack of muscle control and incoordination
g. Shooting pain

Also ask the patient about visceral and somatic structures involvement that receive innervations from the same areas and influence on the each other.

Observations

- Check for the abnormal attitude of the patient (claw hand, foot drop, drooped shoulders, etc.).
- Observe for wasting of the muscles.
- Observe for any skin changes.

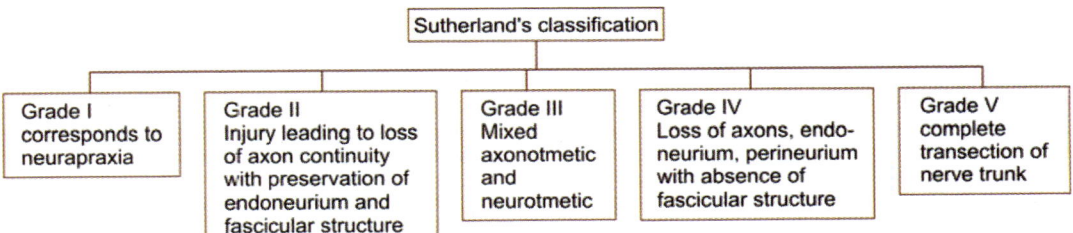

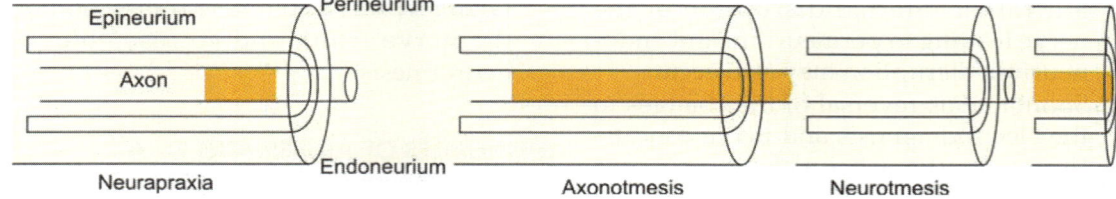

Fig. 15.4: Types of nerve injuries

- Check whether any swelling in the involved area or any gross swelling which may be relevant.
- Observe for any scars or unhealed wounds or skin infections.

Palpation

- Check for the temperature (local) over the area of affection and compare with the normal.
- Palpate the edema, if present and grade it.
- Check for the tenderness over the area of affection.

Examination

- Evaluation of sensory system.
- Localize areas of diminished or absent tactile sensation, increased pain sensations and abnormal sensations. Check for the senses of temperature, vibrations, touch both crude and fine, joint kinesthetic and joint proprioceptive senses.

Note: The evaluation of sensory system should be along the cutaneous distribution of peripheral nerve but not by dermatome.

Common Areas to be Evaluated for Sensory Functions for Various Peripheral Nerves

- The radial nerve sensory area is evaluated over the back of the hand towards the thumb.
- The median nerve sensory area is evaluated over the area occupying the tips of the index and middle fingers.
- The ulnar nerve sensory area is evaluated over the skin of the little finger.
- The musculocutaneous nerve sensory area is evaluated over the medial aspect of the forearm.
- The axillary nerve sensory area is evaluated over the point of the shoulder.
- The femoral nerve sensory area is evaluated over the anterior medial aspect of the thigh.
- The femoral lateral-cutaneous nerve sensory area is evaluated over the distal part of the lateral surface of the thigh.
- The common peroneal nerve sensory area is evaluated over a vertical strip on the front of the ankle.
- The sciatic nerve sensory area is evaluated over the skin occupying the entire foot and ankle except the medial aspect.

Reflex Examination

The deep and superficial reflexes are to be evaluated and elicited and observed for responses. The reflexes are checked only if a particular nerve or its muscular supply is involved in the reflex arc for any specific reflex, e.g. biceps jerk is required to be tested, if there is involvement of musculocutaneous nerve.

Usually, the reflexes shall show a diminished or absent response.

Motor Examination

In peripheral nerve injuries, depending upon the nerve involved, the muscles supplied by that nerve shall show weakness, atrophy, paralysis, and hypotonicity.

The motor examination should involve:
1. Examination of tone
2. Assessment of strength
3. Measurement of muscle volume.

Tone is assessed by quick passive movement which demonstrates hypotonicity in muscles supplied by the nerve.

The strength of the muscle supplied by a particular nerve is assessed by MMT (manual muscle testing) (Fig. 15.5). Take a note that all the muscles individually is checked for strength supplied by a particular nerve. The therapist should have a keen observation skills and techniques to prevent all types of trick movements during the evaluation of the MMT.

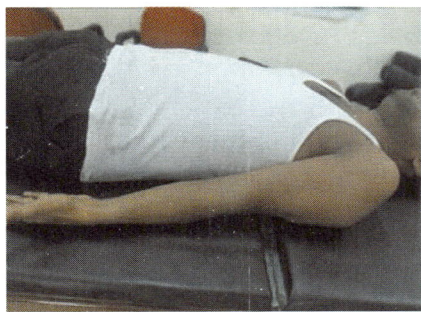

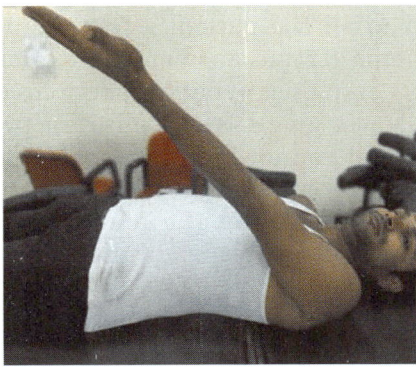

Fig. 15.5: Manual muscle testing

The muscle mass is evaluated by palpation and measuring the muscle girth (Fig. 15.6) and comparing with the normal side (unaffected side). Usually, in peripheral nerve injuries, the mass is decreased because of wasting and atrophy.

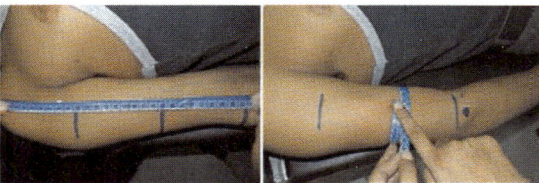

Fig. 15.6: Measurement of limb girth

SPECIAL TESTS RECOMMENDED FOR PERIPHERAL NERVE INJURIES

Sweat Function Test

Quantitative Sudomotor Axon Reflex Test (Fig. 15.7)

This test evaluates how the nerves that regulate your sweat glands respond to stimulation.

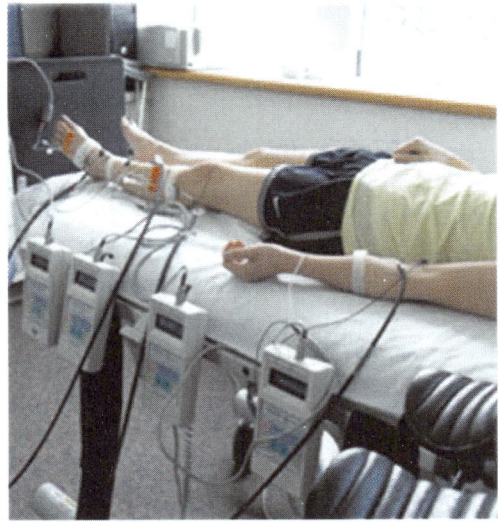

Fig. 15.7: Quantitative sudomotor axon reflex test

A small electrical current passes through four capsules placed on your forearm, foot and leg, while a computer analyzes the response of your nerves and sweat glands.

The patient may feel warmth or a tingling sensation during the test.

Ninhydrin Test

Make the patient assume a comfortable position. Simulate an environment to stimulate sweat glands. Now dust Ninhydrin powder over the skin on area to be checked. If there is normal sweat gland function, there will be change of color, if the color is unchanged, then it means the area is affected.

Galvanic Skin Resistance Test

The skin resistance for galvanic current is decreased by sweating, if the skin resistance is more than the other areas, then the area has sweat dysfunction.

Tinnel Sign

- **Elicitation:** Tap over the nerve throughout the course.
- **Positive response:** A sensation of tingling in the distribution of nerve will be experienced.

Common Investigations in Peripheral Nerve Injuries

- **Strength duration curves:** The SD curves shall show a normal response till Wallerian degeneration is complete but later will show signs of denervation.
- **FG test:** FG test will be galvanic positive.
- **EMG:** Shall demonstrate the typical neurogenic picture.
- **NCV:** There will be decrease in the nerve conduction velocity

Apart from these, CT, MRI and some blood investigations would be necessary.

PRINCIPLES OF PHYSIOTHERAPY MANAGEMENT IN PERIPHERAL NERVE INJURIES

Short-term Goals

1. To provide psychological support and develop good rapport with the patient
2. To maintain muscle properties in the muscles supplied by the affected nerve
3. To relieve radiating pain in the patient
4. To maintain normal range of motion at all joints
5. To prevent contractures and deformities
6. To take care of anesthetized skin and to manage tropical changes
7. To maintain circulation to the affected area
8. To prevent accumulation of edema.

Long-term Goals

1. To improve strength of the muscles affected.
2. To re-educate the muscles that have recovered or after tendon transfers
3. To improve functional capacity of the patient
4. To prevent secondary complications.

To Provide Psychological Support and Develop Good Rapport with the Patient

Discuss with the patient and explain his condition and importance of physiotherapy and early intervention. Develop rapport with the patient and counsel him regularly. The importance of effective communication with the patient is proven in physiotherapy care for better prognosis and improving the rapport with the patient. Have patience to listen to all the patient's complaints and problems. Try to answer and explain the doubts or queries of patient about his condition. Never give false promises or promise goals that cannot be achievable. Effective channel of communication and coordination is required among the members of the multidisciplinary team and also with patient and guardians. In a research by Barnes et al, 2012, it was believed that facilitating a successful intervention might require a health care professional to enhance their communication skills.

To Maintain Muscle Properties in the Muscles Supplied by the Affected Nerve

Interrupted galvanic stimulation of the denervated muscles will ensure a good blood supply as well as help in maintenance of excitation, contraction and coupling. Apart from IG stimulation, cryotherapy in the form of brisk stroking over the affected muscles, hacking, vibrations would maintain the muscle properties.

To Relieve Radiating Pain in the Patient

Radiating pain in the course of the nerve in entrapment neuropathies is very common. Transcutaneous electric nerve stimulation (TENS) can be given to relieve radiating pains (Fig. 15.8).

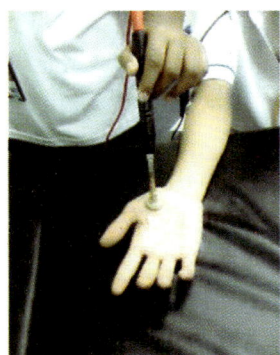

Fig. 15.8: ENMS

Neural mobilization techniques (Fig. 15.10) would be beneficial in relieving the symptoms of radiating pain like in sciatica. Neurodynamics would be another option in relieving the symptoms. Cold laser therapy has a beneficial effect on the radiating pain. Cold lasers are useful in decreasing nerve pain as well, given that in the visible red spectrum at 635 nm.

To Maintain Normal Range of Motion at all Joints

Gentle sustained stretching, full range repetitive rhythmical passive movements of all the joints would help to maintain the full ROM at joints. These maneuvers shall also maintain the flexibility of the muscles affected.

To Prevent Contractures and Deformities

Splinting in the functional positions, gentle passive stretching, and passive movements shall prevent the contractures and deformities. Positioning of the patient will also help in prevention of contractures. Static splints should be used. Pillows or supporting pads can also be used to prevent the contractures.

To Take Care of Anesthetized Skin and to Manage Tropical Changes

Care of the skin which is having sensory loss is an important preventive measure for ulcerization. Advise the patient not to handle any sharp edges including forks and knives with the affected side. Avoid extreme temperatures. Always check temperature with the normal side. Avoid hot saunas. The skin must be kept moist but not wet. Advise the patient to wear soft gloves and stockings. Microcellular insoles can be advised in the footwear. Ask the patient to inspect the anesthetized skin regularly for any bruises, color changes.

To Maintain Circulation to the Affected Area

Massage therapy would be beneficial to improve the circulation of blood to the affected area. Kneading, and efflurage would be beneficial. Passive movements, ENMS, mild stretching, and skin rolling would promote the circulation to the affected areas.

To Prevent Accumulation of Edema

Edema tend to accumulate in gross injuries like brachial plexus injuries where the whole limb is paralyzed. The patient may develop dependent edema due to gravity and lack of muscle tone. Gentle massage, passive movements, elastic crepe bandaging, elevation of the limbs, ENMS would be beneficial to prevent the accumulation of dependent edema.

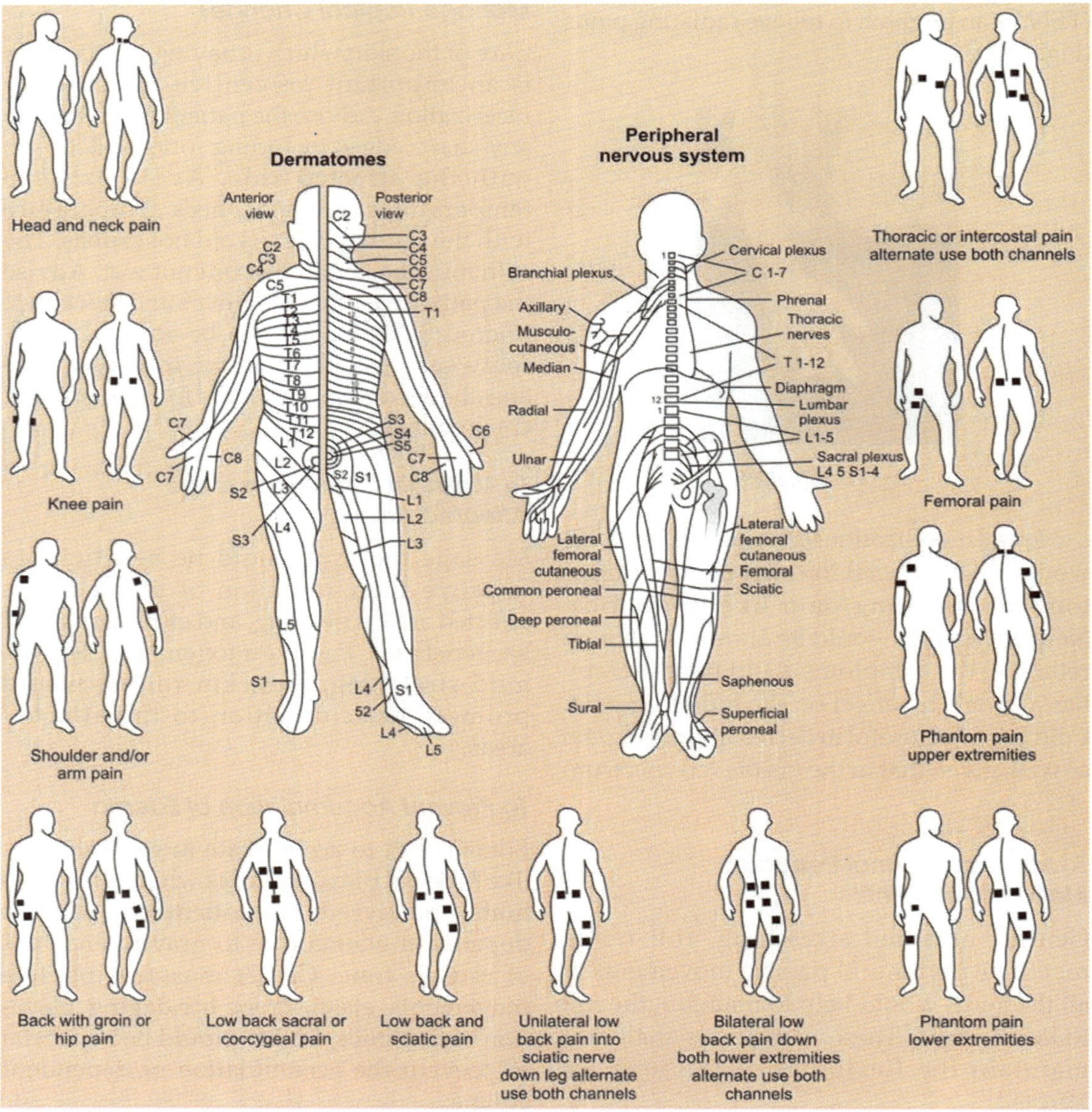

212 Physiotherapy in Neurological Conditions

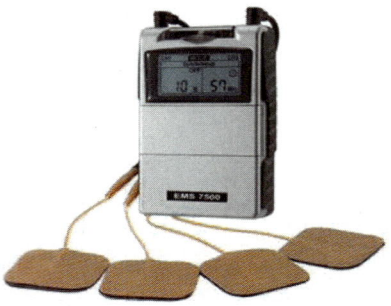

Fig. 15.9: Electrode placement chart for TENS therapy

Peripheral Nerve Injuries

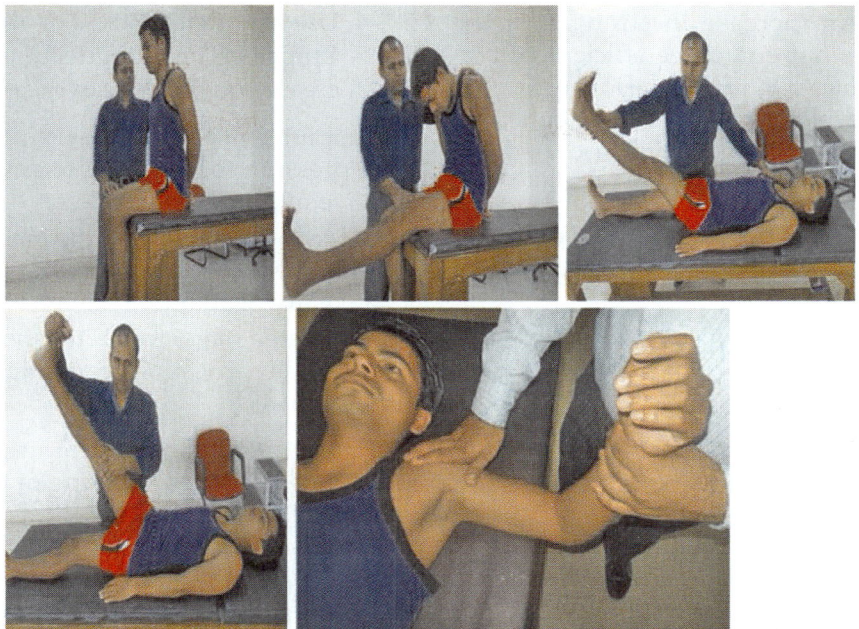

Fig. 15.10: Neural mobilization

Long-term Plans of Management

To Improve Strength of the Muscles Affected

Grade	Program
Grade 0–1	Passive movements Electric nerve muscle stimulation Cryotherapy in the form of brisk stroking Hacking over the muscle Weight bearing
Grade 1–2	Faradic re-education Suspension therapy Hydrotherapy Cryotherapy in the form of brisk stroking Hacking over the muscle Weight bearing
Grade 2–3	Suspension therapy Re-education board exercises Gravity eliminated exercises
Grade 3–5	Progressive resistance exercises

To Re-educate the Muscles that have Recovered or After Tendon Transfers

Muscle re-education is the process by which the muscles are taught how to act like they were before the injury and to perform a new action after they are transferred. It should be noted that there is no ability to the muscle to have its action in memory.

Biofeedback and Faradic re-education program would be effective to re-educate the muscles that have recovered or after a tendon transfer surgery is done.

Eccentric exercises would be beneficial to re-educate the muscle.

To Improve Functional Capacity of the Patient

It is important to improve the muscle strength by exercise program to meet the needs of activities of daily living. In some patients, there may be need of modification of environment and providing the necessary support (orthosis) to meet the ADL demands. It is very much essential to incorporate functional activity into the rehabilitation program. An increase in strength of the muscle does not guarantee efficient performance in the ADL. Thus various gripping exercises, level walking and other function-related exercises should be incorporated and encouraged to perform by the patient.

To Prevent Secondary Complications

The possible complications due to prolonged bedrest would be:

1. Contractures and deformities
2. Pressure sores
3. Deep vein thrombosis
4. Psychological depression
5. Respiratory complications
6. Cardiovascular deconditioning
7. Disturbed bone metabolism

The contractures and deformities are prevented by proper positioning of the patient, passive movements, use of orthotic supports, passive mobilization techniques and stretching.

Pressure sores are deep wounds due to constant and continuous pressure on the skin overlying the bony prominences in the body. It is a very common, and painful complication of the patients who are on the prolonged bed rest. The pressure sores can aggravate the existing disability. It should be noted that the development of pressure sores is soley due to negligence of the caregivers and health providers but not accidental. Hence, the health care providers and caretakers should take all necessary precautions to prevent the pressure sores.

Precautions to prevent pressure sores:

1. Closely monitor all the areas of risk daily.
2. Frequent change in the position of the patient.
3. Small rubbing friction massage with a moisturizer for a small duration (15 sec) over the areas of risk shall improve the skin elasticity and pliability.
4. Use of water mattress or air bed.
5. Proper bed making technique, avoid creases on the bed cover.

Even after taking necessary precautions, if the pressure sores appear to develop, then curative measures to be taken:

1. At redness stage: Stop the massage, start changing the positions frequently and take use of adaptive devices.
2. At blister stage, frequent dressing to be done. Hydrocolloid dressing would be helpful.

If an ulcer has already formed, debridement of dead tissues, dressing of the wound is done. UVR therapy, and ultrasound have been found useful to promote healing.

Deep vein thrombosis is another major complication of immobility. It can be prevented by leg lifts, ankle pumping movements, toe curls, stretching and taking plenty of fluids would be helpful.

The respiratory complications could be avoided by teaching deep breathing exercises, postural drainage, suctioning, good coughing and huffing techniques. Chest muscle stretching exercises would be beneficial.

Cardiovascular deconditioning is prevented by regular endurance exercises.

COMMON PERIPHERAL NERVE INJURIES OF CLINICAL INTEREST

Brachial Plexus Injuries (Table 15.1)

The brachial plexus (Fig. 15.11) is a network of nerves that conduct signals from the spine to the shoulder, arm and hand.

Causes of Brachila Plexus Injuries

a. Birth trauma
b. Traction injuries
c. Penetrating wounds
d. Gunshot wounds
e. Motor vehicle accidents
f. Falls and assaults

Table 15.1: Brachial plexus injuries

Injury	Causes	Sensory symptoms	Motor symptoms	Reflexes	Deformities and functional disability	Principles of management
Erb's palsy injury to the upper trunk tearing the C5 and C6 roots	• Obstetric brachial plexus injury—forceps delivery • Fall on the point of shoulder • Indirect injuries • During anesthesia • Injection palsies • Undue pressure on the supraclavicular area	Loss of sensations over the deltoid insertion and lateral aspects of the forearm of hand	• Supraspinatus, infraspinatus and teres minor, deltoid, biceps, brachii, brachialis, brachioradialis, rhomboids shall have complete motor loss (paralysis) • Triceps, latissimus dorsi, serratus anterior, pectoralis major and extensor carpi radialis—motor weakness (paresis)	Loss/diminished reflexes in biceps and brachioradialis jerks	• Policeman's tip or waiter tip hand deformity • Flexion movement of the shoulder and elbow is lost • Patient shall have difficulty in ADL activities like feeding self, combing, washing, dressing, etc.	• Positioning of the patient and passive range of motion • Airplane splint, cheese splint are popular choices, a simple shoulder sling can be used • IG stimulation • Other principles same as general management of PNI
Klumpke's palsy lesion to the lower trunk of branchial plexus or to the roots C8, C1	• Breech delivery • Fall from a height and clutching for an object during fall • Tumors of apical lobes of lung • Cervical rib	Loss of sensation over the medial aspect of the arm, forearm, hand and over the hypothenar eminence	• Paralysis of intrinsic muscle of hand (interossei and lumbricals) wrist and finger flexors	—	• Claw hand deformity, lumbrical grip is lost making the patient difficult to hold things in the affected hand • The skin will become dry, scaly and nails turn brittle	• Knuckle bender splint is given • Follow general principles of management of PNI

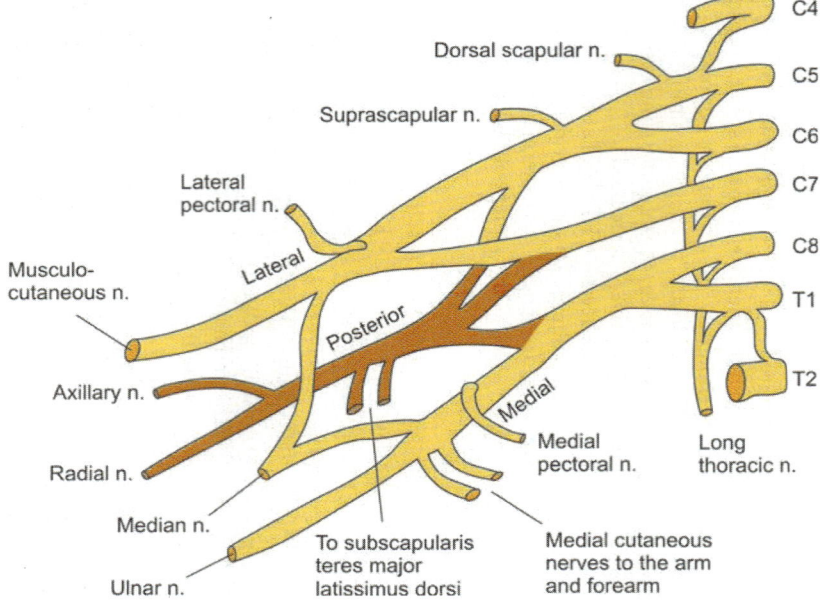

Fig. 15.11: Brachial plexus

Prevalence of Brachial Plexus Injuries[9-15]

- 70% injuries in adults are caused by Motor vehicle accidents
- Men and boys between 15 and 25 years
- Rate of incidence is 1.75/100000/year.
- 62% of the lesions were supraclavicular
- 60% cases occurred due to traction injuries, 25% to gunshot wounds, 8.5% to compression and 5.7% due to perforation/laceration.

g. Fractures of scapula, clavicle and humerus
h. Shoulder girdle neuritis
i. Malignancy in the brachiocervical region
j. Congenital cervical spine abnormalities
k. Apical lung tumors
l. Cervical rib.
m. Radiation

Types of Brachial Plexus Injuries

The types of brachial plexus injuries are classified based on the site of injury.

Preganglionic injuries (Fig. 15.12):
- Caused by avulsion of roots from the spinal cord.
- Both sensory and motor dysfunction are present.
- Axon reflex remains intact in the initial stage of injury.
- Sensory axon remains intact but motor axon is lost.
- Prognosis of this type of injury is very poor.

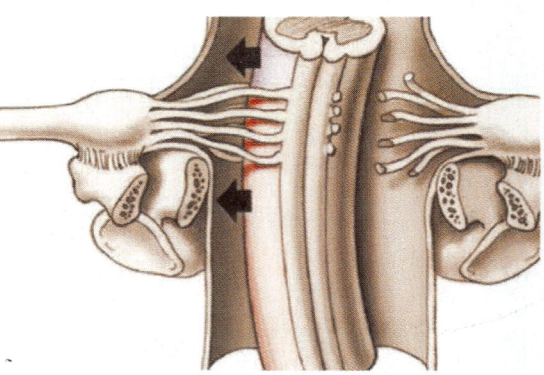

Fig. 15.12: Preganglionic injuries

Table 15.2: Summary of clinical manifestations of radial nerve injury

Sensory symptoms	Motor symptoms	Reflex	Deformities and functional disability	Principles of management
Loss of sensation over the following areas but solely depends on the level of lesion Posterior aspects of upper arm Lower lateral aspect of arm Posterior aspect of the forearm Posterior part of the hand and the fingers up to nail beds Paresthesia Sharp or burning pain along the course of nerve Numbness or tingling	• Paralysis of the following muscles depending upon the level of lesion • Triceps, brachioradialis, extensor carpi radialis longus and brevis, extensor carpi ulnaris, extensor digitorum, extensor digiti minimi, supinator, anconeus, abductor pollicis longus, extensor pollicis longus and brevis, extensor indicis	Loss/diminished reflexes of triceps and brachiradialis jerks	• Wrist drop • Thumb palmar abduction and slight flexion Disabilities: Trouble in straightening the arm Trouble in moving the wrist and finger Weak hand grips Patients will have difficulty in wearing on and off eye glasses or putting the cups or other similar things flat on the table	• Avoid thumb-index web space contracture • Static or dynamic cock-up splints • Reverse knuckle bender splint • Other principles are same as general principles of management of PNI

For surgical management of radial nerve injuries, please refer to neurosurgery chapter.

Postganglionic injuries:
- These are injuries distal to the dorsal root ganglion.
- There is sensory and motor loss.
- The axon reflex is also lost.
- There is Wallerian degeneration and hence the prognosis is expected to be good.

Total plexus injuries:
- These injuries are very close to the vertebral column.
- All the muscles supplied by the brachial plexus are paralyzed.
- Tendon reflexes are lost.

Radial Nerve Injury
(Fig. 15.13 and Table 15.2)

Root value: C5–C8 and T1.

Etiology
a. Crutch palsy
b. Penetrating injuries in the axilla

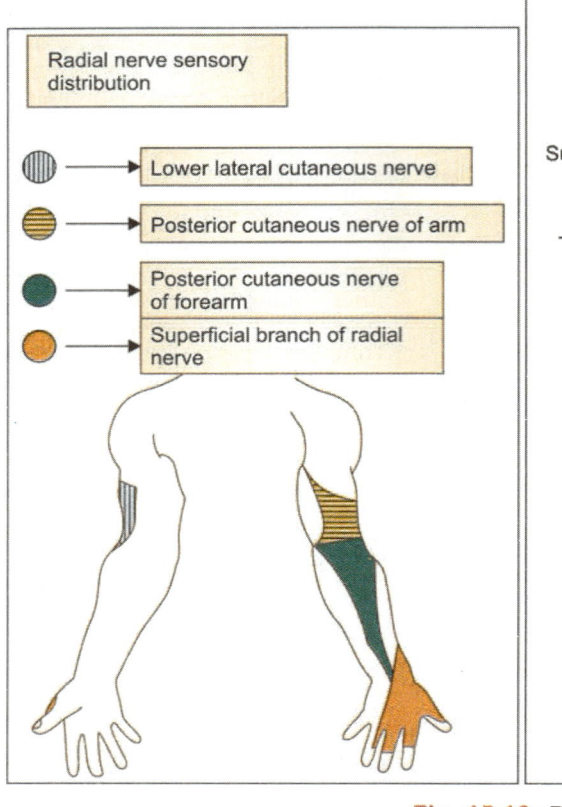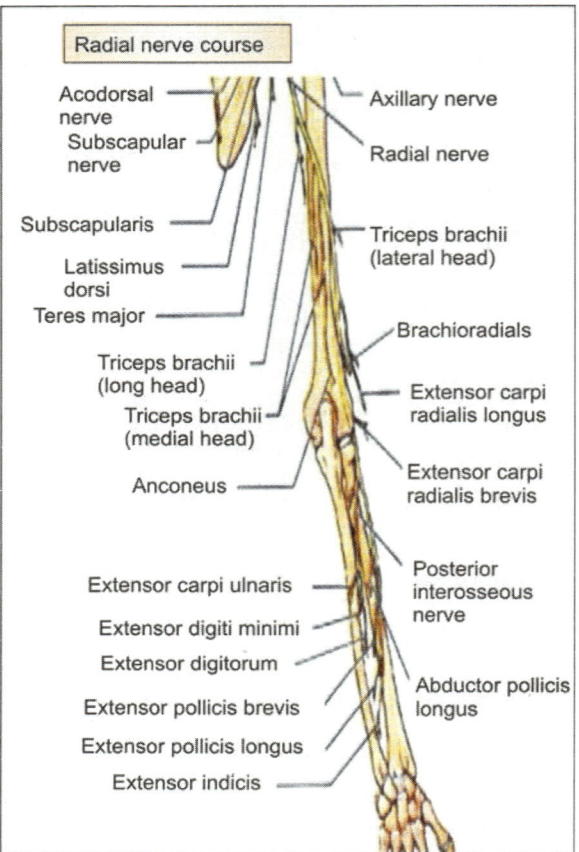

Fig. 15.13: Radial nerve

c. Infections, e.g. diphtheria
d. Neurotoxins, e.g. lead
e. Saturday night palsy
f. Tourniquet's palsy
g. Fracture shaft humerus
h. Injection palsy
i. Supracondylar palsy
j. Gunshot wounds
k. Lacerated wounds or glass cut injuries
l. Compression injury by triceps
m. Tennis elbow
n. Fibrosis of common extensor tendon leading to compression
o. Upper end both bone fractures
p. Supinator muscle spasm
q. Neoplasms

Median Nerve Injury (Fig. 15.14, Table 15.3)

Root value: C6–C8, T1.

Etiology

- Deep and penetrating injuries to the arm, forearm or wrist
- Blunt force trauma or neuropathy
- Axillary aneurysms
- Traction injuries
- Hansen's disease
- Compression underneath Struther's ligament, bicipital aponeurosis, two heads of pronator teres
- Carpal tunnel syndrome
- Tourniquet's plasy
- Golfer's elbow
- Supracondylar fracture
- Glass cut injuries
- Volkmann's ischemic contracture

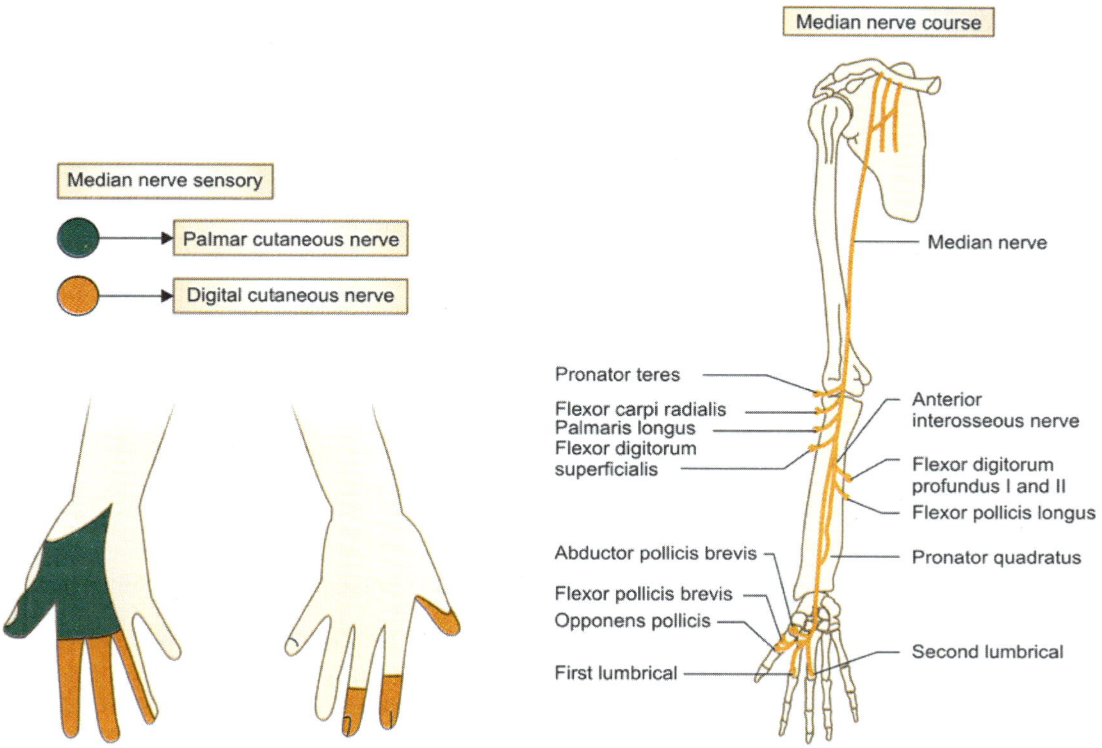

Peripheral Nerve Injuries

Fig. 15.14: Median nerve

Table 15.3: Median nerve injuries

Sensory symptoms	Motor symptoms	Deformities and functional disability	Principles of management
Loss of sensation over the volar aspect of lateral 3½ fingers up to distal phalanx on the dorsal side and the skin over the thenar eminence	Motor paralysis LMN type is observed in: • Pronator teres • Flexor carpi radialis • Flexor digitorum superficialis • Palmaris longus • Flexor digitorum profundus (lateral half) • Pronator quadratus • Flexor pollicis longus • Thenar muscles and 1st and 2nd lumbricals of the hand	• Ape hand deformity • Partial claw hand • Pointing index finger • Tear drop tip to tip pinch The patient have difficulty in holding objects, writing and any activity involving thumb	• Opposition splints to maintain first web space • Cock-up splint (Fig. 15.15) • C-bar (Fig. 15.15) Other principles are same as general principles of management of PNI

Note: The sensory over the autonomous zone of median nerve over the pulp of the thumb

The thenar eminence of the hand is wasted in median nerve injuries as a result there shall be atrophy of the thenar muscles. If patient tries to make a fist, only the little and ring fingers can flex completely. This results in a characteristic shape of the hand, known as *"Hand of Benediction"*.

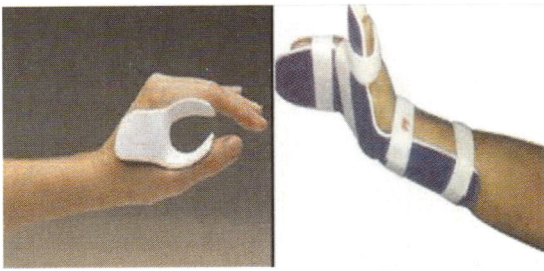

Fig. 15.15: C-bar splint and cock-up splint

Carpal Tunnel Syndrome

It is a medical condition where the median nerve is compressed along its course in the carpal tunnel in the wrist formed by the carpal bones as floor of the tunnel and flexor retinaculum as the roof.

Etiology[16]

- In most of the cases, it is unknown.
- Obesity, oral contraceptives, hypothyroidism, arthritis, diabetes, trauma would be a cause in precipitating the CTS.
- Lipomas, ganglion's, vascular malformations
- Often seen in IT professionals due to faulty biomechanics and in-ergonomic use of computers
- Genetic susceptibility
- Mostly menial workers, hairstylists, musicians, writers, drivers, restaurant servers, jack-hammer and chain saw operators are affected.

Symptoms (Fig. 15.16)

- Numbness, burning sensation, pain and tingling sensation in the hand.
- Decreased sensations in thumb, index finger and middle finger.
- Atrophy of the thumb may be present.
- Loss of grip strength and manual dexterity are lost.
- Phalen's maneuver—positive.
- Tinel's sign—positive.
- Durkan test—positive.

Instructions to Patient for Self-care in CTS

- Perform some gentle exercises regularly like rotating the wrist clock and anticlockwise for at least 2 min frequently.
- Take foods rich in vitamin B_6 or supplements.
- The bromelain in pineapples found to reduce swelling and pain, so have at least half piece.
- Reduce the mechanical tasks of your wrist as far as possible.
- Sleep with arms close to your body.
- Try using night splints during sleep.
- Avoid forcible flexion of wrists.
- Ergonomic use of office tools like desktop and keyboard.
- If you are an IT professional, take a break every hour in your work and shake out your hands frequently.
- TENS, ultrasound therapy and LASER would be effective measures to relieve inflammation and pain. Ask your therapist for more details.

- Hand elevation test—positive.
- Electrodiagnosis-combined sensory index-Robinson index—positive.

Ulnar Nerve Injuries (Fig. 15.17 and Table 15.4)

Root value: C8–T1.

Etiology

- Prolapse intervertebral disk (cervical)
- Cervical spondylosis with ulnar radiculopathy
- Secondary to rheumatoid arthritis
- Cervical rib
- Thoracic outlet syndrome
- Scalenus anticus syndrome
- Crutch palsy
- Tourniquet palsy
- Supracondylar/medial epicondylar fracture of Humerus
- Hansen's disease

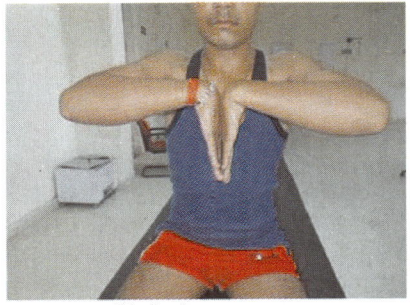
Phalen's maneuver

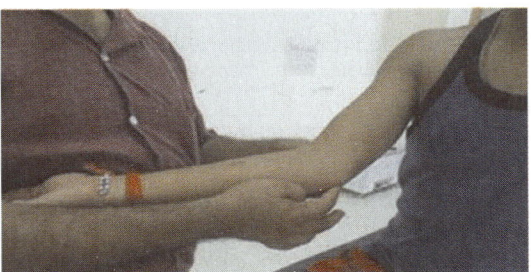
Tinnel's sign

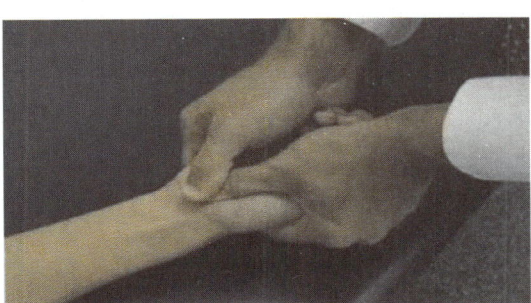
Durkan test

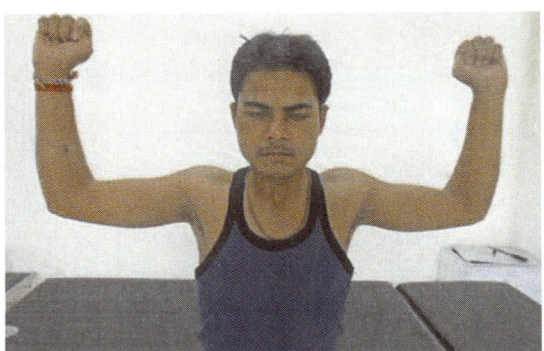
Hand-elevation test

Fig. 15.16: Clinical tests for carpal tunnel syndrome

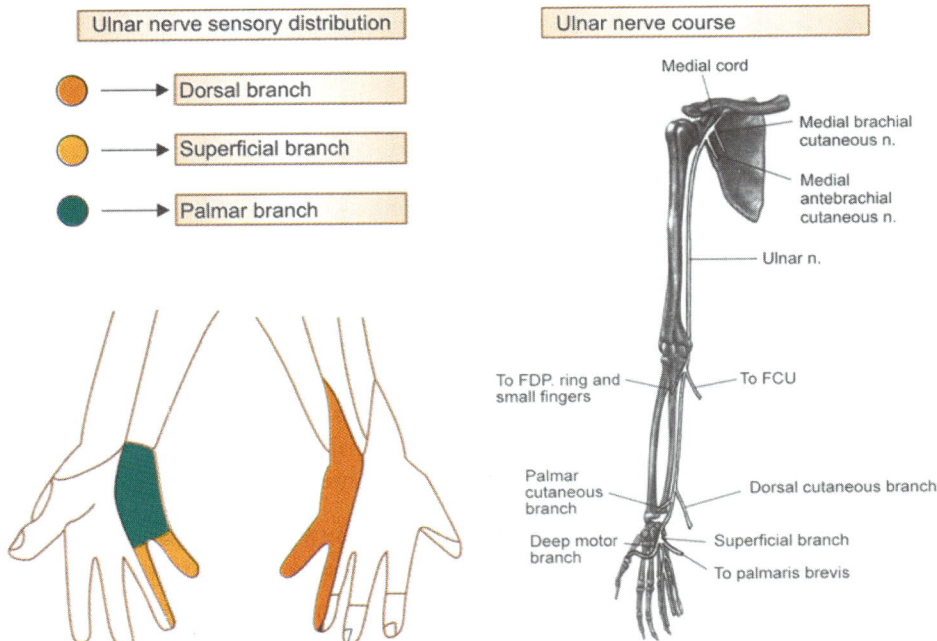

Fig. 15.17: Ulnar nerve

Table 15.4: Summary of ulnar nerve palsy

Sensory symptoms	Motor symptoms	Deformities and functional disability	Principles of management
Patient shall have loss of sensation on the skin over hypothenar eminence Medial 1½ finger up to the nail beds.	LMN type motor paralysis is observed in • Flexor carpi ulnaris • Medial half of flexor digitorum profundus • Hypothenar muscles • Medial two lumbricals and interossei of the hand • Adductor pollicis	• Ulnar claw hand • The patient shall have difficulty in gripping • The classical sign is the patient cannot grip the paper held between the fingers • Disturbed pinch grips • Spherical grip is lost	Knuckle bender splint is advised Follow the general principles of management of PNI

- Tardy ulnar nerve palsy
- Volkmann's ischemic contracture
- Tight POP/plaster
- Glasscut injuries
- Fracture of carpal bones
- Tumors

Thoracic Outlet Syndrome

It is the collection of variety of conditions attributed to the neurovascular structures which transverse the thoracic outlet.

These conditions are brought about by the abnormal compression of the neurovascular structures that pass through the outlet by bony, ligamentous or muscular obstacles between cervical spine and the lower border of the axilla.

Common nerve roots involved are C8 and T1.

Who are at Risk for TOS?
- Jack hammer operators[17]
- Cash register operators, students
- Railway coolies
- Dental hygienists
- Weightlifters
- Pregnancy

Clinical Manifestations

Vascular symptoms:
- Swelling of arm or hand
- Cyanosis of hand
- Pulsating lump above the clavicle
- Easily fatigued arms or hands
- Distension of superficial veins in the hand
- Deep, boring toothache which increases in night

Neurological symptoms:
- Burning, tingling numbness, casualgia, pins and needles sensations (paresthesia) along the medial border of the forearm and the palm.
- Muscle weakness and atrophy of hypothenar, medial two lumbricals, interossei, flexor carpi ulnaris, flexor digitorum profundus, adductor pollicis.
- Difficulty in doing fine motor activities.
- Cramps of the muscles in the forearm and hand.

Common Causes of Neurovascular Compression at the Thoracic outlet
- Anterior scalene tightness
- Costoclavicular approximation
- Pectoralis minor tightness

Special Tests
- East-test—positive
- Adson test—positive
- Costoclavicular maneuver—positive
- Allen's test—positive
- Provocative elevation test—positive

Physiotherapy Management
a. Local heat therapy (moist heat/dry heat)
b. Postural retraining exercises
c. Shoulder strengthening program
d. Self-stretching and passive stretching of the back and neck, chest muscles
e. Ultrasound therapy
f. TENS/IFT for pain relief
g. Scapular bracing exercises
h. Intermittent cervical traction (mechanical)
i. Mobilization of 1st rib, sternoclavicular joint, acromioclavicular joint, glenohumeral joint, cervical.
j. Taping to brace the shoulder

Obturator Nerve Injury[18] (Fig. 15.18 and Table 15.5)
Root value: L2–L4.

Etiology
- Hernia
- Hip joint dislocation
- Tumor or fetus compressing the nerve
- Fracture pelvis
- Sacroiliac joint pathology
- Prolonged difficult labor
- As a postoperative complication in hip arthroplasty

Sciatic Nerve Injury (Fig. 15.19 and Table 15.6)
Root value: L4-S3.

Etiology
a. Penetrating wounds around pelvis
b. Fractures/dislocations around pelvis
c. Injection palsy at gluteal region
d. Neoplasm compressions
e. Piriformis syndrome
f. Lumbar spinal stenosis
g. Degenerative disk disease
h. Spondylolisthesis
i. Pregnancy
j. Ankle jerk is diminished or lost.

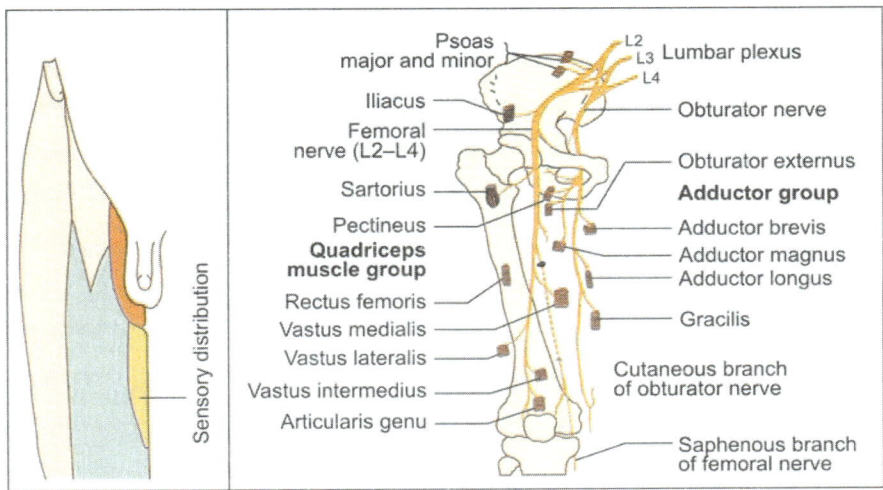

Fig. 15.18: Obturator nerve

Table 15.5: Summary of obturator nerve injuries

Sensory symptoms	Motor symptoms	Deformities and functional disability	Principles of management
Loss of sensations/paresthesias over the distal medial aspect of the thigh and medial aspect of the knee	LMN type of paralysis of the following muscles: • Adductor longus • Adductor brevis • Gracilis • Pectineus • Adductor magnus • Obturator externus • Adductor brevis	Hip flexion-abduction deformity Loss of adduction and internal rotation and have difficulty in gait Wide based or circumductory gait is present	IG stimulation TFL stretching exercises Strengthening of weak muscles Gait training

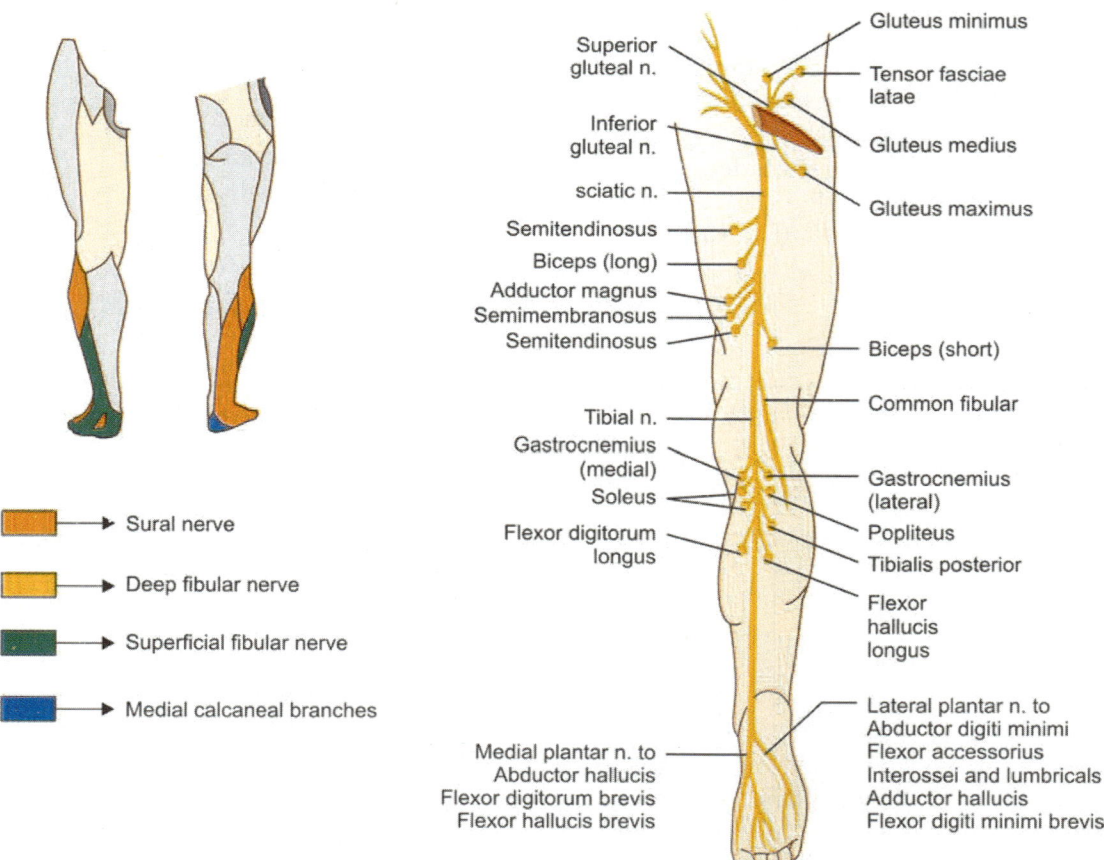

Fig. 15.19: Sciatic nerve sensory innervation

Peripheral Nerve Injuries

Table 15.6: Summary of sciatic nerve injuries

Sensory symptoms	Motor symptoms	Deformities and functional disability	Principles of management
Loss of sensations/paresthesias below the knee except the area that is supplied by femoral nerve	LMN type motor loss is observed in: • Biceps femoris • Semimembranous • Semitendinosus • Hamstring part of adductor magnus • Tibial nerve muscles—calf muscles, intrinsic muscles of the foot • Common fibular nerve muscles—anterior leg, lateral leg and remaining intrinsic foot muscles	• Final leg with foot drop • Clawing of toes with tropic ulceration • Patient shall have difficulty in gait	• Night splints like "L" orthosis • Ankle foot orthosis • Padded footwear or microcellular insoles • Metatarsal bar to the foot • Other principles same as general principles of management of PNI

Sciatica

Sciatica is a set of symptoms caused by the irritation or compression of sciatic nerve leading to tingling numbness and/or weakness of the muscles of the back of the thigh, anterior and lateral leg muscles and intrinsic foot muscles.

Clinical Manifestations

- Lower back pain usually affects one side of the body.
- Pain in the back of the leg—radiating type usually originates in the low back or buttock and continues along the course of sciatic nerve.
- Pain is relieved when patient lie down or walking and becomes worst in standing or sitting.
- Burning, tingling numbness along the back of the thigh and leg.
- Shooting pain.
- Cramps on prolonged standing—neural claudication.
- Ankle jerk is lost or diminished.

Special Tests

Test name	Position of the patient	Position of the therapist	Technique	Observation	Remark
1. Slump test	Sitting at the edge of table	Standing at the side of the table	Patient is asked to bend without neck flexion. Later on asked to extend the knee with dorsiflexion while therapist applies pressure over the neck and shoulders	Pain	Stretch in meninges of the spinal cord
2. Straight leg raise test (SLR) or Lasegu's test	Supine with medial rotation and adduction of hip along with extended knee	Standing at the side of the table	Patient is asked to raise the leg with extended knee. Then instructed to lower the leg in knee extension, thereby relieving the pain. Dorsiflexion of ankle provokes the pain	Excruciating pain at 70	Disc herniation resulting in lower back pain
3. Brudzinski's sign	Supine with medical rotation and adduction of hip along	Standing at the side of the table	Same as SLR test but neck is flexed passively	1. Pain 2. If pain does	Dura mater stretch or lesion in spinal cord Hamstring

(Contd.)

Peripheral Nerve Injuries 227

Test name	Position of the patient	Position of the therapist	Technique	Observation	Remark
	with extended knee			not elicited	tightness
4. Modified SLR test	Side lying with test leg positioned superiorly along with hip and knee flexed to 90°	Therapist stands at the side of the patient	By supporting the pelvis, therapist slowly extends the knee	Pain in lower back	Disc herniation resulting in lower back pain
5. Lhermitt's test or crossover sign	Patient in supine lying position	Therapist stands by the table	Leg of the unaffected side lifted gradually and pain is elicited in opposite leg	Pain	Disc herniation resulting in pain
6. Prone knee bending test	Prone lying	Therapist stands by the table	Flex the knee to the maximum ensuring no hip rotation	Pain in lumbar region, buttock or posterior thigh	L2, L3 nerve root lesion
7. Bowstring test	Supine lying	Stands towards foot end	Therapist places subject's affected leg on his shoulder girdle	Pain radiating lumbar area	Sciatic nerve compression

(Contd.)

Test name	Position of the patient	Position of the therapist	Technique	Observation	Remark
8. One leg standing lumbar extension test (stork standing)	Stands on one leg	Stands behind the patient	Subject extends his/her back while standing on one leg	Pain in back	Stress fracture of pars interarticularis (spondylolisthesis) Scottish dog appearance
9. Stoop's test	Standing or walking		Patient perform brisk walk for hardly 50 m	Pain in gluteal region and lower limbs	Intermittent ckayducation

Management

Medical management includes administrations of NSAIDs like ibrufen, epidural steroids, oral steroids, vitamin B_6 injections.

Physiotherapy includes neural mobilization, self-stretching exercises, back strengthening exercises, myofacial release.

Dexamethasone iontophoresis[19] combined with strong surged Faradic (SSF) current in case of piriformis syndrome would relieve the symptoms from piriformis syndrome and improve functional abilities of the patient (Potturi et al. 2014)

Ultrasound therapy, TENS, SWD, LASER, MWD would reduce the pain and spasms.

FEMORAL NERVE INJURY[20-22]
(Fig. 15.20 and Table 15.7)

Root value: L2–L4.

Etiology
- Psoas abscess
- Pelvic neoplasm
- Fracture of the pelvis/femur
- Hip dislocation
- As a complication to spinal anesthesia
- Intervertebral disk proplapse/lumbar canal stenosis
- Diabetic neuropathy

Guidelines for Patient Self-care in Sciatica
- Avoid any activities that trigger the pain
- Avoid prolonged bedrest
- Sleep on a flat bed
- Avoid wearing a wallet in the back pocket and driving
- Avoid lifting objects

- Post-surgical complication in OBG surgeries.
- Penetrating wounds in lower abdomen.

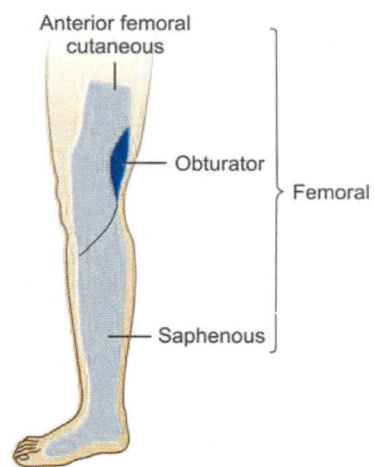

Fig. 15.20: Femoral nerve

Peripheral Nerve Injuries

Table 15.7: Summary of femoral nerve injuries

Sensory symptoms	Motor symptoms	Deformities and functional disability	Reflex lost	Principles of management
Loss of sensations/paresthesias in the anterior and medial aspect of the thigh for anterior division lesions Medial aspect of the leg and foot right up to the ball of the great toe, if posterior division is involved The autonomous zone is small area superior and medial to the patella	LMN type of paralysis is seen in the following muscles • Sartorius and pectineus, if anterior division is involved • Quadriceps, if posterior division is involved	• Genu recurvatum patient shall have difficulty in locking the knee, difficulty in gait and balance	Quadriceps jerk is lost or diminished	Anterior knee guard Knee ankle orthosis Other principles of management are same as PNI

> **Stripping of Femoral Nerve**
>
> It is a surgical procedure in individuals with varicosity of saphenous vein, where the long saphenous vein is stripped. During this procedure, there may be damage to the saphenous nerve as its course is closely related to the saphenous vein resulting in pain, paresthesia or complete loss of sensation on the medial aspect of lower leg.

Common Peroneal Nerve Palsy
(Fig. 15.21 and Table 15.8)

Root value: L4–S3.

Etiology
a. Fracture neck of fibula
b. Compression of the nerve by tight plaster or a splint
c. Fracture dislocation of head of fibula
d. Hansen's disease
e. Trauma
f. Entrapment neuropathy in the fibrous arch at fibula
g. Diabetic neuropathy
h. As a complication of skeletal traction
i. Prolonged immobilization in leg external rotation position.

Tibial Nerve palsy (Medial Popliteal Nerve)
(Fig. 15.22 and Table 15.9)

Root value: L4–S3

Etiology
a. Injection palsy
b. Tarsal tunnel palsy
c. Deep penetrating trauma
d. Dislocation of knee joint
e. Diabetic neuropathy
f. Hansen's disease
g. Arthritis
h. Post-traumatic ankle deformities

Table 15.8: Summary of common peroneal nerve palsy

Sensory symptoms	Motor symptoms	Deformities and functional disability	Principles of Management
Loss of sensations/paresthesias in the sensory distribution of the nerve as follows: Common peroneal nerve palsy: • Skin along the lateral aspect of the knee in the proximal third of the calf muscle • Skin over posterolateral aspect of the calf and over the lateral melleolus, lateral aspect of foot and 4th, 5th toes Deep peroneal nerve palsy • 1st web space of foot, i.e. between great and second toe • Lateral aspect of the dorsum of the great toe • Medial aspect of dorsum of the second toe Superficial peroneal nerve palsy: • Anterior and lateral area of leg • Dorsum of the foot and toes except in the area of distribution of deep peroneal nerve	Motor loss in the form of LMN type muscle paralysis of (in deep peroneal nerve palsy) • Tibialis anterior • Extensor hallucis longus • Extensor digitorum longus and brevis • Peroneus tertius In superficial peroneal nerve palsy • Peroneus longus and brevis	Foot drop Equinovarus deformity Patient shall have difficulty in gait	Foot drop splint Calipers with dorsiflexion top or plastic ankle foot orthosis is advised Other principles shall be similar to the general principles of management of PNI

Peripheral Nerve Injuries

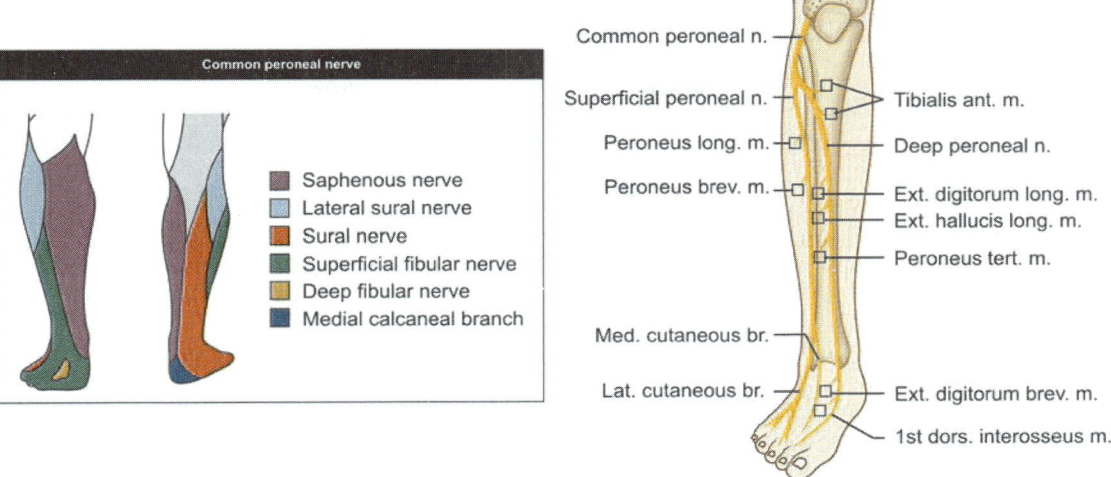

Fig. 15.21: Common peroneal nerve

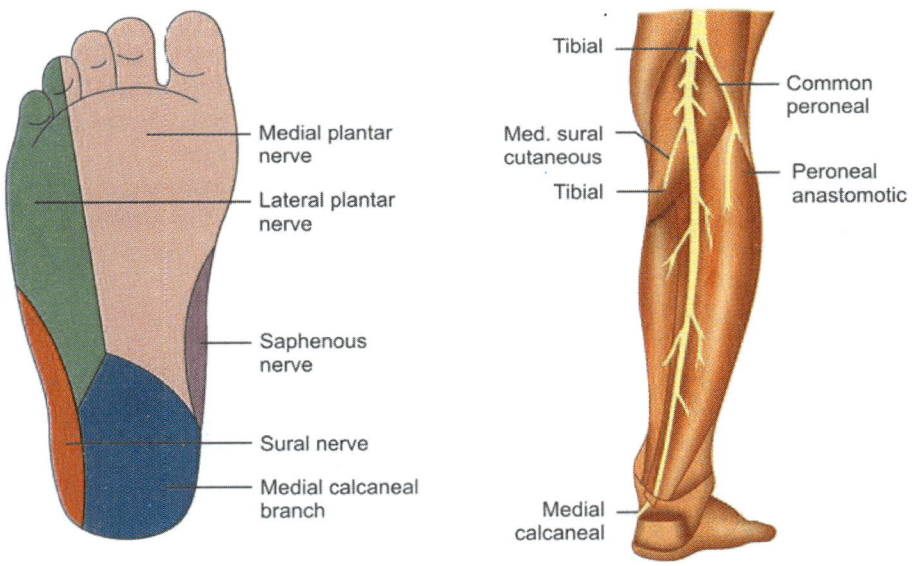

Fig. 15.22: Tibial nerve

Tarsal Tunnel Syndrome

It is a clinical condition where the tibial nerve within the tarsal tunnel which is posterior to the medial malleolus is compressed.
Common causes are osteoarthritis, rheumatoid arthritis, post-traumatic ankle deformities.
Paresthesia or loss of sensations in the sensory innervations of tibial nerve may be experienced which are usually aggravated by activity and relieved by rest
Conservatively treated by NSAIDs, footwear modifications and physiotherapy.
If the conservative management is failed, then the nerve is released surgically by cutting the flexor retinaculum.

Table 15.9: Summary of tibial nerve injuries

Sensory symptoms	Motor symptoms	Deformities and functional disability	Reflex lost	Principles of management
Loss of sensation or paresthesia in the sensory distribution of the nerve as follows: • Sole of the foot • Skin over the medial aspect of heel Autonomous zone is sole of the foot except the medial border of foot, lateral surface of the heel and plantar surface of the toes	Motor loss in the form of LMN type Paralysis of the following muscles: • Calf muscles • Flexor hallucis longus • Flexor digitorum longus • Tibialis posterior • Popliteus	Talipes calcaneovalgus deformity Dorsiflexion deformity	Ankle jerk is lost or diminished Plantar reflex No response	Below knee caliper with dorsiflexion stopper Other principles of management is same as general principles of PNI

Facial Palsy

It is the UMN type lesion of the facial nerve characterized by the paralysis of the lower half of the face.

Etiology

- Supranuclear lesions involving the corticospinal fibers
- Frontal lobe tumors
- Nuclear and infranuclear lesions
- Pontine lesions
- Facial canal lesions
- Skull fractures
- Spread of infections in the facial canal
- Surgical operations of the ear
- Herpes zoster infections
- Mumps
- Inflammations of the facial nerve within the stylomastoid foramen
- Cerebrovascular accident.
- Uncontrolled high blood pressure
- Lyme disease
- Ramsay–Hunt syndrome

Signs and Symptoms

1. Hemiparalysis or hemiparesis of the contralateral muscles of facial expression on the lower quadrant of the face
2. Facial asymmetry
3. Atrophy of muscles of lower portion of the face on affected side
4. Unlike Bell's palsy there is no difficulty in closing the eyes
5. No eyebrow drop
6. Intact folds on the forehead
7. Smoothing of nasolabial folds on affected side
8. Lips cannot be held tightly together or pursed
9. Difficulty keeping food in mouth while chewing on the affected side
10. Muscle twitching
11. Dryness of the eye and mouth
12. Drooling of saliva
13. Facial pain
14. Headache or dizziness

Bell's Palsy (Fig. 15.23)

Bell's palsy is an LMN type paralysis of the facial nerve resulting in inability to control facial muscles on the affected side.

Named after Scottish anatomist Charles Bell, who first described it.

Bell's palsy is the most common acute mononeuropathy and is the most common cause of acute facial nerve paralysis.

Peripheral Nerve Injuries

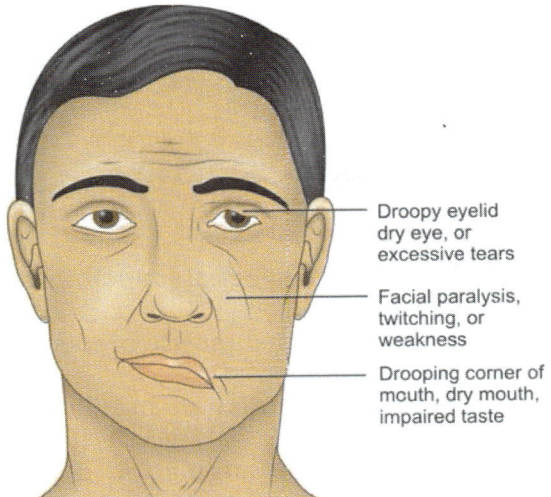

Fig. 15.23: Bell's palsy symptoms

Epidemiology[23,24]

- 0.2% of world population suffer from Bell's plasy, 1 in every 5000 people are effected.
- Right side is more affected (63% cases).
- In winters, the prevalence is more than in summers.
- Recurrence rate being 4–14%.
- Usually unilateral but bilateral is not uncommon.
- Diabetic patients have found to be at more risk.
- High incidence in patients of age above 60, but can affect any age group.
- Men and women are equally affected but women in ages 10–19 are more affected than men of the same age group.

Etiology

a. Most of the cases are of unknown cause—idiopathic
b. Postoperative complications of ear surgeries
c. Infections of the ear
d. Post-herpes simplex virus infection
e. Upper respiratory tract infections
f. Prolonged exposure to cold winds
g. Water retention in pregnancy

Pathology

- It is thought that an inflammatory condition leads to swelling of the facial nerve. The nerve travels through the skull in a narrow bone canal beneath the ear. Nerve swelling and compression in the narrow bone canal are thought to lead to nerve inhibition, damage or death.
- No readily identifiable cause for Bell's palsy has been found, but clinical and experimental evidence suggests herpes simplex type 1 infection may play a role.

Clinical Manifestations

1. Symmetry of the face is lost.
2. Patient cannot raise the eyebrow on the affected side.
3. The patient cannot close the eyelid on the affected side.
4. Blowing, smiling, speech, pursing of lips, retraction of mouth are affected.
5. Taste is impaired due to cauda tympanic branch of facial nerve involvement.
6. Water drools out when the patient tries to drink water.
7. **Bell's phenomenon:** On attempt of closing the eyelid, the affected side eyeball rolls upward.
8. Angle of mouth is deviated to the normal side.
9. Rarely some patients may experience mild pain at the TMJ.

Anomalous Regeneration of Facial Nerve

- Fibers that innervate orbicularis occuli after regeneration can get connected with orbicularis oris, as a result, there will be retraction of the corner of the mouth with closure of eyelid.
- The viscera motor fibers which are innervating salivary gland may get connected to lacrimal glands leading to tears from eyes whenever the patient salivates. This phenomenon is called "Crocodile tears".

10. **Ectoprism:** Eversion of lower eyelid impairs absorption of tears which tend to overflow the lower eyelid.
11. **Synkinesis:** Attempt to move one group of facial muscles results in contraction of all of them.
12. **Hemifacial spasm:** Any initiation of facial muscle movement results in the spasm of the facial muscles on the affected side.

Diagnosis

- The Bell's palsy is diagnosed based on clinical symptoms and by exclusion, i.e. eliminating the other reasonable possibilities. Bell's palsy is commonly referred as idiopathic or cryptogenic which means "cause not known".
- Strength duration curves usually shows a typical graph indicating denervation of the muscles.
- FG test, galvanic shall be positive due to denervated muscles.
- NCV is decreased.
- ESR is elevated indicating the inflammation.
- X-ray/CT scan can be used to rule out suspected tumors.

Management

1. Antiviral drugs, like acyclovir, famcyclovir, are prescribed for underlying viral infection.
2. Antibiotics are prescribed accordingly for a bacterial infection.
3. For faster resolution of inflammation, steroids, like prednisolone, are prescribed.
4. For early nerve regeneration, methylcobalamin, alpha lipoiec acids are prescribed.

Then the patient is referred to physiotherapy for further management.

Surgical management:

- Very rarely in unsuccessful cases, skin grafting is done to maintain the facial symmetry.
- Skin is taken from thigh region for the purpose.
- In cases of nil nerve regeneration, nerve graft is indicated.

PHYSIOTHERAPY MANAGEMENT

Aims of Management

1. To provide psychological support to the patient.
2. To maintain normal muscle properties in the muscles affected.
3. To relieve inflammation and muscle spasm.
4. To maintain the angulation of the mouth.
5. To relieve pain.
6. To improve functional capacity of the patient.
7. To improve circulation over the face and aid in venous and lymphatic drain-age.
8. To prevent infections/irritation to the affected eye

Plans of the Treatment

In order to achieve the above aims, the following physiotherapy interventions are incorporated.

- Counseling of the patient
- Exercise therapy
- Massage therapy
- Supportive measures
- Electrodiagnosis and therapy
- Home regime/advices

Counseling of the patient: *Face is the index of the mind, everyone wants an esthetic face and any deviation would be an insulting and embarrassing situation.* The patient with Bell's palsy would show social withdrawal symptoms leading to depression. Hence, the therapist should counsel the patient regularly. Explain the importance of physiotherapy, the outcomes of treatment and motivate the patient to participate actively in the treatment sessions.

Exercise therapy (Fig. 15.24):
1. Always exercise before a mirror
2. Raise eyebrows
3. Frown
4. Flare nostrils
5. Try closing and opening eyes
6. Smile with showing teeth
7. Smile without showing teeth
8. Grin the cheeks
9. Try moving tongue to both sides of the cheek
10. Munching movements of mouth

Massage therapy: Massage therapy is indicated to aid in lymphatic circulation and to reduce edema, if any. The strokes of the massage should be directed upwards and the lymph is drained into subauricular lymph nodes and cervical lymph nodes.

Techniques include:
1. Stroking
2. Effleurage
3. Soft-padded kneading
4. Reverse hacking, if tolerated.
5. Static vibrations at the TMJ

Supportive measures:
- Taping will help to maintain the angle of the mouth and helps in eating and drinking for the patient.

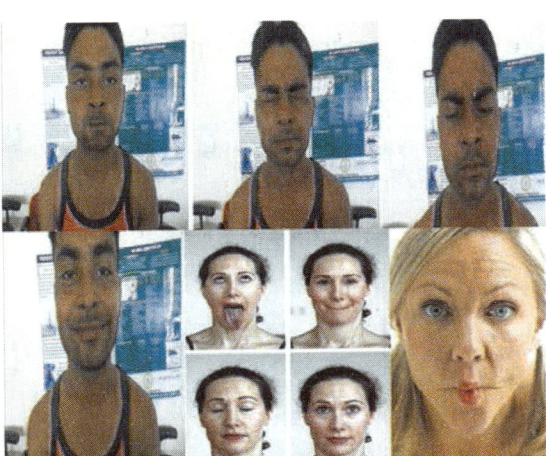

Fig. 15.24: Facial exercises

- Protective eye goggles should be worn to prevent dust entering the eye

Electrotherapy: To relieve inflammation, if the patient immediately visits the therapist following the paralysis, IRR radiation would be beneficial.

IG stimulation would be beneficial to stimulate the muscle to maintain the normal muscle properties.

- Face muscles are small and tend to fatigue easily.
- Hence galvanic stimulation is used with a pad and pen electrode.
- Explain the procedure to the patient
- Clean the skin before starting the treatment.
- Place the pad electrode under the neck.
- Stimulate each muscle on the respective motor point to have maximum efficiency.
- Number of stimulations should not exceed more than 30 contractions.
- After the treatment, mild redness on the skin is noticed as a sign of vasodilatation.

Cryotherapy in the form of brisk stroking would be beneficial to improve the muscle tone.

EMG biofeedback/Faradic re-education (if patient is tolerable) would be beneficial for regaining the muscles power.

Home regime/advices:
- Exercise regularly
- Blow balloons
- Use straw to drink water or fluids
- Chew gum
- Goggle the mouth with water frequently.
- Self-massage once a day.
- Use goggles to protect the eyes
- Splint used to correct the mouth deviation

REFERENCES

1. Blumenfeld H. Neuroanatomy through Clinical Cases 2nd ed.
2. Snell RS. Clinical Neuroanatomy 7th ed.
3. Blumenfeld H. Neuroanatomy through Clinical Cases. Sinauer Associates 2002.
4. Brazis PW, Masdeu J, Biller J. Localization in Clinical Neurology 6th ed. Lippincott Williams & Wilkins.
5. Hart AM, Terenghi G, Kellerth JO, Wiberg M. Sensory neuroprotection, mitochondrial preservation and therapeutic potential of N-acetyl-cysteine after nerve injury. Neuroscience 2004;125:91.
6. Tuncel U, Turan A, Kostakoglu N. Acute closed radial nerve injury. Asian J Neurosurg 2011;6:106–109.
7. Daroff RB, Fenichel GM, Jankovic J, Mazzlotta JC. Bradley's Neurology in Clinical Practice, 6th ed. Philadelphia, PA: Saunders 2012.
8. Noble J, Munro CA, Midha R. Analysis of upper and lower extremity peripheral nerve injuries in a population of patients with multiple injuries. J Trauma Inj Infect Crit Care 1998;45:116–122.
9. Y Allieu and P Cenac. "Is surgical intervention justifiable for total paralysis secondary to multiple avulsion injuries of the brachial plexus? Hand Clinics 1988;4(4):609–618.
10. Azze RJ, Mattar R Jr, Ferreira MC, Strack R and Canedo AC. "Extraplexural neurotization of brachial plexus". Microsurgery 1993;15(1):28–32.
11. Brandt KE, Mackinnon SE. "A technique for maximizing biceps recovery in brachial plexus reconstruction". Journal of Hand Surgery 1993;18(4):726–733.
12. Brunelli G and Monini L. "Direct muscular neurotization". Journal of Hand Surgery 1985;10(6): 993–997.
13. Doi K, Muramatsu K, Hattori Y, et al. "Restoration of prehension with the double free muscle technique following complete avulsion of the brachial plexus. Indications and long-term results, Journal of Bone and Joint Surgery A 2000;82(5):652–666.
14. Doi K, Kuwata N, Muramatsu K, Hottori Y and Kawai S. "Double muscle transfer for upper extremity reconstruction following complete avulsion of the brachial plexus". Hand Clinics 1999;1(4):757–767.
15. Dubuisson AS, Kline DG, Amar AP, Gruen JP, Kliot M and Yamada S. "Brachial plexus injury: a survey of 100 consecutive cases from a single service." Neurosurgery 2002;51(3):673–683.
16. Derebery J. "Work-related carpal tunnel syndrome: the facts and the myths". Clinics in occupational and environmental medicine 2006;5(2):353-367, viii.
17. Sara A Neal, Karl B Fields, Moses Cone Health System. Greensboro, North Carolina, Peripheral Nerve Entrapment and Injury in the Upper Extremity. Am Fam Phys 2010;17:81(2):147–155.
18. John Sison Tipton, Obturator neuropathy. Curr Rev Musculoskelet Med 2008;1(3/4):234–237.
19. Potturi G et al. (2014). Effect of Dexamethasone Iontophoresis Combined with Strong Surged Faradic Current on Piriformis Syndrome—A Simple Randomized Control Clinical Trail Indian Jour. Physhitherapy and Occupational therapy 2014;8(4).
20. Al Hakim M, Katirji B. Femoral mononeuropathy induced by the lithotomy position: A report of 5 cases with a review of literature. Muscle Nerve 1993;16(9):891–895.
21. Peirce C, O'Brien C, O'Herlihy C. Postpartum femoral neuropathy following spontaneous vaginal delivery. J Obstet Gynaecol 2010;30(2):203–204.
22. Wong CA. Nerve injuries after neuraxial anaeshesia and their medicolegal implications. Best Pract Res Clin Obstet Gynaecol 2010.

16

Polyneuropathies

LEARNING OBJECTIVES

At the end of this chapter, you will be able to:
1. Describe various etiological factors and types of polyneuropathies.
2. Identify various clinical manifestations of the patients with various polyneuropathies.
3. Identify the various investigations done in patients with polyneuropathies.
4. Demonstrate the skills to assess the patients with polyneuropathies.
5. Design treatment protocols for rehabilitating the patients with polyneuropathies.

INTRODUCTION

It is a clinical condition, in which there is a diffuse peripheral nerve disorders and causes impairment of function of many peripheral nerves simultaneously.[1-3] This will result in the symmetrical distribution of flaccid muscle weakness and also sensory disturbances usually affecting more distal segments than the proximal segments of the limbs and may even cranial nerves would be involved.

The term polyneuritis used to be synonym for **polyneuropathy** but now it is less commonly used. Pathologically, polyneuropathy refers to those conditions where there is primary degeneration in the nerve parenchyma, whereas **polyneuritis** refers to inflammation of the peripheral nerves.

Definition

It is a disease of **peripheral nerves** featuring weakness, numbness, paresthesia involving many nerves simultaneously. The spinal nerves or cranial nerves may be involved.

Polyneuropathies could be caused by several agencies including numerous endogenous and exogenous toxins, acute infections, vitamin deficiencies which would directly affect the nerves.

Types

Polyneuropathies are classified based on cause, speed of progression, or the primary involvement of myelin or axon or cell body.

Distal Axonopathy

It is due to metabolic or toxic derangement of neurons.

Usually caused by the metabolic diseases such as diabetes, kidney failure, connective tissue damage, alcoholism or effects of toxins or drugs such as chemotherapy.

These may be divided further according to the type of axon affected as large fiber, small fiber or both.

The distal most portions of axons would be affected and later the degeneration advances slowly towards the nerve cell bodies.

Regneration is only possible, if the cause is removed.

Myelinopathy

Also known as demyelinating polyneuropathy.[4]

It is mainly due to loss of myelin and may completely block the conduction of action potentials through the axon of the nerve cell.

Guillain-Barré syndrome (GBS) is acute inflammatory demyelinating polyneuropathy and some genetic metabolic disorders like leukodystrophy are common examples of myelinopathies.

Neuronopathy

It is the result of destruction of peripheral nervous system neurons.[5–7]

Mostly caused by motor neuron diseases, sensory neuronopathies (herpes zoster), toxins.

Neurotoxins such as vincristine which is used in chemotherapy may cause neuronopathies.

Pathology

1. Each nerve fiber consists of two types of cells, cell body with the axon and the satellite cell—Schwann cell.
2. In myelinated fibers, myelin is formed by the invagination of the Schwann cell surface membrane round the axon. It supports a length of myelin sheath between two nodes of Ranvier which may be as much as 1 mm in length.
3. There are two types of degeneration occurs in the neuropathies, one where

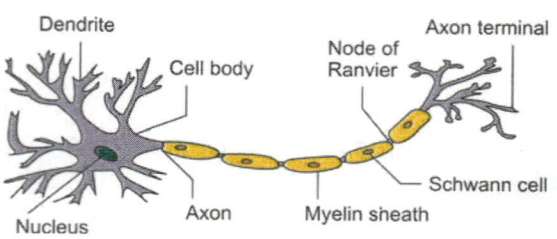

Fig. 16.1: Structure of a typical neuron

Axonal Degeneration

- In axonal degeneration, degeneration of axon begins towards the nerve cell.
- The cell body often undergoes chromatolysis.
- All other changes are similar to Wallerian degeneration, but the regeneration reaction is limited or absent.

Wallerian Degeneration

- This type of degeneration is seen in peripheral nervous system.
- It occurs after transection of the axon.
- The reasons for this is knife wounds compression, traction and ischemia.
- Following transection, initially there is accumulation of organelles in proximal and distal ends of transection sites.
- Later, the axon and myelin sheath distal to transection site undergo disintegration up to next node of Ranvier, followed by phagocytosis.
- The regeneration occurs by sprouting of axons and proliferation of Schwann cells from proximal end.

the nerve cell body or axon or both are affected which is called *axonal degeneration* and *Wallerian degeneration*.

4. The second where the Schwann cell may be affected, causing demyelination without the involvement of axon which is known as segmental demyelination.
5. In chronic neuropathies, there is evidence of partial regeneration of nerves.

Etiology[8]

Infective causes:
  a. Viral—herpes zoster
  b. Bacteria—*Mycobacterium leprae*
  c. *Corynebacterium diphtheriae* indirectly with its endotoxins

Autoimmune

Caused by immunologically activated cells after sensitization associated with infection, e.g. GBS.

Metabolic Causes

  a. Vitamin B_1 deficiency
  b. Poisons like heavy metals, arsenic, copper, thallium, mercury, lead and gold, organic compounds like triorthocresylphosphate, acrylamide, etc.
  c. Drugs like isonizid, sulphonamides, etc.
  d. Uramia—increasing use of dialysis in the treatment of renal failure
  e. Diabetes

Vascular Disorders

  a. Atheroma
  b. Beurger's disease
  c. Collagen disorders

Others

- Systemic diseases like rheumatism, amyloid disease, carcinoma
- Genetically determined disorders.

GUILLAIN–BARRÉ SYNDROME

Definition

It is acutely or subacutely generalized ascending type of polyneuropathy.[9]

Synonyms of GBS

- Acute idiopathic demyelinating polyneuropathy (AIDP)
- Acute infective polyneuropathy (AIP)
- Landry–Guillain–Barré syndrome
- Acute idiopathic polyneuropathy (AIP)

It is thought to be an autoimmune disorder and in some cases, viral infection could be a cause.

It is predominantly a motor paralysis, often may be associated with **paresthesia**.

Other clinical conditions with same onset in which there is purely motor involvement with an ascending form of paralysis were previously called **Landry syndrome**. Later these syndromes are called *Landry-Guillain-Barré syndrome*.

In some cases, there is evidence of involvement of spinal cord or even brainstem in addition to the nerve roots and peripheral nerves.

Often cranial nerves are also involved. The disease reaches peak of disability by 4 weeks.

Etiology[9,10]

- Predominantly idiopathic
- Common in females
- Infections by viruses (Epstein-Barr virus), bacteria (*Mycoplasma pneumoniae*)
- Post-vaccination against rabies, typhoid, tetanus or influenza can precipitate GBS.
- In some post-surgical cases, due to release of neural antigens, surgical stress, blood transfusions.
- Some drugs on prolonged usage can precipitate GBS, e.g. antidepressant drugs.
- Autoimmune due to presence of antigen CD+ T cells.

Clinical Manifestations[10-12]

  a. The onset is usually acute or subacute and occasionally febrile.
  b. Motor weakness of muscles of LMN type which is symmetrical and usually proximal muscles are more affected than the distal.
  c. Cranial nerve involvement can lead to dysphagia, diplopia and respiratory failure.

d. Shoulder elevators and neck flexors are in parallel with diaphragm weakness and respiratory failure (Cooper et al).
e. Incordination of trunk and limb movements which can be attributed to muscle weakness.
f. Sensory changes may be severe, slight or sometimes absent. Symptoms range from mild pain to paresthesia. Loss of complete sensations is not common.
g. The areflexia (absent of reflexes) may be not because of muscle weakness alone but may be because of asyncronization in the firing by the motor axon.
h. Myalgia occurs because of release of cytokinins released by the macrophages.
i. Autonomic disturbances: Involvement of sphincters leading to retention or overflow incontinence, orthostatic hypotension, anhydrosis or hyperhydrosis.

Investigations

CSF analysis shows raised levels of proteins (>2.0–3.0 g/l), raised pressure.

Electrophysiological studies show neurogenic type of presentation with an increase in the amplitude, increase in the duration of motor unit potential with **polyphasia**.

Progression of Disease

The motor paralysis spreads within 30 min to 4 weeks time to peak. Once reached peak, a plateau phase may be seen for 15–20 days and then recovery phase for 4–6 months and may sometimes even extend to 2 years.

Some remissions and relapses are also observed. In most cases, the recovery has good outcome even though the improvement is slow.

Medical Management

a. Plasmapheresis
b. Steroid therapy
c. IgG injections
d. Good nursing care in:
  1. Fluid intake
  2. Maintenance of TPR charts
  3. Maintenance of input–output flowchart
  4. Mouth care, skin and pressure care
  5. Administration of drugs
  6. Dietary needs

Complications of GBS

1. Lower respiratory tract infections
2. Deep vein thrombosis
3. Retention of urine
4. Cardiac arrhythmias

ALCOHOLIC NEUROPATHY[14,15]

a. Also known as nutritional or alcoholic neuropathy, beriberi or vitamin B deficiency syndrome.
b. It may be because of consumption of spirits or wine over long periods.
c. Most patients are middle aged and males are more affected than females.
d. Alcoholism leads to malabsoprtion of vitamin B_{12} leading to neuropathy.
e. There is gross degeneration of the axons as well as the myelin sheath. In extreme cases, anterior and posterior roots of spinal nerves and vagus and phrenic nerves may be involved.

Clinical Manifestations

Sensory disturbances: Pain and paresthesia may be felt by the patient at the feet or hand. The patient can also complain of cold and burning feet sensations which is known as *dysesthesia*.

The patient may exhibit the symptoms of increased sensitivity in which mild touch can be perceived as a severe painful sensation which is known as *hyperpathia*.

Motor disturbances: The motor disturbances can vary from case-to-case and usually exhibits LMN type weakness which may be evenly distributed or distal muscles more involved than proximal muscles.

Patient complains of muscle tenderness which can be elicited by squeezing the calf muscles.

Contractures and deformities: Due to muscular imbalances, there may be contractures and deformities.

Reflexes: Because of LMN type of weakness, the reflexes may be diminished or absent.

Autonomic disturbances: Hyperhydrosis in the palm and sole is seen. Orthostatic hypotension is observed. Involvement of the vagus nerve can lead to dysphasia and hoarseness of voice.

Respiratory problems: Due to involvement of phrenic nerve, there may be diaphragm weakness and respiratory problems.

Medical Management

1. Psychological counseling for alcohol de-addiction
2. Parenteral injection of vitamin B_{12}.
3. Balanced diet.

Other symptomatic medical management like aspirin for pain relief, etc., could be used.

OTHER NEUROPATHIES

Diphtheritic Polyneuropathy

a. Caused by *Corynebacterium diphtheriae*.
b. Paralysis of palate would be the earliest nervous symptom and may develop during third or fourth week after infection.
c. Nasal voice and regurgitation of fluid through the nose during swallowing may be observed.
d. Accommodation of eyes is affected leading to blurring of eyes.
e. Generalized polyneuritis may be present.
f. Postural sensitivity is affected.
g. Gross ataxia is present which is known as *pseudodiabetic form of ataxia*.
h. Sphincters are not involved.
i. Respiratory muscle weakness may lead to serious complications.
j. Segmental weakness of the muscles supplied by same spinal segment may occur when the organism infects a skin wound.

Lead Neuropathy

a. Usually associated by lead poisoning and is seen in lead workers such as plumbers, painters, etc.
b. Can be due to contamination of lead in water, beer or cider passed through lead pipes.
c. Lead usually affects the extensor muscles of wrist and fingers bilaterally leading to wrist and finger drop.
d. An upper arm type of paralysis involving deltoid, biceps, brachialis and brachioradialis muscles. Lower limb muscle involvement is rare.
e. There is no sensory involvement.
f. Investigations show anemia with basophilic stippling of red cells and a high blood lead levels.

> Patients should not be tested to exhaustion since the recovery from fatigue can take time and will delay the rehabilitation. Note the patient, if he started slowly the signs of fatigue, and if so, adjust or discontinue the activity accordingly.

PHYSIOTHERAPY

Physiotherapy Assessment

The physiotherapy assessment usually includes history taking of present illness, past history, personal history and other medical history.

Observation

Observe for the general attitude of the limbs, signs of malnutrition, contractures, deformities, wasting of muscles, clubbing or cyanosis for respiratory complications.

Take the patient consent and check the skin for lesions or pressure sores. The skin over the bony prominences is to checked regularly for any changes.

Palpation

Palpate for checking muscle tenderness and grade it:
1. **Severe tenderness:** Patient shows grimace on the face or withdraws or screams to touch of the therapist.
2. **Moderate tenderness:** Patient shows grimace to hold by the therapist.
3. **Mild tenderness:** Patient shows grimace to press or pressure by the therapist.

Check for the temperature changes (local) by comparing with proximal and distal segments, upper and lower limbs involved and normal areas.

Check whether any edema is present. Grade the edema whether it is localized or generalized, indurate or non-indurated, pitting or non-pitting type.

Check for cranial nerves (Fig. 16.2) by checking on both sides and compare.

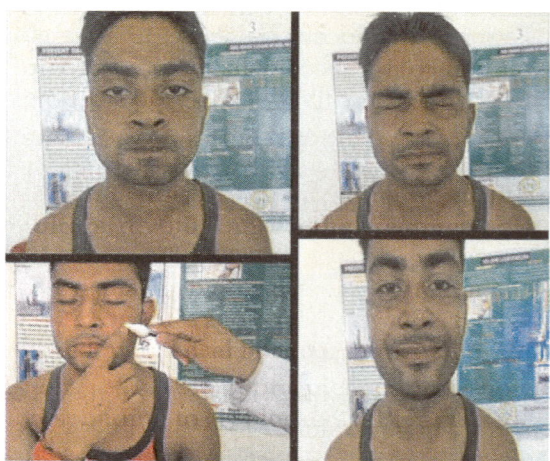

Fig. 16.2: Cranial nerve assessment

Sensory Examination (Fig. 16.3)

Check for the dermatomes involved and uninvolved and compare all the superficial, deep and cortical sensations.

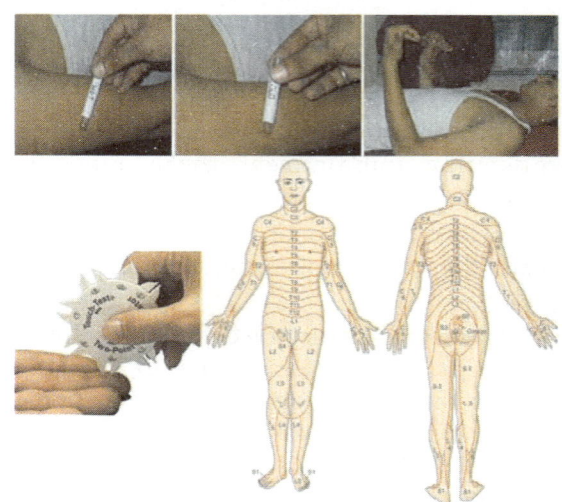

Fig. 16.3: Sensory examination

Reflex Examinations (Fig. 16.4)

Check for superficial and deep reflexes and compare.

Usually there will be dimished or absent responses.

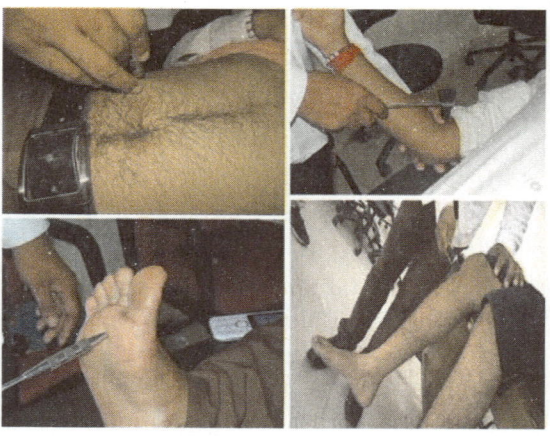

Fig. 16.4: Reflex examination

Motor Examination

Perform the tone examination. Flaccidity is observed in muscles affected.

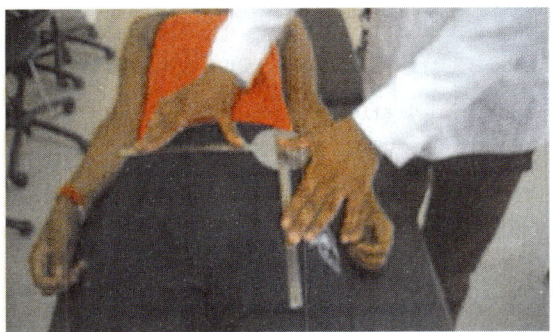

Fig. 16.5: Therapist measuring joint range of motion

The MMT grading is done for the individual muscles.

Coordination and balance: May be affected due to muscle weakness.

Joint range of motion (Fig. 16.5): Assess the patient's joint range of motion by goniometry, paying particular attention on the ankles, knees, hips would be considerative especially if a patient spents most of the time in sitting posture or in wheelchair.

Gait Assessment

The gait parameters are evaluated and observed for changes. If a patient is functionally mobile, a brief gait assessment and if required wheelchair assessment is carried out.

Functional Assessment

Check for the functional capabilities of the patient which shall guide for modifications to be done in rehabilitation. Depending upon the assessment done, mat mobility, transfers, self-care activities like grooming, feeding, dressing and other activities of daily living, leisure activities should be assessed.

Respiratory Assessment

It is very important to assess the respiratory functions of the patient. Check the breathing pattern, use of accessory muscles of respiration, $VO_{2\,max}$, auscultate for breath-sounds. On percussion, there may be diminished sounds.

Autonomic Dysfunction

Check the patient for postural hypotension, pulse in lying, standing and sitting postures and check whether the patient has any dizziness or any symptom related to postural hypotension.

PHYSIOTHERAPY MANAGEMENT

Short-term Goals

1. To provide psychological support
2. To maintain clear airways
3. To remove secretions that lead to lung infection
4. To maintain normal joint range of motion
5. To prevent contractures and deformity
6. To prevent pressure sores
7. To maintain normal muscle properties
8. To improve ventilatory capacity.

Long-term Goals

1. To prevent secondary complications
2. To improve functional capacity of the patient
3. To improve the ambulatory functions
4. To improve the strength of the affected muscles
5. To improve respiratory and cardiovascular capacity.

Plans of Management

To Provide Psychological Support

To gain maximum co-operation from patient and to prevent him from going into depression, psychological counseling is very important. Without encouraging negative thoughts, therapist should explain the patient about his condition. Without giving false hope, therapist should explain the importance of physiotherapy for his condition.

Motivate the patient all through the course of the patient. Explain briefly the condition and importance of physiotherapy for his rehabilitation. Develop good rapport with the patient.

To Maintain Clear Airways

The airways must be clear from secretions to prevent respiratory complications. Frequent change of positions to drain out the secretions and also assist in proper air entry to all the lobes of the lung. Postural drainage with active coughing techniques should be carried out, if patient's cough is not productive, then nebulization is done with a broncholytic and bronchodilator agent and followed by suctioning. Manual techniques like percussion, shaking, manual mobilization can be carryout to loosen the secretions.

If the patient is not on ventilator, deep breathing exercises, and external tracheal stimulation may be given for good bronchial hygiene.

Deep breathing exercises, frequent change of positions would assist in clearing bronchial secretions and decrease paradoxical movement of the rib cage (Doughlas et al. 1977). Nebulizations, postural drainage with manual techniques like clapping, vibrations to remove the accumulated secretions would be useful. Spirometric and chest mobility exercises are to be encouraged to improve ventilator capabilities. Coughing and huffing tech-niques are taught to the patient and he/she is encouraged for effective coughing.

A manual abdominal thrust to assist the patient with coughing has been shown to be effective in increasing the expiratory force to the patient and can generate aid in clearing secretions (Kirby et al. 1966). Glossopharyngeal breathing would be helpful to increase vital capacity and enhance the effectiveness of a patient's cough, provide the patient with ability to sigh and improve lung and chest wall compliance. Diaphragmatic breathing exercises would improve strength and endurance, decrease the work of breathing and increase vital capacity of the patient. Manual stretching of chest wall and use of air shift maneuvers assist in maintaining the

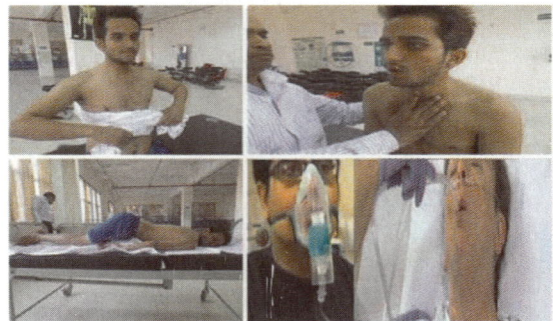

Fig. 16.6: Chest physiotherapy techniques

mobility and compliance of chest wall (Fig. 16.6). A use of abdominal corset is found useful in sitting position to improve the patient's tidal volume and decrease shortness of breath (Alvarez et al. 1981).

To Maintain Normal Joint Range of Motion

The range of motion is maintained by passive movements. Slow gradual stretching of the muscles is very important to prevent tightness, contractures and deformities. Passive movements should be given at least 3 times a day involving all the movements and at all joints at least 10 repetitions per movement. Teach the caretaker also the techniques of passive movement and advice her to continue the passive movements (Fig. 16.7).

At joints where the active or active assisted movements are possible, encourage the patient to perform free exercises at least 3 times a day.

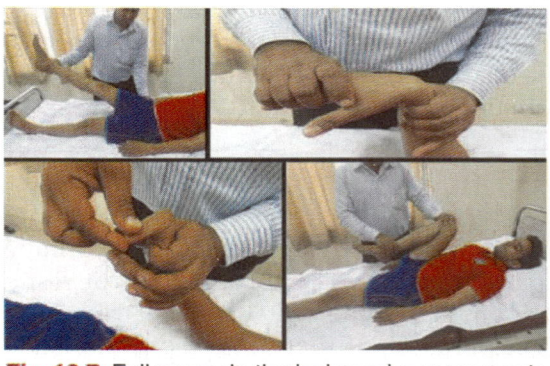

Fig. 16.7: Full range rhythmical passive movements

To Prevent Contractures and Deformity

Passive sustained stretching of the muscles should be done periodically to prevent tightness and contractures. Positioning of the patient will also help in prevention of contractures. Static splints should be used. Pillows or supporting pads can also be used to prevent the contractures (Fig. 16.8).

Fig. 16.8: Splints to prevent contractures and deformities

To Prevent Pressure Sores

Pressure sores are deep wounds due to constant and continuous pressure on the skin overlying the bony prominences in the body. It is the very common, painful complication of patients who are on the prolonged bedrest. The pressure sores can aggravate the existing disability. It should be noted that the development of pressure sores is solely due to negligence of the caregivers and health providers but not accidental. Hence, the health care providers and caretakers should take all necessary precautions to prevent the pressure sores.

Precautions to prevent pressure sores:
1. Closely monitor all the areas of risk daily
2. Frequent change in the position of the patient
3. Small rubbing friction massage with a moisturizer for a small duration (15 sec) over the areas of risk shall improve the skin elasticity and pliability
4. Use of water mattress or air bed
5. Proper bed making technique, avoid creases on the bed cover.

Even after taking necessary precautions, if the pressure sores appear to develop, then curative measures to be taken:

1. **At redness stage:** Stop the massage, start changing the positions frequently and take use of adaptive devices.
2. **At blister stage**, frequent dressing to be done. Hydrocolloid dressing would be helpful.

If an ulcer has already formed, debridement of dead tissues, and dressing of the wound is done. UVR therapy, and ultrasound have been found useful to promote healing.

Deep vein thrombosis is another major complication of immobility. It can be prevented by leg lifts, ankle pumping movements, toe curls, stretching and taking plenty of fluids would be helpful.

The respiratory complications could be avoided by teaching deep breathing exercises, postural drainage, suctioning, good coughing and huffing techniques. Chest muscle stretching exercises would be beneficial.

To Maintain Normal Muscle Properties

Muscles begin to degenerate very rapidly in polyneuropathies. To maintain the normal muscle properties and prevent atrophy, Electrical muscle stimulation of all muscles should be performed (Fig. 16.9). Functional electrical stimulation ameliorates muscle loss, if the muscles are allowed to produce significant forces during the stimulation.

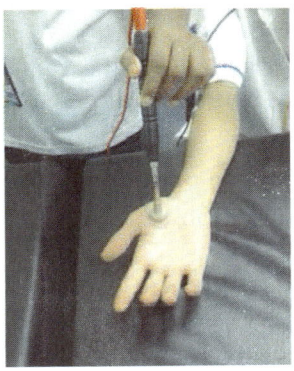

Fig. 16.9: ENMS

To Improve Ventilatory Capacity

Respiratory muscle training with spirometry, assisted treadmill walking, frequent stretching exercise, active and passive mobility exercises, and deep breathing exercises would be beneficial to improve the ventilator capacity of the patients.

Long-term Plans

To Prevent Secondary Complications

The possible complications due to prolonged bedrest would be:
1. Contractures and deformities
2. Pressure sores
3. Deep vein thrombosis
4. Psychological depression
5. Respiratory complications
6. Cardiovascular deconditioning
7. Disturbed bone metabolism

The contractures and deformities are prevented by proper positioning of the patient, passive movements, use of orthotic supports, passive mobilization techniques and stretching.

Pressure sores are deep wounds due to constant and continuous pressure on the skin overlying the bony prominences in the body. It is the very common, painful complication of patients who are on the prolonged bedrest. The pressure sores can aggravate the existing disability. It should be noted that the development of pressure sores is solely due to negligence of the caregivers and health providers but not accidental. Hence, the health care providers and caretakers should take all necessary precautions to prevent the pressure sores.

Precautions to prevent pressure sores:
1. Closely monitor all the areas of risk daily.
2. Frequent change in the position of the patient.
3. Small rubbing friction massage with a moisturizer for a small duration (15 sec) over the areas of risk shall improve the skin elasticity and pliability.
4. Use of water mattress or air bed.
5. Proper bed making technique, avoid creases on the bed cover.

Even after taking necessary precautions, if the pressure sores appear to develop, then curative measures to be taken:

1. **At redness stage:** Stop the massage, start changing the positions frequently and take use of adaptive devices.
2. **At blister stage**, frequent dressing to be done. Hydrocolloid dressing would be helpful.

If an ulcer has already formed, debridement of dead tissues, and dressing of the wound is done. UVR therapy, and ultrasound have been found useful to promote healing.

Deep vein thrombosis is another major complication of immobility. It can be prevented by leg lifts, ankle pumping movements, toe curls, stretching and taking plenty of fluids would be helpful.

The respiratory complications could be avoided by teaching deep breathing exercises, postural drainage, suctioning, good coughing and huffing techniques. Chest muscle stretching exercises would be beneficial.

Cardiovascular deconditioning is prevented by regular endurance exercises.

To Improve Functional Capacity of the Patient

It is important to improve the muscle strength by exercise program to meet the needs of activities of daily living. In some patients, there may be need of modification of environment and providing the necessary support (orthosis) to meet the ADL demands.

To Improve the Ambulatory Functions

Gait training is initiated in the parallel bars and then progressed to walking on a plane surface. If required, walking aids can be given. Once the patient is comfortable and confident to walk on the even surfaces, the challenge can be increased by walking on the uneven surfaces. Later stair climbing ascending and descending the stairs could be taught.

To Improve the Strength of the Affected Muscles

Grade	Program
Grade 0–1	Passive movements Electric nerve muscle stimulation Cryotherapy in the form of brisk stroking Hacking over the muscle Weight bearing
Grade 1–2	Faradic re-education Suspension therapy Hydrotherapy Cryotherapy in the form of brisk stroking Hacking over the muscle Weight bearing
Grade 2–3	Suspension therapy Re-education board exercises Gravity eliminated exercises
Grade 3–5	Progressive resistance exercises

To Improve Respiratory and Cardiovascular Capacity

Respiratory muscle training with spirometry, assisted treadmill walking, frequent stretching exercise, active and passive mobility exercises, and deep breathing exercises would be beneficial to improve the aerobic capacity of the patients.

The postural hypotension can be prevented by stimulation of the vascular reflex by getting the patient into the erect position very gradually using a tilt table and by using elastic crepe bandage. Aerobic training would be useful to improve the cardiovascular deconditioning.

Interval training: It is repeated bouts of high and low intensity exercise. The length and duration of the high intensity work bout will depend on the needs of the patient.

High intensity training can also be incorporated once the patient is fit for the program.

DIABETIC NEUROPATHY

These are nerve disorders caused because of diabetes. It is usually a polyneuropathy involving peripheral nerves.

Many people with diabetes can over time, develop neural damage throughout the body. It is estimated that about 60–70% of people with diabetes suffer from neuropathies. At any time, people with diabetes can develop nerve problems at any time, but the risk of developing the disease rises with age and longer duration of diabetes. Researches have revealed that the highest rates of neuropathy can occur in people who have diabetes for at least 25 years.

Types of Diabetic Neuropathies

The neuropathy in diabetes can take several forms.

Peripheral Neuropathy

The most common type of diabetic neuropathy. It is a mild chronic symmetrical motor and sensory neuropathy affecting particularly the lower extremities. Severe burning sensation would be the common complaint of the patient. The sensory symptoms can be associated with the tropic lesions of the skin and joints which are usually distal.

What causes Diabetic Neuropathies?

- **Metabolic:** High blood glucose, abnormal body fat composition, low levels of insulin
- Neurovascular factors
- Autoimmune factors
- Mechanical injury to nerves
- Lifestyle factors: Smoking, alcoholism.

Autonomic Neuropathy

This occurs in diabetic patients and may lead to digestion, bowel, bladder, and sexual functional disturbances. Postural hypotension can also occur. There is reduced heart rate due to vagal fibers involvement and also can lead to sudden unexplained deaths.

Diabetic amyotrophy: This type of neuropathy is occasional and can be restricted to the proximal muscles of the lower limb, proximal leg pain and diminished reflexes.

Focal neuropathy: It results in the sudden weakness of one nerve or a group of nerves, causing muscle weakness or pain.

Clinical Manifestations

- Symptoms depend on the type of neuropathy and nerves affected.
- Some cases may be asymptomatic.
- Symptoms can involve sensory, motor and autonomic.
- Numbness, tingling, or pain in the toes, feet, legs, arms and fingers.
- LMN type of muscle weakness leading to the wasting of the muscles.
- Indigestion, nausea and vomiting.
- Diarrhea or constipation.
- Dizziness or faintness.
- Bladder and bowel incontinence or urgency.
- Erectile dysfunction or vaginal dryness
- Malaise.

Peripheral Neuropathy

- Numbness or insensitivity to pain or temperature
- Tingling, burning or pricking sensations
- Sharp pains or cramps which are usually worse on night
- Loss of balance or coordination
- Muscle weakness especially in the distal muscles of lower limbs
- Foot deformities like hammer toes, collapse of mid-foot
- Loss of reflexes
- Blisters and sores

Autonomic Neuropathy

- Shakiness, sweating and palpitations
- Hypoglycemic unawareness
- Postural hypotension
- Abnormal heart rates
- Urinary incontinence
- Sexual dysfunction
- Profuse sweating at night or while eating
- Pupillary reflex may be lost

Proximal Neurotherapy

- Pain in the thighs, hips, buttocks or legs usually on one side of body
- Weakness of the muscles of the lower limbs
- Inability to raise from sitting position
- Length of recovery period varies

Focal Neuropathy

- Inability to focus the eye
- Double vision
- Pain in the eyes
- Paralysis of the muscles of the face (Bell's palsy)
- Pain in the front of thigh, chest, stomach, outside the shin, chest or abdomen
- Entrapment neuropathies

Diagnosis of Diabetic Neuropathies

1. Clinical manifestations
2. **Foot examination:** A comprehensive examination of foot including assessment of skin, muscles, bones, circulation and sensation of the feet.
3. Nerve conduction studies and electromyography

4. Clinical examination to check the variability of heart rate to deep breathing and posture.
5. Ultrasound of bladder may be required.

Medical Management

1. The line of management is to control the blood glucose levels which helps to prevent further nerve drainage.
2. Blood glucose monitoring, diet planning, physical activity and drugs can control the diabetes.
3. Pain relief can be done by NSAIDs.
4. Antidepressants to prevent psychological stress.
5. Antibiotics to prevent UTI.
6. Oral medications are available for erectile dysfunction and vaginal lubricants can be used.
7. Foot care.

Foot Care in Diabetic Patients

1. Clean the feet daily by using warm water and a mild soap. Dry the feet with soft cloth, do not soak the feet
2. Inspect your toes regularly for any cuts, blisters, redness, swelling, calluses
3. Use moisturizers, but avoid in between the toes
4. Cut the nails regularly
5. Always wear the socks and footwear, do not walk on barefoot
6. Look shoes over carefully before putting them on and make sure the shoes are free of tears, sharp edges or objects that might injure the feet
7. Use microcellular insoles to prevent abnormal pressures.
8. Avoid smoking and alcoholism.

REFERENCES

1. Richard AC Hughes. "Clinical review: Peripheral neuropathy". British Medical Journal 2002;324:466.
2. Janet M Torpy, Jennifer L Kincaid, Richard M Glass. "Patient page: Peripheral neuropathy". Journal of the American Medical Association 2010;303(15).
3. "Peripheral neuropathy face sheet". National Institute of Neurological Disorders and Stroke 2012.
4. Dr Sara J Cuccurullo. Physical Medicine and Rehabilitation Board Review (3rd ed). Demos Medical Publishing p 434.
5. McCane Kathryn L, Huether Sue E. Pathophysiology: The Biologic Basis for Disease in Adults and Children. Elsevier Health Sciences, 635.
6. Moloney Elizabeth B, de Winter Fred, Verhaagen Joost. "ALS as a distal axonopathy: molecular mechanisms affecting neuromuscular junction stability in the presymptomatic stages of the disease". Frontiers in Neuroscience 2014.
7. Barohn Richard J, Amato Anthony A: "Pattern-Recognition Approach to Neuropathy and Neuropathy". Neurologic Clinics 2013;3(2):343–361.
8. Mahdi-Rogers, Mohamed, Rajabally, Yusuf A. "Overview of the pathogenesis and treatment of chronic inflammatory demyelinating polyneuropathy with intravenous immunoglobulins". Biologics: Target and Therapy 2010;4:45–49.
9. van den Berg, Bianca, Walgaard Christa, Drenthen Judith, Fokke Christiaan, Jacobs Bart C, van Doorn Pieter A. "Guillain-Barre syndrome: pathogenesis, diagnosis, treatment and prognosis". Nature Reviews Neurology 2014;10(8):469–482.
10. Rinaldi Simon. "Update on Guillain-Barre syndrome". Journal of the Peripheral Nervous System 2013;18(2):99–112.

11. Yuki Nobuhiro, Hartung Hans-Peter. "Guillain-Barre Syndrome". New England Journal of Medicine 2012;366(24):2294–2304.doi:10.1056/NEJMra1114525.
12. Ryan Monique M. "Pediatric Guillain-Barre syndrome". Current Opinion in Pediatrics 2013;25(6): 689–693.
13. Cooper AB, Ferguson ND, Hanly PJ, et al. Long-term follow-up of survivors of acute lung injury: lack of effect of a ventilation strategy to prevent barotrauma. Crit Care Med 1999;27:2616–2621.
14. Mawdsley C, Meyer RF. "Nerve Conduction in Alcoholic Polyneuropathy". Brain, a journal of neurology 1965;88(2):335–336.
15. Aminoff Michael J, Brown William A, Bolton Charles Francis. Neuromuscular function and disease: basic, clinical and electrodiagnostic aspects. Philadelphia: WB Saunders, 2002;1112–1115.
16. Preedy Victor R, Watson Ronald R. Nutrition and alcohol: linking nutrient interaction and dietary intake. Boca Raton: CRC Press 2004;7–13.

17

Pediatric Neurology

LEARNING OBJECTIVES

At the end of this chapter, you will be able to:
- Describe the various clinical manifestations of common pediatric disorders related to nervous system.
- Demonstrate the skills in assessing the children with spina bifida, hydrocephalus, cerebral palsy, muscular dystrophy.
- Design the treatment protocols for various disabilities for these children.
- Identify the prognosis of the disease and redesign the protocols.

SPINA BIFIDA

Spina bifida is a general term for any neural tube defect (also called an NTD) that involves the brain, spinal cord, and/or meninges. Spina bifida occurs when the neural tube (area around the spinal cord) does not close during a baby's development.[3]

Definition

It is a developmental abnormality in the vertebral column resulting in a lack of fusion of vertebral arches as a result the vertebral canal is not closed (Fig. 17.1).

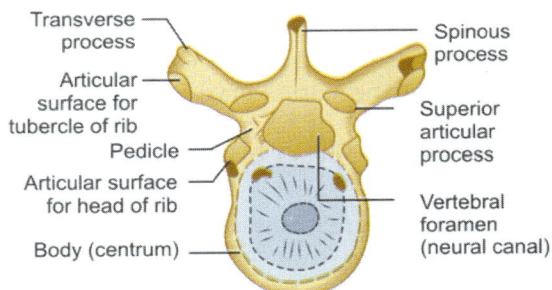

Fig. 17.1: Structure of vertebra

The common regions involved in order of familiarity are thoracolumbar region, lumbosacral, thoracic and cervical.

Classification (Fig. 17.2)

What causes the spina bifida? The exact cause of spina bifida is unknown. The cause for neural tube defect and causing a malformation is not yet known. Scientists suspect that the cause is multifactorial and involves genetic, nutritional and environmental factors.

Many researches have found that insufficient intake of folic acid in the mother's diet is a key factor in causing spina bifida and other neural defects.[1,2]

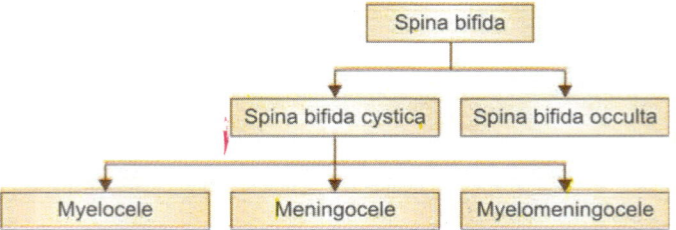

Fig. 17.2: Spina bifida classification

Spina Bifida Occulta

This is the mildest form of defect. The name *'occulta'* which means *"hidden"* indicates the malformation or opening in the spine is covered by a layer of skin. This is present in 10–20% of population and rarely causes disability or symptoms. There is a defect in the fusion of laminal arch. A bond of fibrous tissue between the skin and vertebra causes dimple over the skin. Nervous tissue is not affected. A tuft of hair is present at the depression. It appears like a dimple on the back (Fig. 17.3).

The common features seen in spina bifida occulta are:
- A hairy patch in the lower back
- A fatty lump over the bottom of the spine
- A hemangioma (a reddish or purple spot) on the skin
- A dimple or sinus (hole) above the level of the crease in the buttocks
- A pigmented area or birthmark over the bottom of the spine.
- A small tail

Spina Bifida Cystica

It is a more complicated form of spina bifida. There will be a visible sign of a sac or cyst.

The spina bifida takes different forms.[5]

Meningocele: In this defect, the vertebral arch is not fused. The sac contains meninges and cerebrospinal fluid. Meningocele is the least found type of spina bifida. There may be no signs of neurological deficit. This is commonly found in the lumbar region. The skin over the lesion is usually intact (Fig. 17.4).

Myelomeningocele: Most severe type of spina bifida. There is severe neurological

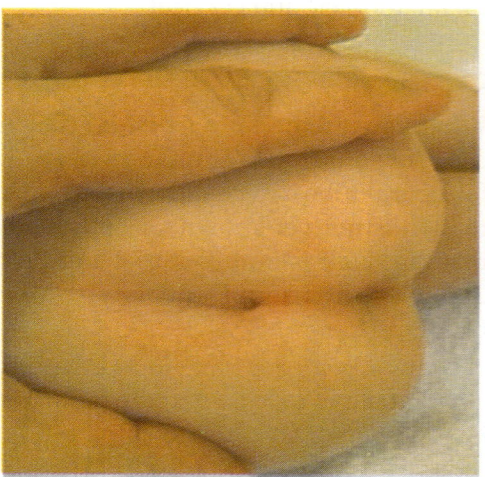 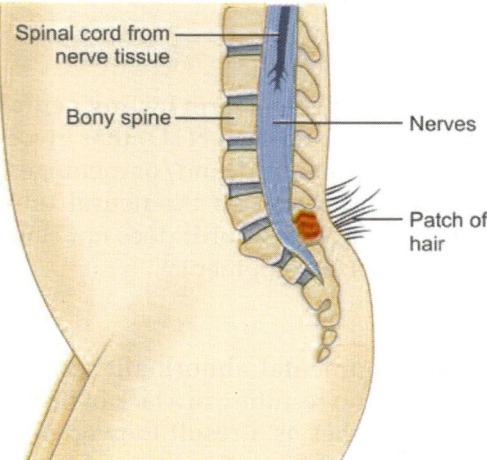

Fig. 17.3: Dimple and tuft of hair in case of spina bifida occulta

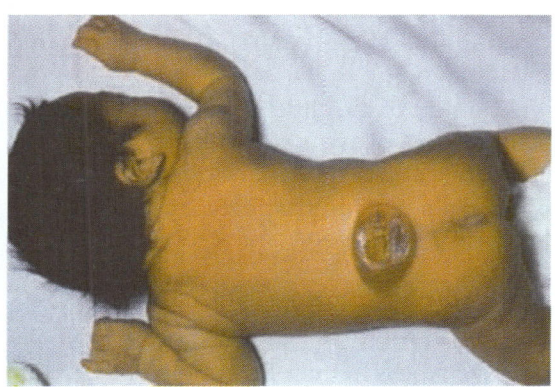

Fig. 17.4: Meningocele

damage. It is exhibited as a sac containing spinal cord. The cauda equina protrudes out. Infection may cause further damage.

Myelocele: The neural plate lies exposed on the surface as a reddish, vascular granulating mass, usually lying in the middle and cranial part of the defect. The lesion may be dry or if the central canal is patent, there may be a continuous leak of CSF.

Clinical Features of Spina Bifida

- Muscle paralysis or weakness depending upon the level of lesion
- Loss of senses below the level of lesion
- Rectal and bladder incontinence
- Mental retardation
- Hydrocephalus associated with spinabifida is called **Arnold-Chiari malformation**

Physiotherapy Assessment

1. Take the demographic data and history from the parents or medical records.
2. **Observation:** Observe for the following:
   a. Tuft of hair, subcutaneous lipoma or dimple over the back of the child
   b. Localized sac in cystic type
   c. Increased head circumference
   d. Deformities of the lower limbs
   e. Deformities of the spine
   f. Skin ulcerations and soft tissue injuries

On palpation: Palpate for bony defects, subcutaneous lipoma, temperature and vital signs.

Examination

- **Sensory assessment:** Assess the superficial, deep and cortical sensations.
- Measurements of head circumference, chest circumference and muscle girths.
- Assessment of primitive reflexes.
- Assessment of milestones.
- Assessment of muscle tone, voluntary control and power.

Management of Spina Bifida

The management of spina bifida is under 3 headings.

Surgical Management[7,8]

- It is necessary to repair the spinal defect and prevent further damage.
- If associated with hydrocephalus, then shunt is done to drain the CSF
- Ureteroileostomy in girls with urinary incontinence (Fig. 17.5).

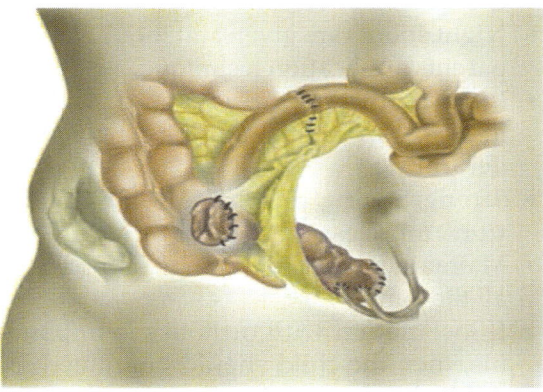

Fig. 17.5: Ureteroileostomy

Medical and Nursing Management

- Counsel the parents and give psychological support.
- Advice braces to avoid deformities.
- Prophylactic antibiotics to prevent infections.
- Sterile dressing for surgical wounds.

PHYSIOTHERAPY AND REHABILITATION

Aims
- To provide psychological support to parents
- To develop good rapport with the child and parents
- To prevent abnormal movement pattern
- To prevent contractures
- To take care of the anesthetized skin
- To improve and maintain functional independence

Plans of Physiotherapy[10-13]
- Give general encouragement and support to the child and family.
- Parents of the child are explained regarding the child's condition, its outcome and coping strategies.
- Proper care and prevention of complications is also explained to them.
- Deformities can be corrected by strapping, splinting, serial plaster casts and passive mobilization techniques.
- Circulation of the limb is constantly checked when strapping.
- Night splints are given and thought to the parents application technique.
- Proper sitting posture is attained with the help of orthoses to prevent spinal deformity.
- Strengthening of arm, lower limb and trunk muscles.
- Maintain normal joint range of motion at all the joints.
- If locomotion with orthotics is impossible, then the child should be trained for wheelchair ambulation
- Intermittent catheterization is initially taught to the parents and later to the child.
- The loop operation where the ureters are diverted through the body wall in special bag makes the child independent of nappies and he becomes socially acceptable.
- Hygiene is maintained at inguinal and anal area to prevent skin infections and damage.
- The skin of the child is well-protected and child is kept away from objects of extreme temperatures and sharp objects.
- Child's skin is inspected daily for any injury.

Upper Limb Exercises
Children with spina bifida needs to compensate weakness in their legs and trunk. Strong arms assist for ADL such as:
- Helping in sitting without trunk stability.
- Using hand-operated mobility aids—wheelchair, crutches.
- Periodically rise from chair to relieve pressure symptoms on skin.
- Transferring from seat to toilet, standing up from chair.
- Extending the arms to lift the seat off the floor.
- Press-up with pillows under the knees and feet.
  Exercises for improving sitting balance:
  - Head control exercises.
  - Strengthening exercises of back muscles and balance exercises (rolling, getting from prone to sit).
  - Special seats to provide adequate support.
  - Regular relieve of the pressure is necessary.

Sensory integration techniques and other principles of management are same as cerebral plasy.

HYDROCEPHALUS

The word **hydrocephalus** is derived from Greek word which literally means water on the brain.

It is a condition in which there is increased cerebrospinal fluid or excessive accumulation of fluid in the brain leading to increase in the head circumference.[14]

Etiology

Congenital/acquired: Hydrocephalus can be congenital or acquired. The causes are usually genetic but can also be acquired and usually occur within first few months of life. The main causes may be:
- Intraventricular matrix hemorrhages which are common in preterm babies.
- Infections.
- Type II Arnold-Chiari malformations.
- Aqueduct atresia and stenosis.
- Dandy-Walker malformations.
- Brain tumors, head injuries or intracranial hemorrhages.

Types of Hydrocephalus

a. **Non-communicating hydrocephalus:** In non-communicating hydrocephalus, the CSF in the ventricles cannot reach the subarachnoid space. This results from obstruction of interventricular foramina, cerebral aqueduct or the outflow foramina of the fourth ventricle.
The most common obstruction is in the cerebral aqueduct. A block at these sites leads to rapid dilatation of one or more ventricles. In young children, the skull is pliable resulting in increased head circumference.

b. **Communicating hydrocephalus:** In communicating hydrocephalus, the obstruction of CSF flow is in the subarachnoid space leading to thickening of the arachnoid matter. This thickening causes blockage of the return-flow channels.

c. **Internal hydrocephalus:** If the foramina of the fourth ventricle or cerebral aqueduct are blocked, CSF accumulates within the ventricles. This is called internal hydrocephalus and it results in increased CSF pressure.

d. **External hydrocephalus:** It is usually benign and is a condition generally seen in infants and involving enlarged fluid spaces or subarachnoid spaces around the outside of the brain.

e. **Normal pressure hydrocephalus (NPH):** This form of hydrocephalus usually begins slowly and is more common in adults over the age of 60. One of the easiest signs is falling suddenly without losing consciousnesss. Common symptoms include changes in the gait, incontinence of bowel and bladder, and headaches.

Pathophysiology

Hydrocephalus is usually due to blockage of cerebrospinal fluid outflow in the ventricles or in the subarachnoid space over the brain.

Normally, CSF continuously circulates through the brain, its ventricles and spinal cord and is continuously drained thereafter.

There may be increase in the production of the CSF resulting in hydrocephalus.

The accumulating fluid can compress the brain tissue resulting in neurological dysfunction.

The signs of neurological dysfunction will occur soon in adults, whose skulls are no longer able to expand.

In infants and young children, there will be enlarged head excluding the face because the pressure of fluid causes the individual bones which are not fused to bulge outward at their juncture points.

Clinical Manifestations

In infants with hydrocephalus, CSF builds pressure in the central nervous system, causing the fontanelle to bulge and head circumference increases. The symptoms shall be:

a. In infants, there will be fixed downward gaze with the sclera (white eye) showing the above iris, as though the infant is trying to look its own lower eyelids.
b. Irritability and seizures
c. Separated suture joints of the skull bones
d. Drowsy and sleepiness
e. Nausea and vomiting

f. Brief, shrill, high pitched cry
g. Uncontrolled eye movements and squints
h. Altered feeding behavior
i. Headaches and muscle spasm
j. Delayed milestones
k. Bowel and bladder incontinence.

Investigations (Fig. 17.6)

Magnetic resonance imaging:[15] MRI scans can be used to look for the signs of excess CSF.

Computerized tomography (CT) scans can be used to look for the signs of excess CSF and enlarged ventricles.

CSF tests to predict shunt responsiveness and/or determine shunt pressure.

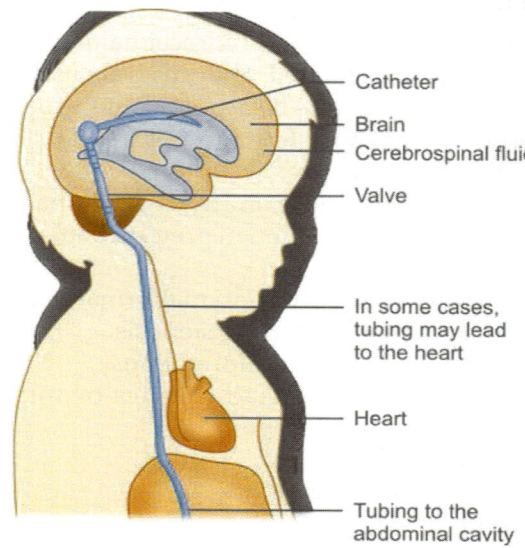

Fig. 17.7: Shunt system

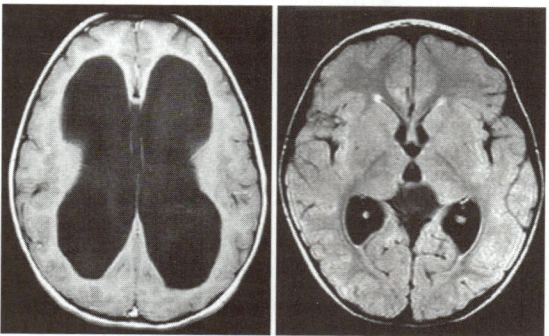

Fig. 17.6: Radiological manifestations of hydrocephalus

Management of Hydrocephalus

There is no known way to prevent or cure hydrocephalus and the only treatment option today requires brain surgery. With early diagnosis and interventions, there will be good prognosis.

The most common surgical management of hydrocephalus is shunt system[16,17] (Fig. 17.7).

In shunt operations, the flow of CSF is diverted from a site of blockage within the CNS to another area of the body where it can be absorbed as part of circulatory process.

A shunt is a flexible but sturdy silastic tube. A shunt system consists of a shunt, a catheter and a valve (Fig. 17.8). One end of the catheter is placed in the CNS (usually within a ventricle of brain), or a cyst or in a site close to the spinal cord. The other end of the catheter is commonly placed within the abdominal cavity or a chamber of the heart or a cavity in the lung where the CSF can drain and be absorbed. A valve located along the catheter maintains one-way flow and regulates the rate of CSF.

Third ventriculostomy:[18] It is rarely done, where a small hole is made in the floor of the third ventricle by using fiberoptic technology—neuroendoscope—allowing the CSF to bypass the obstruction and flow toward the site of reabsorption around the surface of the brain.

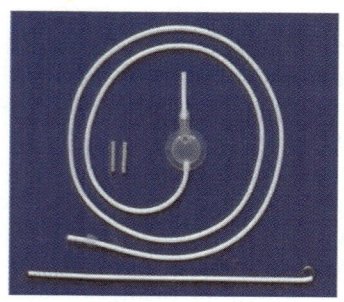

Fig. 17.8: Apparatus for shunt system

S. no.	Shunt type	Location of fluid drain
1.	Ventriculoperitoneal shunt (VP shunt)	Peritoneal cavity
2.	Ventriculoatrial shunt (VA shunt)	Right atrium of the heart
3.	Ventriculopleural shunt (VPL shunt)	Pleural cavity
4.	Ventriculocisternal shunt (VC shunt)	Cistern magna
5.	Ventriculosubgaleal shunt (SG shunt)	Subgaleal space (it is a temporary measure used in infants where a pocket beneath the epicranial aponeurosis is made and CSF is drained)
6.	Lumbar-peritoneal shunt (LP shunt)	Peritoneal cavity

Choroid plexus cauterization: Along with third ventriculostomy, the surgeon uses a device to burn or cauterize tissue from the choroid plexus, thus reducing the rate of production of CSF.

Complications of Shunt Operations

1. Shunt malfunction which is a partial or complete blockage of the shunt.
2. Shunt infections (*Staphylococcus epidermidis*): The infections are caused by child's own bacterial organisms and is not acquired.
3. Over drainage leading to increase in the size of the ventricles.
4. Subdural hematoma.
5. Multiloculated hydrocephalus.
6. Abdominal complications in VP shunts.

Complications of Other Surgical Procedures

1. Life-threatening situations due to sudden closure of the pathways created by third ventriculostomy.
2. Infections.
3. Fever and bleeding.
4. Short-term memory loss.

Physiotherapy Management of Hydrocephalus

Excessive pressure on the brain can result in physical problems in children and may affect their physical development, achievement of milestones, balance, coordination or mobility.

Aims of Physiotherapy

1. To develop good rapport with the child and parents
2. To promote achievement of milestones
3. To improve circulation
4. To prevent contractures and deformities
5. To improve balance and coordination
6. To improve functional capabilities of the child
7. To improve confidence and quality of life

For plans of physiotherapy, please refer to *cerebral palsy*.

CEREBRAL PALSY

Introduction

Cerebral palsy[19,20] is a term used to describe a broad-spectrum of motor disability which is caused by non-progressive damage to developing brain at or around birth. It is a common developmental disability first described by William Little in the 1840s. It is one of the three most common lifelong developmental disabilities, the other two being autism and mental retardation causing considerable hardship to the affected individuals and their families.

Cerebral palsy can be described as, "umbrella term covering a group of non-progressive, but often changing, motor impairment syndromes secondary to lesions or anomalies of the brain arising in the early stages of its development" (Mutch et al, 1992).[20]

Definition

- It is a neuromotar disorder resulting from non-progressive damage to the developing brain.
- It is a persistant but not unchanging disorder of movement and posture.

Etiology

Figure 17.9 depicts causes of cerebral palsy.

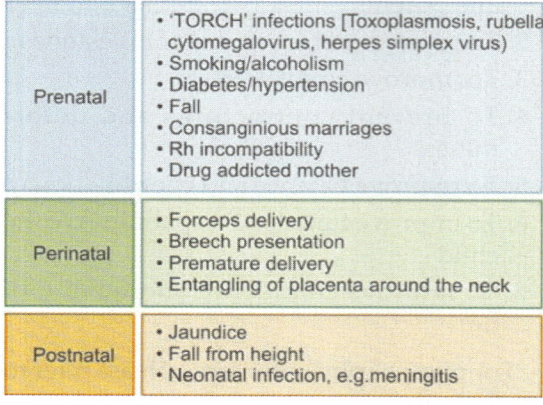

Fig. 17.9: Causes of cerebral palsy

Incidence

CP is a common problem, the worldwide incidence being 2–2.5 per 1000 live births.[21] 75–80% of the cases are due to prenatal injury, less than 10% being due to significant birth trauma or asphyxia. The most important risk factor seems to be prematurity and low birth weight.[22]

Classification of Cerebral Palsy

Clinical Classification

1. Spastic
2. Athetoid
3. Ataxic
4. Flaccid
5. Mixed

Topographical Classification (Fig. 17.10)

1. **Quadriplegic:** All four limbs are involved.
2. **Hemiplegic:** One-half of the body is involved.
3. **Paraplegic:** Both the the lower limbs are involved.
4. **Monoplegic:** Single limb is involved.
5. **Diplegic:** All four limbs are involved, but lower limbs are more involved than upper limbs.

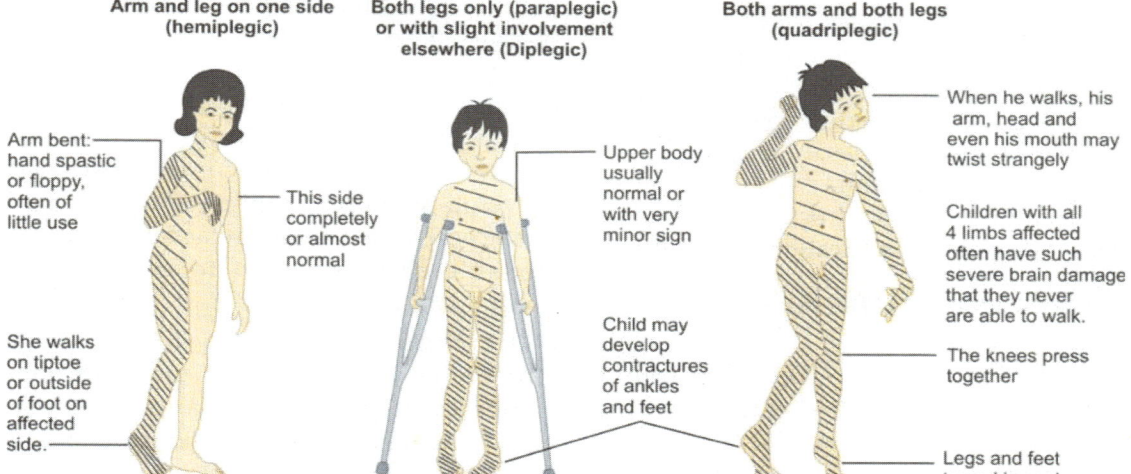

Fig. 17.10: Topographical types of cerebral palsy

Types of Cerebral Palsy
(Figs 17.11 and 17.12)

Spastic Cerebral Palsy (Fig. 17.13)

This is the most common type of cerebral palsy. It is characterized by increased tone in the muscles.

The kids with spastic type of cerebral palsy are known as hypertonic kids.

The spasticity may be present in all the limbs or half of the body.

These kids have fear of fall so should never be treated on beds or couches.

Increased tone results in faulty postures resulting in contractures and deformities.

The movements in these kids are slow and awkward. A change in the position of the head triggers abnormal positions of the whole body.

These kids are more irritable and spasticity increases when a child is upset or excited.

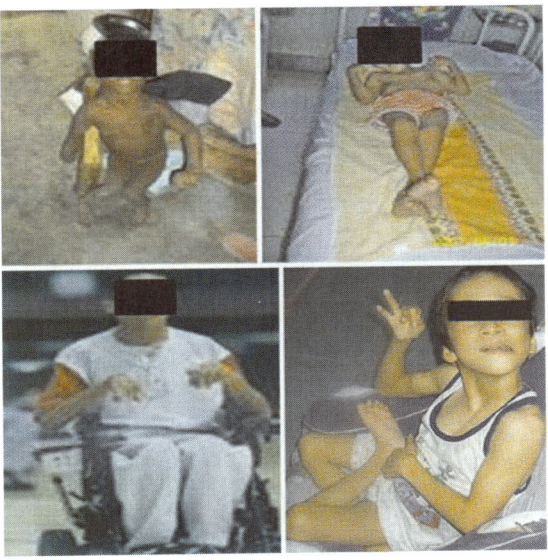

Fig. 17.13: Spastic cerebral palsy

These kids usually suffer from mental retardation and cognitive problems, difficulty in motor learning and delayed milestones.

Athetoid Cerebral Palsy (Fig. 17.14)

There will be irrhthymical, irregular, jerky, purposeless, involuntary writhing movements.

These movements are present at rest, increases on activity and decreases on fatigue.

These kids may have normal IQ and usually go to normal school.

When the kid moves, body parts move too fast and too far. The balance is very poor and falls frequently.

The voluntary movements are incoordinated. A few kids suffer from mood disorders and auditory problems.

Ataxic Cerebral Palsy (Fig. 17.15)

These kids have ataxia and have poor balance and coordination. This is because of the damage to the cerebellum and its pathways.

Signs of hypotonia, intentional tremors, dysarthria and sometimes nystagmus is seen.

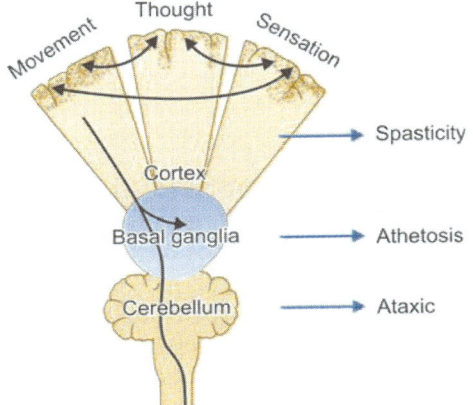

Fig. 17.11: Level of lesions—types of cerebral palsy

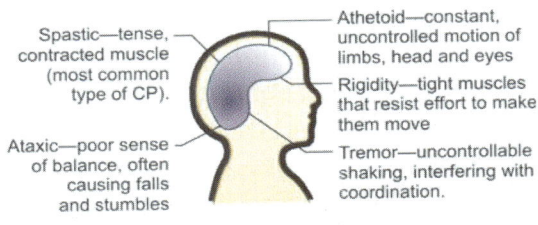

Fig. 17.12: Types of cerebral palsy

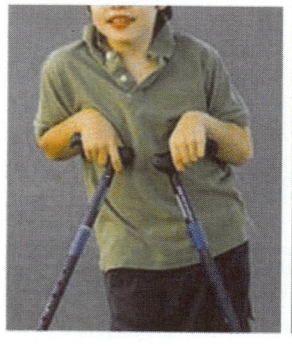 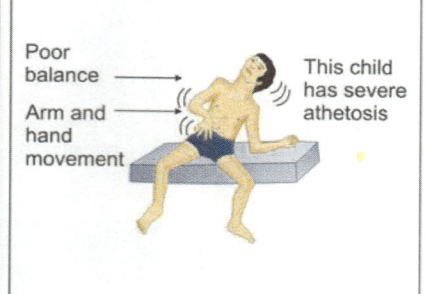

Fig. 17.14: Athetoid CP

 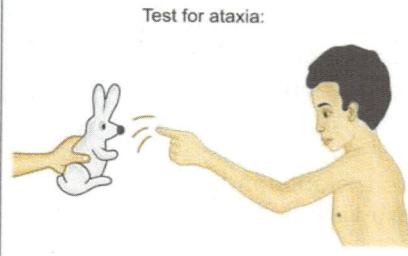 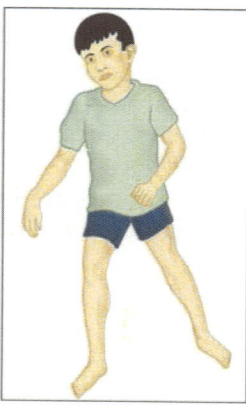

Fig. 17.15: Ataxic cerebral palsy

The voluntary movements will be clumsy and mental retardation is common.

Proximal fixation of head, shoulder girdle, trunk and pelvic girdle is poor. Lack of balance is compensated by excessive balance saving reaction of the upper limbs. A wide based gait is seen due to imbalance and fear of fall.

These kids will also have poor IQ levels, with visual and auditory defects.

Flaccid Cerebral Palsy

These kids are also called **floppy kids**. The kids are having low muscle tone [hypotonic] and are usually mentally retarded.

Joint subluxations are common due to decreased muscle tone. The reflexes are poor and arms, legs hang down like a "Rag doll". These kids will have difficulty in swallowing and breathing. Speech is difficult and intelligence may be unaffected.

Mixed Type of Cerebral Palsy

It is a developmental disorder with combination of movement problems. It has the traits of spastic, athetoid and ataxic cerebral palsies.

Children with this condition may have issues with movement including spasticity, involuntary movements, imbalance and lack of co-ordination.

Common Clinical Features of Cerebral Palsy

- Abnormal tone
- Abnormal reflexes
- Disturbed higher functions
- Sensory disturbances

- Delayed milestones
- Contractures and deformities
- Respiratory and oromotor dysfunction
- Dysmorphic features:
    a. Low set eyes and ears
    b. Frontal bossing
    c. Delayed closure of anterior fontanelle
    d. Cleft lip/cleft palate
    e. Excessive drooling of saliva
    f. Irregular dentition
- Mental retardation

Multiple Associated Deficits

- Mental retardation
- Convulsions
- Visual deficits
- Hearing defects
- Perceptual problems
- Learning disabilities
- Feeding problem
- Emotional and behavioral problems
- Speech and language disorders

Early Intervention of Cerebral Palsy

a. It is always a known fact that 'early intervention—better prognosis'.
b. As the age at which diagnosis is made goes on increasing, secondary complications of developmental delay come into picture.
c. Therefore, the CP child should receive therapeutic intervention as early as possible.
d. The earliest intervention is immediately after birth.
e. The neonate is seen by the therapist earliest in NICU where baby is admitted for medical complications.
f. When neonate is referred to physiotherapist before starting the therapeutic intervention, assessment of the infant has to be carried out.

Assessment of Cerebral Palsy

Assessment starts with history. Detailed history of prenatal/perinatal and postnatal risk factors has to be obtained either from mother or from medical records.

Apgar Scoring[23,24]

- It is a quantitative method for assessing[23,24] infants respiratory, circulatory and neurological status immediately after the birth.
- Timing: 1 min, 5 min, 10–20 min after the birth.

As in newborn, extremities are always blue immediately after birth, ideal score is never 10 at 1 min but 9.

Score	Effect
8–10	Normal
5–7	Moderate asphyxia
Less than 4	Severe distress

Illingworth Scale (Table 17.1)

- Along with birth asphyxia, preterm babies also form a major group in cerebral palsy children.
- Therefore, a preterm infant should be identified from normal term infant.
- Illingworth scale differentiates a preterm baby [risk baby] from fullterm baby.

Other Factors Regarding General Condition of the Baby

S. No.	Factor	Value
1.	Height of the baby	50 cm
2.	Head circumference	34–35 cm
3.	Chest circumference	Usually 3–4 cm less than head circumference
4.	Respiratory status	30–40/min
5.	Heart rate	120–140 beats/min
6.	Birth weight	2.5–3.5 kg

Table 17.1: Illingworth scale

	Factor	Preterm	Fullterm
1.	Sleep	Disturbed small sleep cycles	Sound sleep
2.	Movements	Faster/bizzare/uncoordinated	Co-ordinated
3.	Cry	Cry is infrequent/feeble/not prolonged	Prolonged vigorous cry
4.	Feeding behavior	Cannot relied upon to demand feeds may be unable to suck and swallow regurgitation—cyanotic attacks	Can be relied upon for feeds rooting/sucking/swallowing—normal
5.	Muscle tone	Less flexor tone	Good flexor tone
6.	Posture of baby	Prone: Flat pelvis and knees at the side of abdomen Acute flexion at hips Supine: Lower limbs externally rotated and abducted Head turned to side	Prone: Pelvis high knees drawn up under abdomen Supine: Limbs are strongly flexed. Head aligned to trunk
7.	Head rotation	Head can be rotated so far that chin is well beyond acromion	Chin can be rotated only as far as acromion
8.	Scarf sign	Hand reaches beyond opposite acromion	Hand does not go beyond opposite acromion
9.	Wrist flexion	Wrist flexion is incomplete There is a window between hand and forearm	Complete wrist flexion. No gap between palm and forearm
10.	Grasp	Less than 28 weeks, it is weak	Strong palmar grasp
11.	Knee extension	When hip is flexed completely knee can be fully extended	After complete hip flexion, knee extension is short of 20°
12.	Dorsiflexion of foot	Dorsiflexion of foot is incomplete	Complete dorsiflexion such that the dorsum of foot touches shin of tibia
13.	Automatic walking	28 weeks: Feeble 32 weeks: Walks on toes 40 weeks: Walks with foot flat	Normal walk
14.	Horizontal suspension	Hangs limply no flexion of limbs	Flexes upper and lower limbs strongly

Vojta's Reactions (Fig. 17.16)

- These are useful for diagnosis of brain damage in infants.
- Dr Vojta, a German pediatric neurologist, standardized 7 postural reflexes along with neurological and behavioral assessment technique to diagnose the development of cerebral palsy in the neonate.[25,26]

The 7 reactions are as follows:
1. Traction
2. Landau
3. Axillary suspension
4. Vojta's side tilt reaction
5. Colli's horizontal suspension reaction
6. Pieper and Isbert's reaction
7. Colli's vertical suspension reaction

S. No.	Factor	Score = 0	Score = 1	Score = 2
1.	Heart rate	Absent	Less than 100 beats/min	More than 100 beats/min
2.	Respiratory effort	Absent	Slow, irregular cry	Good cry
3.	Muscle tone	Limp	Some flexion in extremities	Active good flexor tone
4.	Response to catheter	No response	Grimace	Cough/sneeze
5.	Color of baby	Blue/pale	Body pink and extremities blue	Complementary pink

Apgar score immediately after birth

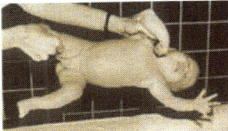
Colli's horizontal suspension

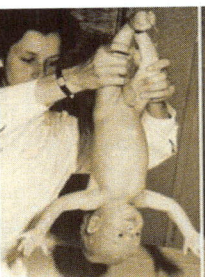
Pieper and Isberth reaction

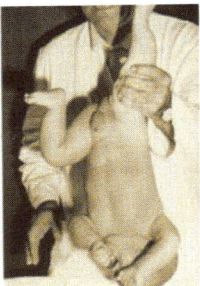
Colli's vertical suspension

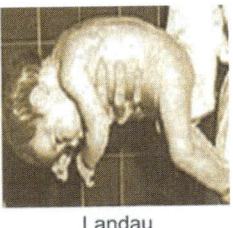
Landau

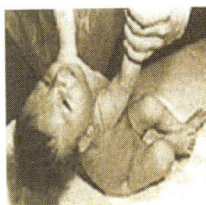
Traction response

Axillary suspension

Fig. 17.16: Vojta's reactions

These reactions develop which are dependent on the age of infant from 0 to 12 months.

Abnormal postural reactions indicate *disturbed central coordination* (DCC). The development of cerebral palsy depends upon the severity of DCC. It is scaled as follows:
- Mild DCC: 3 or less than 3 abnormal reactions
- Moderate DCC: 4–5 abnormal reactions
- Severe DCC: 6–7 abnormal reactions

At birth (0 months), the child will show 7 reactions as shown in Table 17.2.

Reflex Examination

A reflex is a stereotypic response to stimulus. Reflex examination is used for early intervention, level of functional independence identification, and to plan the treatment. Primitive reflexes should be lost before the corresponding voluntary control is achieved.

The various reflexes to be checked in risk babies are listed in Table 17.3.

Cortical Reactions

Equilibrium: It is tested on equilibrium board in all the functional positions or by pushing the baby from static posture.

Equilibrium	Age attained
Prone	6 months
Supine	8 months
Quadriped	8–10 months
Sitting	8–10 months
Kneeling	15 months
Standing	15–18 months

Assessment of Milestones

The therapist should record the milestones and the age they have attained by interviewing the parents.

In order to assess the milestones, the therapist should have keen knowledge on normal development and the age at which normally a milestone is attained.

Fetal growth extends from 0 to 12 weeks during which organogenesis takes place. First trimester consists of:
- Hazard of faulty implantation
- Embryonic defects
- Maternal infections

Second trimester is relatively safe and the fetus gains about 70% of height and 20% of birth weight.

Third trimester consists of hazards such as:
- Hazards of abortion and infection
- Rapid weight gain

Table 17.2: Seven reactions of child at birth

S. no.	Reaction	Elicitation and body part to be observed	Normal response
1.	Traction	Infant is slowly pulled up from supine to an angle of 45° head and lower limbs are observed	Complete head lag, but head does not fall on one side. Head remains in center lower limbs in mild flexion
2.	Landau	Prone infant is held in horizontal suspension, head, spine, upper and lower limbs are observed	Head hangs in center, spine, upper and lower limbs are in flexion
3.	Axillary suspension	Infant is lifted in vertical suspension holding just below the axilla, lower limbs are to be observed	Mildly flexed
4.	Vojta's side tilt	Vertically held infant suddenly tilted to lateral horizontal position, overlying upper and lower limbs are to be observed	Overlying upper extremity Moro-response, lower limb flexed
5.	Colli's horizontal suspension	Infant is suddenly suspended by ipsilateral upper and lower limbs free upper and lower limbs are to be observed	Free upper limb Moro-response, free lower limb flexion
6.	Pieper and Isbert's vertical suspension	The infant is held by its thighs and lifted suddenly head down in vertical position, head, spine and upper limb are observed	Head hangs in the center Upper limb–Moro-response No response in spine
7.	Colli's vertical suspension	Infant is lifted up with one thigh, head down, free lower limb to be observed	Flexion of lower limb

Principles of Normal Development

1. Development is a continuous process, rate of development in each field is different though sequence is same.
2. Development is related to maturation of nervous system which is cephalocaudal in direction and proximal to distal.
3. General mass activity is replaced by specific individual response.
4. Primitive reflexes should be lost before the corresponding voluntary control is achieved.
5. At first, brainstem, thalamus and basal ganglia are dominant in development. The rapid growth of cerebral cortex and cerebellum taking place later.
6. The blood supply during development from subependymal plate to the cerebral cortex. During third trimester, circulation shifts from a central to cortical and white matter orientation.
7. Early development is directed towards decrease of flexor tone.

Note: In fullterm newborn baby, there is predominance of extensor tone in neck and flexor tone in limbs.

Prone Development

Attitude at birth: Reflexes present are:
- Neonatal and spinal
- Positive supporting
- Neck righting

At 1–3 months:
- Reflexes present are labyrinthine righting reflexes

At 3 months:
1. Reflexes:
   a. ATNR decreased
   b. STNR increased
2. Galant's reflex decreases.

At 6 months:
1. Palmar grasp integrates.
2. Chest off the supporting surface.
3. Weight-bearing on open hands.
4. Visual field improved.

Table 17.3: Various reflexes to be checked in risk babies

S. no.	Reflex	Age of normal	Stimulus	Response
Neonatal reflexes				
1.	Doll's eye reflex	Birth–10 days	Baby head is turned to one side	Eyes lag behind
2.	Rooting reflex	Birth–3–4 months	Light touch around lips	Turning of head, lower lip and tongue on the side of stimulus
3.	Sucking reflex and swallowing	Birth–3–4 months	Place a finger on baby's lips	Sucking movement of lips and swallows
4.	Palmar grasp reflex	Birth–4 months	Pressure on palm of hand from ulnar side	Finger flexion with strong grip that persists and resists removal of stimulus
5.	Plantar grasp	Birth–10–11 months	Strong pressure on ball foot	Flexion of toes
6.	Placing of upper extremity	Birth–6 months	Brush the dorsum of one of baby's hand against edge of the table	Flexion of upper limb with placement of hand on the table
7.	Placing of lower extremity	Birth–1½ months	Brush the dorsum of the foot against the under edge of the table	Flexion of the lower limb with placement of foot on the table top
8.	Moro's reflex	Birth–3–4 months	Dropping the baby head backwards from semi-sitting position	Abduction, external rotation, extension of arms and extension of fingers followed by adduction of arm to midline
9.	Automatic standing and walking	Birth–1½ months	Place the baby in the vertical suspension near to supporting surface and touch the feet to the ground	Extension of lower limbs as if baby is standing. If pelvis is rotated forwards, then child will automatically put steps forward
10.	Gallant's reflex	Birth–3–6 months	In horizontal suspension, stroke unilateral lumbar region with blunt object	Lateral flexion of trunk on the same side
Spinal level reflexes				
1.	Flexor withdrawal	Birth–2 months	Quick tactile stimulus applied to the sole of the foot	Uncontrolled flexion of hip and knee
2.	Extensor thrust	Birth–2 months	One leg in extension and other fully flexion apply pressure on the ball of the foot of flexed leg	Uncontrolled extension of same leg
3.	Crossed extensor	Birth–2 months	One leg in flexion and other in extension. Give pressure on the ball of the foot of extended leg without allowing flexion of the same leg	The flexed leg extends
Brainstem reflexes				
1.	Asymmetric tonic neck reflex [ATNR]	Birth–4 months	Passively turn the head 90°	Increase in the extensor tone on face side and increase in flexor tone of limbs on occipital side

(Contd.)

Table 17.3: Various reflexes to be checked in risk babies (Contd.)

S. no.	Reflex	Age of normal	Stimulus	Response
2.	Symmetrical tonic neck reflex [STNR]	Birth to 4–5 months	Sti1: Flex the child head bringing his chin towards chest Sti2: Extension of baby's head	Res1: Flexion of upper extremities and extension of lower extremities Res2: Extension of upper extremities and flexion of lower extremities
3.	Tonic labyrinthine reflex	Birth to 3–4 months	Patient in supine and prone position	Increase in flexor tone in prone position and extensor tone in supine
4.	Positive supporting reactions	Birth to 6 months	Patient upright standing, firm contact on ball of foot to floor	Rigid extension of lower limbs resulting from co-contraction of flexors and extensors
Midbrain reaction				
1.	Neck righting reflex	Birth to 4–6 months	In supine position, turn the baby's head to one side and hold it in that position	Body rotates on the same side as a whole [log rolling]
2.	Labyrinthine righting	2 months–lifelong	Baby is blind folded suspended in space by holding at pelvis the baby is tipped sideways so that head is laterally flexed	Head brought into horizontal position
3.	Body righting on head	6 months–5 years	Baby is blind folded and first placed in supine then in prone	The head is brought back to vertical position
4.	Body on body righting	6 months to 4–5 years	Baby in supine, passively turn the head to one side	Segmental rolling on turned side
5.	Parachute reaction	6 months–lifelong	Baby is held in prone suspension at pelvis, push baby to the side with sufficient surprise and force that he/she believes his head will contact the supporting surface	Extension of all the four limbs

At 7–10 months:
1. Learns equilibrium and weight shifting
2. Walks sideways with support

At 12 months:
1. Confident to walk forward
2. Keeps hands out in abduction
3. Uses broad 'base of support' (BOS) for walking
4. A few steps alone possible. Walks if held by hands

Supine Development

At birth:
1. Doll's eye reflex is positive.
2. Pelvis off the support.
3. Rooting, sucking, swallowing reflexes are positive.
4. Weight-bearing on head and upper trunk.

At 3rd month:
1. Trunk flat on support.
2. Hands come to midline.
3. Eye-hand regard positive.

At 4th–5th month:
1. Isolated arm movements
   Able to take hands across midline.
2. The baby exhibits segmental rolling

At 6th month:
1. Child becomes aware of his body parts.
2. Good hip flexion and extension.
3. Transfer of objects from hand-to-hand, palmar grasp integrates.
4. Parachute reactions are absent.

At 7th month:
1. Eye-hand-foot co-ordination.
2. Developments of parachute reactions.
   a. Forward: 7 months
   b. Sideways: 8 months
   c. Backwards: 9 months.

Advantages and disadvantages of supine development:

Advantages:
1. For neck flexion and stability—static posture
2. Hands free for play—no mobility.
3. Development of eye-hand coordination.

Disadvantages:
1. Static posture
2. No mobility.

Development of Hand Function

0–8 months: Initially hands are opening from fisted position and reflex present is palmar grasp which will be getting weaker.

3–5 months: Child brings hands to midline and grasps objects placed in his hands.

5–7 months: Child is now sitting up and develops normal postural reactions.

Hands are free for activity. Bilateral activity of hands develops.

7–9 months: Child transfers objects from one hand to another. His/her radial grasp is developing. When a block of one inch is given, he is not able to release it normally but releases by pressing down palm on hard surface.

9–12 months: Individual finger movements develop and there will be development of crude to fine pinch develop, development of pronation and supination release and gross opening is learnt by throwing and looking for fallen objects.

12–15 months: Since he learned release, built a tower of two blocks.

12–18 months: Delicate pincer grasp develop. He can only take off shoes and socks.

2 years: Fine movements are increasing.

3 years: Able to take off clothes, feed himself and wash alone.

4 years: Able to dress alone except for putting buttons and laces.

5 Years: Dress and undress independently.

Development of Vision

At birth by 3 weeks, focus on mother's face, Doll's eye view fixing within 8–10 inches.

1 month: Range of vision, midline to 45°.
2 months: Midline—90°
3 months: Midline—180°

Development of Speech

Initially cooing and gurgling is present. After 3 months, if it stops, then it can indicate some hearing defect.

28 weeks: Starts speaking monosyllables.
44 weeks: Uses one word—dog, cat
1 year: Makes sentences of 2–3 words without pronouns, i.e. give toy, give box.
2 years: Uses pronouns—give me box.

Note: If a child does not speak monosyllables till 2 years needs investigations.

Aims:
- To enable the baby to use his/her potential to maximum extent
- To enable the baby to have some kind of locomotion and interact with environment

- To enable him to have some kind of communication.

These aims cannot be achieved by physiotherapists alone. Treatment of cerebral palsy involves team work.

Team work includes the following professionals:
- Physiotherapist
- Occupational therapist
- Audio-speech therapist
- Orthotist
- Neurologist
- Orthopedic surgeon
- Clinical psychologist
- Special teacher
- Social worker
- Mother of the baby

Aims of Physiotherapy

- Developing rapport with parents and baby
- Management of abnormal tone
- Maintaining the length of muscle
- Developing postural reactions and improve balance and coordination
- Sensory integration
- Training the respiratory and oromotor functions
- Training for functional independence

There are various approaches to rehabilitate the cerebral palsy kid. All these techniques or approaches are showed different insight into the conditions. A combination of these approaches would be beneficial in rehabilitating the cerebral palsy kid.

Neurofacilitation Approaches for Cerebral Palsy Rehabilitation

Vojta Method of Therapy

Vojta established 18 points in the body for stimulation and used the positions of reflex crawling and reflex rolling. He proposed that placing the child in these positions and stimulation of key points in the body would enhance CNS development. In this way, the child is presumed to learn normal movement patterns in place of abnormal motion.

According to Vojta, reflex locomotion is activated from three main positions—prone, supine and sidelying.

Two coordination complexes in reflex locomotion: In the practical use of reflex locomotion, there are two coordinated complexes:

1. Reflex creeping
2. Reflex rolling

The movement sequences of reflex locomotion are retrievable at all times.

The three main positions—prone, supine and side lying and have more than 30 variations.

By combining and varying stimulation zones and resistances, as well as making changes in directions of pressure and joint angles in the starting position, therapy can be adapted to the patient's individual treatment goal and condition.

1. Reflex Creeping (Fig. 17.17)

Reflex creeping is a movement sequence that include the most fundamental components of locomotion:
- Specific postural control

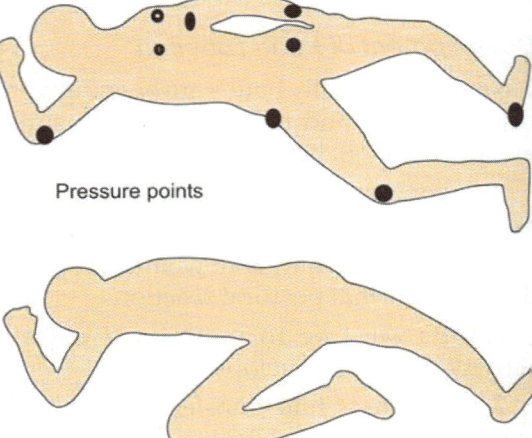

Fig. 17.17: Reflex creeping

- Upright posture or extension against gravity

Goal: Directed stepping movements of arms and legs.

The main position is prone lying with the head resting on the bed rotated to one side.

In the newborn babies, reflex creeping can be fully activated from one zone; in adults, a combination of several pressure points is necessary.

Movement predominantly ensures in so-called cross-pattern, in which the right leg and left arm, or vice versa, move stimultaneously. A leg and its contralateral arm support the body and move the trunk forwards.

2. Reflex rolling (Fig. 17.18)

Reflex rolling transitions from supine to side lying and leads to crawling.

Therapeutically, the reflex rolling is used in different phases of supine and side lying.

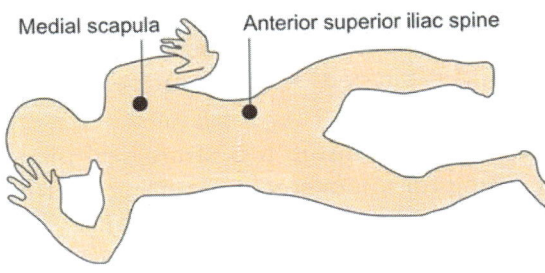

Fig. 17.18: Reflex rolling

First phase: The first phase starts in supine lying with upper and lower limbs extended.

Now stimulate the breast zone in the intercostals space beneath the nipple on the mammillary line, rotation to the side is achieved.

The rotation of the head is resisted by the therapist.

The fundamental reactions are:

1. Extension of spine and flexion of hip, knee and ankle joints.
2. Maintenance of lower limbs in this position against gravity.
3. Preparation of upper limbs for the support function.
4. Lateral eye movements and initiation of swallowing.
5. Increase in depth of breathing.
6. Coordinated activation of abdominal muscles.

Second phase: As now the child has obtained side lying, the second phase shall start from side lying.

The body is supported by the underlying upper and lower limbs which move upward and forward against gravity.

Fundamental reactions:

- Contrary flexion and extension movements of the over and underlying arms and legs with increase in support function on the underlying shoulder progressing to the hand and underlying pelvis progression to leg.
- Extension of the spine during the entire rolling sequence.
- Maintenance of the head in side lying.

Bobath Concept

It is an approach to neurological rehabilitation based on normal movement or neuro-developmental approach. The main principle is to promote motor learning for efficient motor control in various environments, thereby improving participation and function. The Bobath approach mainly concentrates to prevent the synergic patterns and facilitates normal movement. It also promotes to learn how to control postures and movements.

The main strategies of this concept are: *Therapeutic handling* whereby the patient's movements are influenced by facilitation and inhibition techniques. *Facilitation* techniques promote motor learning by using sensory information to reinforce weak movement patterns and discourage overactive patterns. *Inhibition* techniques reduce the abnormal influences on movement or posture that interfere the normal pattern of movement. *Key points of control* generally refer to parts of the body that are advantageous when facilitating or inhibiting the movements and postures.

Brunnstrom Approach

The Brunnstrom approach is based on using the reflexes that represent normal stages of development, and be used in functional rehabilitation.

Brunnstrom approach principles:
1. Reflexes should be used to elicit movement when there is no movement (normal developmental sequence).
2. Proprioceptive and extroceptive stimuli also can be used therapeutically to evoke desired movement or tone changes.

Peto Approach

The concept of conductive education (CE) is an educational system, based on the work of Hungarian professor Andras Peto. This approach has six elements—group, facilitation, daily routine, rhythmic intention, task series and conductor. The patient is encouraged to verbalize the activites as they perform them and focuses on function.

Johnstone Approach

Margatet Johnstone pionerred the use of air splints in active training of the hemiplegic limb in the severely impaired stroke patients. The treatment is based on reflex inhibition with special attention to inhibiting the tonic neck reflexes through use of air splints and positioning. The optimal position for the affected upper extremity in the air splint at 40 mm Hg pressure inflated by mouth is shoulder external rotation, elbow, wrist and fingers in extension, forearm in supination and thumb in abduction. The lower limb is positioned in protracted pelvis, hip in internal rotation, hip, ankle and knee in flexion.

Motor Relearning Program

This was first described by Carr-Shepherd in the year 1987. The training is based on an understanding of kinematics and kinetics of normal movement, motor control process and motor learning. The major factors in the learning or relearning process identified by Carr and Shepherd are:
- .Identification of a goal
- Inhibition of unnecessary activity
- Ability to cope with the effects of gravity and, therefore, to make balance adjustments while shifting weight
- Appropriate body alignment
- Practice
- Motivation
- Feedback

The motor relearning is aimed at gaining functional independence through learning a specific task-oriented movement. The motor control strategies have five components:

i. **Motor program:** It is asset of preset sequences to get a coordinated voluntary movement by muscle activation and is carried out without peripheral feedback.

ii. **Motor planning:** It is strategical planning for a movement requiring coordination of various motor programs.

iii. **Feedback:** It is the stimulation of control centers in brain by sending the information from peripheral receptors regarding the accuracy of movement and adaptations necessary.

iv. **Feed forward mechanism:** It is the strategy of the musculoskeletal system to adapt or respond in anticipation of a movement or changes of an ongoing movement.

v. **Motor skill acquisition:** It is the training of a goal-oriented problem solving strategy through development of motor programs and integration to form a motor plan and execute.

Proprioceptive Neuromuscular Facilitation

This approach was developed by Knott and Voss in the year 1968. General treatment in PNF includes the use of recapitulation of total patterns of developing motor behavior, spiral and diagonal patterns of movements, coupling voluntary movement with postural and rightning reflexes, appropriate sensory and verbal cues, maximal resistance for maximal excitation and inhibition and repetitive activity for conditioning and training.

Temple Fay and Doman Approach

Temple Fay and Doman place emphasis on the use of more primitive patterning initially and encourage reptilian and amphibian pattern movement before working towards the more sophisticated patterns of the mature human being. They also emphasize, more positively, the need for constant repetition and the importance of stimulating the patient physically and intellectually.

Rood Approach

This was proposed by Margaret S Rood in 1950. It is based on the philosophy of treatment concerned with the interactions of somatic, autonomic, psychologic factors and their interactions with motor activities. Rood has used sensory stimuli by stroking or brushing at a given speed for a given duration for activation of a phasic muscle response. Rood applied cold for visceral stimulation and somatic relaxation and applied pressure/stretch for postural muscle activation.

Plans of Physiotherapy[28-30]

- Developing rapport with the kid is very important as any goal will be difficult to achieve without the cooperation of the baby. The baby has to be motivated well enough to gain the confidence. The goals set for the baby must be challenging at the same time achievable.
- False appreciation must be avoided.
- Initially maximum support and feedback must be given.
- Never give false hope to parents.
- Explain the role of mother and teach the home exercises so that it can be carried at home as treatment of cerebral palsy is whole day management.
- Remember always the therapy should be play therapy. Try to include games or play items into the therapy or else the kid will not show interest in the treatment.

Management of Abnormal Tone

The contractures may develop due to imbalance in the opposing muscle groups either because of wasting or spasticity in a muscle group.

The tone can be maintained by the following techniques.

Hypotonicity

- Ice brisk stroking over the muscle
- Hacking over the muscle
- Weight-bearing over a joint where the muscle is crossing
- Vibrations
- ENMS is found useful in developing the tone, but it is not advised in MND
- Suspension therapy and aquatic exercises
- Functional re-education exercises
- Muscle energy techniques
- EMG biofeedback techniques

Hypertonicity

- Cryotherapy in the form of prolonged icing—ice pack
- PNF techniques
- Passive stretching exercises
- Othotic supports to maintain the length of the muscle

Positioning Techniques

Positioning principles in cerebral palsy: Always position the child in a way that prevent problems. CP chair, logs or bolsters, wedges, walkers, pads, pillow can be used for correct positioning.

The basic principles of correct posture of the CP child are:
1. Always the head is straight up and looking forward
2. Trunk remains straight, shoulders symmetrical
3. Avoid hyperextension lateral rotation of the spine
4. Keep the arms in abduction
5. See that the child weight bears on both sides of the body

Techniques of lifting and carrying, positioning in lying, positioning for function and movements are to be taught to the child and also to the parents.

Lifting and Carrying

The mother is taught how to carry the child correctly. The adult's hands may be under or in front of the child's knee in carrying. She should lift the kid, keeping a straight back and wide base of support, bending knees and holding the child as close as possible. Suitable support should be given.

Lying and Sleeping Positions

The therapist teaches the mother to find out ways the child to be positioned opposite to the abnormal reflex posture such as scissoring of legs in supine posture.

If the child's body often arches backward, try positioning the child to lie and play on side lying. Make the patient lie in Hammock.

If the child's head always turns to the same side, do not have the child to lie so that the child turns head to that side to see instead, have him lie so that he has to turn his had to the other side to see the action.

Sitting Posture

Long sitting with the knees slightly flexed or straight is the position which a normal baby learns to establish his bottom and legs as his base of support. The postural education that takes place in long sitting is obviously important. This is also a position in which the hamstrings and hip adductors can be stretched.

If the child is unable to sit flat on the floor without pelvis rotating posteriorly, he can be placed so that his bottom is slightly raised, such as on a very low stool, or sitting downhill on a wedge.

Sitting cross-legged (tailor sitting) can be promoted but avoid in children who have hips externally rotated and abducted (frog legs) as there is a danger of anterior dislocation of hip.

W-sitting has the possible reinforcement of the flexed, adducted and internally rotated position of the hips. It may also contribute to tibial torsion and foot deformities.

Standing

Standing postures are encouraged in these kids to prevent hip deformities.

Benefits of standing:
1. Prevention of flexion deformity at the trunk, hips and knees
2. Prevents equines deformity in ankle
3. Development of weight-bearing surfaces of the feet
4. A child is often more able to move head, arms and hands in standing than in other positions.

Sensory feedback in standing is important both proprioceptively and perceptually.

The child benefits socially from being at the same height as his standing peers.

The standing posture stimulates cardiovascular, digestive, respiratory and excretory functions.

Maintaining the Length of Muscle

- Appropriate length of the muscle is a prerequisite to the normal control and normal postural adjustments.

- In cerebral palsy, because of delay or absence of normal movements, muscles are usually in shortened state.
- Stretching of the muscle is carried before the exercises.
- Orthotic supports/night splints are given.
- Range of motion exercises would be beneficial to prevent the contractures and maintain normal length of the muscles.
- Reflex inhibiting postures, therapists may use these postures to break the muscle spasms.
- Moist heat, saunas, tub baths, whirlpool baths, aquatic exercises, cryotherapy all promote release of spasticity.

Developing Postural Reactions and Improve Balance and Co-ordination

- Equilibrium exercises are taught with the help of Swiss ball, tilt board and bolster.
- Righting reactions, protective reactions and equilibrium reactions are taught.
- Equilibrium reactions are necessary before the next milestone is achieved.
- Activities to facilitate the postural abilities which include the ability to shift weight onto side of the body which normally bears normal weight, etc.
- Activities that challenge the child's postural abilities are established. Ask the child to lift the head follow an object usually a sounding toy with his eyes focussing on the object.
- Tipping him out of his base of support and asking him to regain the position.
- Make the child to stretch to reach something outside his base of support without falling.
- Activities to improve the child's ability to move from one position to another with control, e.g. rolling from supine to side lying and back again, or moving from sitting to standing.

Sensory Integration

- Perception includes whole of sensory motor experience. Sensory integration is ability to organize the sensory inputs for use.
- Various functional activities incorporating different objects/sizes/colors/textures can be used in therapy, e.g. beading, putting different size objects into respective holes, getting the object under the chair, sandplay, putti-clay, coloring squares circles, obstacle walking.
- Daily activities, intellectual, social and emotional development of the patients.
- Sand and water play, vestibular exercises like swings, rocking toys, scooter boards, music in motion, gliders, see saws, therapy balls.
- Aromatherapy, massage also helps in sensory integration.
- Wilbarger brushing protocols: It is a surgical brush which is used to brush the child at regular intervals.
- Oral toys like whistles, blow toys, textured teething rings or spoons, help the children in developing oromotor activites.

Training the Respiratory and Oromotor Functions

Training for the Oromotor Control

Oromotar function depends on well-controlled head and neck flexion which is dependent upon the active use of supra- and infrahyoid muscles that have the primary action on jaw, tongue and hyoid movements.

Common Oromotor Problems

- Drooling
- Problems in sucking and swallowing
- Body movements associated with speech
- Inadequate tongue movements

Therapy

- Develop good neck control (wedge exercises)
- Develop good trunk control
- Use of nook brush to decrease the drooling

MUSCULAR DYSTROPHY

Introduction

Muscular dystrophy was first described by Dr Charles Bell in 1830.[31] These are the group of disorders characterized by progressive degeneration of group of muscles. Muscular dystrophies are characterized by progressive skeletal muscle weakness, defects in muscle proteins and the death of muscle cells and tissue.

Incidence

Muscular dystrophy is a group of rare hereditary muscle diseases with prevalence of MD is 29 per 100, 000 population.

An observation in eastern India (Bihar, West Bengal) reports a high incidence of DMD cases in Muslims compared to Hindu polulation. A report submitted in 1997 estimated that there were 22,000 DMD cases present in India and more than 2500 cases added every year (Duriaswamy Navaneetham, 2009, USA).

Types of Muscular Dystrophies

a. Becker muscular dystrophy
b. Congenital muscular dystrophy
c. Duchenne muscular dystrophy (DMD)
d. Distal muscular dystrophy
e. Emery-Dreifuss muscular dystrophy
f. Facioscapulohumeral muscular dystrophy
g. Limb-girdle muscular dystrophy
h. Myotonic muscular dystrophy
i. Oculopharyngeal muscular dystrophy.

Although many types of muscular dystrophy are found, but there are only a few which are frequently seen in physiotherapy practice. The most common muscular dystrophy is Duchenne muscular dystrophy.

DMD is caused by a mutation of the gene that is responsible in production of a muscle protein called dystrophin. Dystrophin is required for muscle cells to keep their shape and work properly. An abnormal gene cannot produce this protein and thus the muscle cells collapse and die.

About Dystrophin[32]

It is a protein found in the muscle fibre membrane, its helical nature allows it to act like a spring or shock absorber. Dystrophin links actin in the cytoskeleton and dystroglycans of the muscle cell plasma membrane, known as the sarcolemma. In addition, dystrophin also regulates calcium levels.[37]

Pathophysiology

The exact mechanism of pathogenesis is not clear in DMD, there are some theories which explains the probable pathophysiology of DMD.

1. **Neurogenic theory:** According to this theory, the primary problem is in the nerve supply to the muscles rather than the muscles itself, and this has been suggested by some as the explanation for the high association of intellectual impairment in DMD boys.
2. **Vascular theory:** According to this theory, the problem in the muscles was secondary to an impairment in their blood supply.
3. **Membrane theory:** Most current accepted theory. According to this theory, the cell membranes are genetically altered, causing a compromise in cell integrity. An increase in the activity of muscle proteolytic enzymes may accompany the membrane alteration, leaving the muscle cell vulnerable to degeneration.

Genetic Basis of DMD—Who and How?

DMD is inherited. The abnormal gene for DMD is found in X chromosome, hence it is X-linked disease and is usually passed down from mother.

Boys have one X chromosome inherited from their mother and one Y chromosome inherited from their father.

If a boy inherits an X chromosome with the abnormal gene for DMD from his mother, he will develop this disorder.

Girls have two X chromosomes. If a girl inherits an X chromosome with the abnormal gene for DMD, the other X chromosome has a normal copy of the gene. The normal gene can usually make enough protein to prevent symptoms.

Girls are not usually affected by DMD, they are carriers. A woman carrying the DMD mutation can develop mild muscle weakness and an enlarged heart later in life.

If a woman carries the abnormal gene for DMD, each of her children will have 50% chance of inheriting the X chromosome with the abnormal gene.

Thus, **if a son inherits the abnormal gene, he will develop DMD**.

If a daughter inherits the abnormal gene, she will become carrier.

Clinical Manifestations[33-37]

Onset: There will be no abnormalities in early infancy in 50% of boys with DMD. The disease usually noticed at about 3 years of age. While in the early months of walking, he walks more slowly falls more frequently, does not run normally and climbs stairs only with difficulty.

Progression: With time, certainly at the age of 4 years, there will be slightly lordotic stance and the waddling gait, combined with the perceived locomotor difficulties serve to increase diagnostic suspicions that there may be a significant muscle weakness.

A boy will be unable to jump up even a very small step, climbing stairs will be slow and laborious.

Gower's Sign (Fig. 17.19)

If the patient tries to getting up from the floor, this will always start with the person going

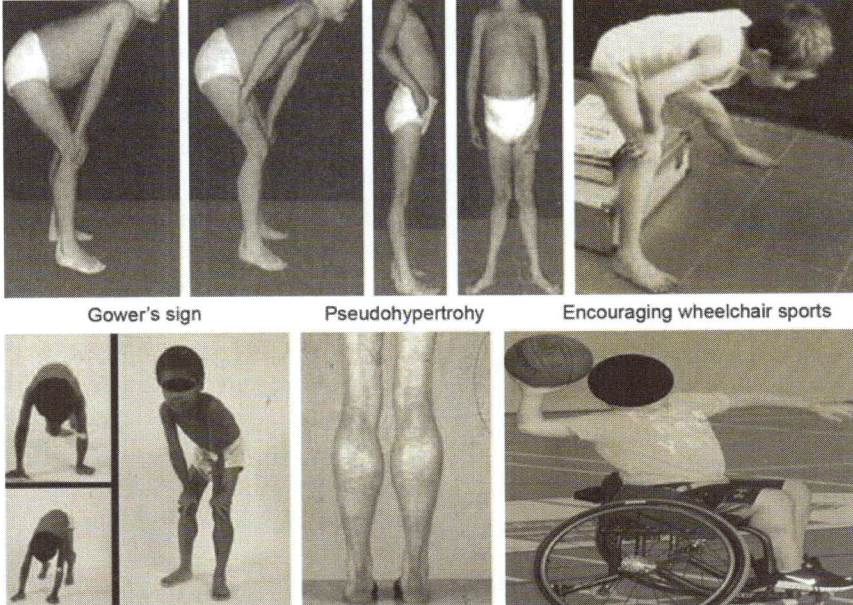

Fig. 17.19: DMD clinical features

Gower's sign   Pseudohypertrohy   Encouraging wheelchair sports

to prone and depending on age, using his hands to some extent to assist in standing up.

By the age of 4, typical Gower's maneuvers are typically seen. A Gower's sign consists of a typical getting up manner by the child using support of various parts in the lower limb and appears as though he is climbing up into standing using support of his lower limbs.

Muscle Weakness

There will be typically LMN type of motor weakness. There will be progressive muscle weakness of plantar flexors, evertors, quadriceps, gluteal muscles, long extensors of hand, biceps, triceps, deltoid, pectoral muscles, lattissimus dorsi and some other shoulder girdle muscles.

Pseudohypertrophy of Muscles (Fig. 17.19)

In contrast with the wasting of muscles as in other LMN type muscle weakness, the child with DMD develops hypertrophy of the muscles. But this hypertrophy is not exactly increase in the muscle bulk, but is due to abnormal deposition of fatty tissue and fibrous tissue in the muscles. Hence it is termed pseudohypertrophy.

The calf muscles, quadriceps, biceps and extensors of wrist and fingers show pseudohypertrophy commonly.

Incoordination

There will be loss of muscle coordination, child may seem unsteady, clumsy, frequently falling or waddle when they walk.

Toe Walking

The muscles become tight around joints, limiting movement. When walking becomes difficult, children may walk on their "tippy-toes" or balls of their feet.

The superficial reflexes are lost later the deep tendon jerks are also lost, although ankle jerk tend to be preserved.

With the progression of the disease, there tends to be better power retained in the distal muscles than in the proximal, in hamstrings than quadriceps, in deltoids than biceps or triceps, in wrist flexors than extensors, in neck extensors than flexors and in plantar flexors of the feet than dorsiflexors.

Joint contractures develop at the hips, knees, elbows and wrists and in majority the lumbar lordosis is replaced by a kyphoscoliosis due to muscular imbalance.

Diaphragm is the only muscle that is not involved in the DMD. But because of inactivity, development of kyphoscoliosis, decreased thoracic mobility, and respiratory complications set in.

By the early teens, muscle weakness is extreme and extensive, dependency will develop for functional activities like dressing, bathing, toileting and turning in bed in night.

Some children have learning disabilities and the IQ is 75–80% and does not deteriorate with age. In very few types of MD, there will be mental retardation.

There will be GIT abnormalities like constipation, vomiting, etc.

Progress

By the early teens, muscle weakness is extreme and extensive. Dependency will develop for functional activities, death will ultimately occur any time from the mid-teens to, in extreme cases, the mid-twenties.

Investigations

Muscle biopsy: The muscle biopsy will show extensive variation in the size of the fibres with increased infiltration of fatty tissues and fibrous tissues.

Serum CPK levels: CPK is an enzyme found in muscles, its levels are elevated.

EMG: It shows typical myogenic pictures with small amplitudes, small duration polyphasic potential. In later stages of disease, there may be presence of fasciculation potentials.

Urine/serum myoglobin levels: When muscle is damaged, the myoglobin is released into the bloodstream, in DMD, myoglobin levels are raised in serum and urine.

Serum creatinine: The serum creatinine levels are raised in muscular dystrophy.

Medical Management

Steroid medications can slow the progress of the disorder.

Prednisolone can help to keep muscles strong and working.

Deflazocort works in the same way as prednisolone.

Certain times, DMD child develops cardiomyopathy with high blood pressure, antihypertensives may be advised by the physician.

Genetic counseling: If any of the family member has DMD, genetic counseling might be helpful.

Orthopedic management: Certain times, a few corrective surgeries may be necessary to make the child functionally independent.

Principles of Physiotherapy Assessment

The demographic data is obtained from the parents or medical records. A detailed history of illness is obtained from the mother.

Observation

- Observe the general built of the patient
- Observe for any abnormal postures like lordosis, kyphoscoliosis to abnormal attitude of the limbs
- Observe for the psuedohypertrophy of the muscles
- Observe for any kind of contractures or deformities
- Observe for the Gower's sign.

Palpation

- Palpate the muscles and check for tenderness
- Palpate for the bony prominences and check whether symmetrical or not

Examinations

As a routine examnation, higher functions are assessed followed by cranial nerve examinations.

In DMD, sensory system is unaffected but the sensory system is evaluated to rule out other causes of presenting symptoms.

Reflex examination: Superficial and deep reflexes are examined. The superficial and deep reflexes are lost, sometimes ankle jerk may be preserved.

Motor examination:

- Tone of the muscle is flaccid, power of the muscles is checked which is usually deteriorating.
- Muscle girth measurements show pseudohypertrophy.
- Limb length measurements to evaluate any deformities.
- Spinal examination to assess for the scoliosis and lordosis or any other spinal deformities.
- Joint range of motion is assessed by goniometry.
- Detailed respiratory assessment by auscultation, percussion, depth and extent of excursion, symmetry of chest mobility.
- Balance and coordination is assessed.
- Examination of bowel and bladder dysfunction and dependency
- Assessment of gait and other functional evaluation.

Management by Physiotherapy

Aims

1. To develop good rapport with the parents and the child
2. To counsel and motivate the child and explain the parents the importance of physiotherapy

3. To prevent contractures and deformities
4. To prevent respiratory complications
5. To improve functional independence of the patient.
6. To prevent all the possible

Plans of Physiotherapy

1. The parents of the child with DMD will be in a great shock as unlike other pediatric problems where the baby is born with disability. The DMD boys are born normal and at later stages develop the problems. It is quite difficult to the parents to accept the truth. The therapist need to talk to the parents, counsel them and explain the importance of physiotherapy. Talk to the child and handle them with care and affection to develop good rapport so that they will actively participate in the physiotherapy
2. **Exercise therapy:** The child is encouraged to lead an active life as long as possible. Incorporate them in all general activities like other peers, in the school like morning PT program (drill in school), games, etc.
3. Endurance exercises and resisted exercise, PNF strengthening techniques to the affected muscles can be beneficial, along with as much as normal acitivity as the child can achieve during the day.
4. Do not be over enthusiastic and do not encourage the child in vigorous exercise therapy program especially when working out on strengthening exercises.
5. Passive mobilization of the spine or corrective exercises would be beneficial.
6. Deep breathing exercises and chest mobility exercises are encouraged.
7. Teach the exercises to the parents and ask them to have sessions even at home.
8. Keep the exercise more interesting and challenging by incorporating play therapy.

Training of ADL Activities

1. All daily living activities such as feeding, dressing and toileting must be watched so that the parents can encourage to let the child do as much as possible by himself.
2. There may be necessary to do some modifications like rasing the height of the bed, chair and toilet seat. Advise the parents to install western toilet seat (Fig. 17.20).
3. Advise to have ramp instead of steps for easy ambulation of the child.
4. Advice the parents to use a mattress which is firm and resilient.
5. In advanced stages, provide the child with an alarm bell to call on for help.

Preventing Contractures and Deformities

1. Passive and active stretching exercises are incorporated to prevent contractures of the affected muscles.
2. General mobility exercises for all the joints are taught.
3. Night splints may be advised in later stages.
4. Serial plaster splinting, orthotic support may be necessary.
5. Advise the patient to lie on prone for some time of the day to prevent hip deformities.
6. Spinal braces may be necessary to prevent and correct spine deformities.

Wheelchair Stage

Transfer techniques are taught from bed to wheelchair, wheelchair to bed, wheelchair to floor and floor to wheelchair.

Using the tilt table the patient is made to stand in upright position to give a prolonged stretch to hip flexors, hamstrings and calf muscles.

Tilt board exercise to train for balance with orthotic supports.

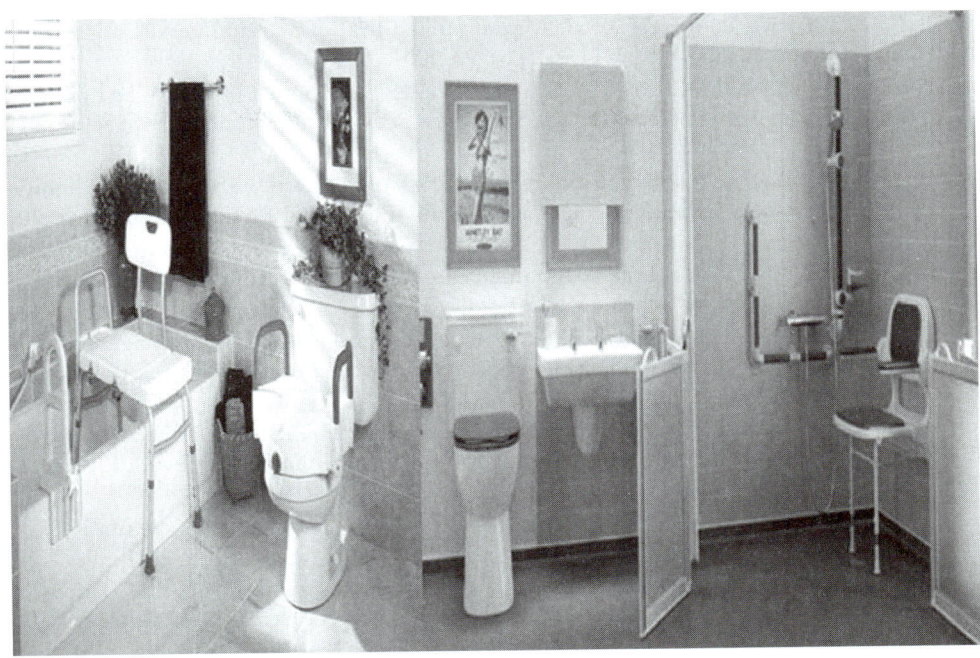

Fig. 17.20: Recommended toilet modifications

Encourage the child to propel the wheelchair and participate in wheelchair sports.

Arrange or modify the home environment for wheelchair accessibility.

The long-term management depends upon the progression of the disease and aims to prevent the secondary complications, encouraging the child to become functionally independent as far as possible.

REFERENCES

1. US Preventive Services Task Force. Folic acid for the prevention of neural tube defects: US Preventive Services Task Force Recommendation Statement. Ann Intern Med 2009;150:626–631.
2. Hark L, Catalano PM. Nutritional management during pregnancy. In: Gabbe SG, Niebyl JR, Simpson JL, et al. (Eds.). Obstetrics: Normal and Problem Pregnancies. 6th ed. Philadelphia, PA: Elsevier Saunders 2012;Chap 7.
3. Vinck A, Nijhuis-van der Sanden MW, Roeleveld NJ et al. Motor profile and cognitive functioning in children with spina bifida. Eur J Paediatr Neurol 2010;14(1):86–92.
4. Thompson DN. Postnatal management and outcome for neural tube defects including spina bifida and encephalocoeles. Prenat Diagn 2009;29(4):412–419.
5. Fletcher JM, Copeland K, Frederick JA, et al. Spinal lesion level in spina bifida: a source of neural and cognitive heterogeneity. J Neurosurg 2005;102(3 Suppl):268–279.
6. Canfield MA, Ramadhani TA, Shaw GM, et al. Anencephaly and spina bifida among Hispanics: maternal, sociodemographic, and acculturation factors in the National Birth Defects Prevention Study. Birth Defects Res A Clin Mol Teratol 2009;85(7):637–646.
7. Fichter MA, Dornseifer U, Henke J, et al. Fetal spina bifida repair—current trends and prospects of intrauterine neurosurgery. Fetal Diagn Ther 2008;23(4):271–286.

8. Adzick NS, Thom EA, Spong CY, et al. A randomized trial of prenatal versus postnatal repair of myelomeningocele. N Engl J Med 2011;364(11):993–1004.
9. Jacobson LA, Tarazi RA, McCurdy MD, Schultz S, Levey E, Mahone EM, et al. The Kennedy Krieger Independence Scales-Spina Bifida Version: a measure of executive components of self-management. Rehabil Psychol 2013;58(1):98–105.
10. Cancel D, Capoor J. Patient safety in the rehabilitation of children with spinal cord injuries, spina bifida, neuromuscular disorders, and amputations. Phys Med Rehabil Clin N Am 2012;23(2):401–422.
11. Houtrow AJ, Maselli JH, Okumura MJ. Inpatient care for children, ages 1–20 years, with spina bifida in the United States. J Pediatr Rehabil Med 2013;6(2):95–101.
12. Stubberud J, Langenbahn D, Levine B, Stanghelle J, Schanke AK. Goal management training of executive functions in patients with spina bifida: a randomized controlled trial. J Int Neuropsychol Soc 201;19(6):672–85.
13. Verhoef M, Barf HA, Post MW, van Asbeck FW, Gooskens RH, Prevo AJ. Functional independence among young adults with spina bifida, in relation to hydrocephalus and level of lesion. Dev Med Child Neurol 2006;48(2):114–119.
14. Rekate HL. A contemporary definition and classification of hydrocephalus. Semin Pediatr Neurol 2009;16(1):9–15.
15. Hattingen E, Jurcoane A, Melber J, Blasel S, Zanella FE, Neumann-Haefelin T. Diffusion tensor imaging in patients with adult chronic idiopathic hydrocephalus. Neurosurgery 2010;66(5):917–924.
16. Hamilton MG. Treatment of hydrocephalus in adults. Semin Pediatr Neurol 2009;16(1):34–41.
17. Woodworth GF, McGirt MJ, Williams MA, Rigamonti D. Cerebrospinal fluid drainage and dynamics in the diagnosis of normal pressure hydrocephalus. Neurosurgery 2009;64(5):919–925; discussion 925–926.
18. Lacy M, Oliveira M, Austria E, Frim MD. Neurocognitive outcome after endoscopic third ventriculocisterostomy in patients with obstructive hydrocephalus. J Int Neuropsychol Soc 2009;15(3):394–398.
19. "Cerebral Palsy: Hope Through Research". National Institute of Neurological Disorders and Stroke. 2015.
20. Rosenbaum P, Paneth N, Leviton A, et al. A report: the definition and classification of cerebral palsy. Dev Med Child Neurol 2007;109:8–14.
21. Mutch L, Alberman E, Hagberg B, Kodama K, Perat MV. Cerebral palsy epidemiology: where are we now and where are we going? Developmental Medicine and Child Neurology 1992;34:547–551
22. John Yarnell. Epidemiology and Disease Prevention: A Global Approach (2nd ed). Oxford University Press, 2013;p 190.
23. Hirtz D, Thurman DJ, Gwinn-Hardy K, Mohamed M, Chaudhuri AR, Zalutsky R. "How common are the "common" neurologic disorders?". Neurology 2007;68(5):326–337.
24. Apgar, Virginia. "A proposal for a new method of evaluation of the newborn infant". Curr. Res. Anesth. Analg. it takes less than 2 seconds and for experienced midwives it would take about less than 1 second 1953;32(4):260–267.
25. Finster M, Wood M. "The Apgar score has survived the test of time". Anesthesiology 2005;102(4):855–857.
26. Zafeiriou DI. Primitive reflexes and postural reactions in the neurodevelopmental examination. Pediatr Neurol 2004;31:1–8.
27. Campbell S, Osten E, Kolobe T, Fisher AG. Development of the test of infant motor performance. Phys Med Rehabil Clin 1993;4:541–550.

28. Graham KH. Mechanisms of deformity. In: Scrutton D, Damiano D, Mayston M (Eds.) Management of the motor disorders of children with cerebral palsy. 2nd edn. MacKieth Press, London 2004; 105–129.
29. Hur JJ. Review of research on therapeutic interventions for children with cerebral palsy. Acta Neurol Scand 1995;91:423–432.
30. Mayston MJ. Physiotherapy management in cerebral palsy: an update on treatment approaches. In: Scrutton D, Damiano D, Mayston M (Eds.) Management of the motor disorders of children with cerebral palsy. 2nd edn, MacKieth Press, London 2004;147–160.
31. "NINDS Muscular Dystrophy Information Page". NINDS 2016. Retrieved 12 September 2016.
32. Lapidos Karen A, Kakkar Rahul, McNally Elizabeth M. "The Dystrophin Glycoprotein Complex Signaling Strength and Integrity for the Sarcolemma". Circulation Research (2004-04-30);94(8): 1023–1031.
33. Donders J, Taneja C. Neurobehavioral Characteristics of Children with Duchenne Muscular Dystrophy. Child Neuropsychol 2009;1–10.
34. Almenrader N, Patel D. Spinal fusion surgery in children with non-idiopathic scoliosis: is there a need for routine postoperative ventilation?. Br J Anaesth 2006;97(6):851–57.
35. Kinali M, Messina S, Mercuri E, et al. Management of scoliosis in Duchenne muscular dystrophy: a large 10-year retrospective study. Dev Med Child Neurol 2006;48(6):513–18.
36. Birnkrant DJ. New challenges in the management of prolonged survivors of pediatric neuromuscular diseases: a pulmonologist's perspective. Pediatr Pulmonol 2006;41(12):1113–1117.
37. Miller RG, Chalmers AC, Dao H, et al. The effect of spine fusion on respiratory function in Duchenne muscular dystrophy. Neurology 1991;41(1):38–40.

18

Vertigo, Dizziness and Vestibular Rehabilitation

LEARNING OBJECTIVES

At the end of this chapter, you will be able to:
1. Define vertigo
2. Classify various types of vertigo
3. Identify the causes of vertigo
4. Describe the principles of assessment of patients with vertigo
5. Describe the principles of vestibular rehabilitation.

INTRODUCTION

Vestibular rehabilitation therapy is used in patients who often experience problems with balance and equilibrium along with movement-related dizziness with inner ear disorders.[1–4]

The decreased strength, loss of range of motion, increased tension leading to muscle fatigue and headaches are the secondary symptoms to inner ear problems.

Jonsson et al. (2004) found that the overall prevalence of problems with balance at age of 70.36% is common in women and 29% in men.

The symptoms affect the patient's ability to change positions or move about without imbalance and vertigo.

Vestibular rehabilitation therapy is an exercise-based program for reducing the symptoms of disequilibrium and dizziness associated with vestibular pathology.

TYPES AND CAUSES OF DIZZINESS

Otologic dizziness: It is the most common type of dizziness in the elderly and is due to BPPV(benign paroxysmal postural vertigo) where brief bouts of vertigo provoked by changing the orientation of the head to gravity.
- Meniere's disease[5] which is clinically presented as spells of rapid decline in hearing, roaring tinnitus, vertigo and monaural fullness.
- Vestibular neuritis[6] which is a monophasic self-limited clinically manifested as vertigo, nausea, ataxia and nystagmus.
- Bilateral vestibular paralysis which is a result of prolonged exposure to some ototoxic medications, e.g. gentamicin, infections of inner ear, autoimmune causes. It is manifested by blurring of vision (oscillopsia), ataxia and vertigo.

Central dizziness:[7] It is secondary to underlying neurological deficits especially secondary to vascular deficits to cerebellum and brainstem. Patient's chief complaints would be ataxia, nausea and illusions of motion which are chronic. A rare condition, where patient complains of spinning sensations with history of seizures is known as *epileptic vertigo*.[8]

Proprioceptive dizziness: It is secondary to diabetes, vitamin B_{12} deficiency, cervical spondylosis or a neuropathy which results in vertigo and unsteadiness especially found in older population.

Medical dizziness: It is manifested as postural hypotension and may result in syncope. Cardiac arrhythmias, hypoglycemia and some drugs and induce medical dizziness.

Psychogenic dizziness: It is common in anxiety disorders, panic attacks, agarophobia, somatization syndrome.

Unlocalized dizziness: In older patients, dizziness is common, but cannot be attributed secondary to any etiological cause described above. In such case, it is termed unlocalized dizziness.

PHYSIOTHERAPY ASSESSMENT OF PATIENTS WITH DIZZINESS AND VERTIGO

Histroy Taking

The therapist should take a detailed history of the patient either by directly interviewing the patient and family members or taking from medical records.

The history taking should mainly focus on the:
- Type and intensity of symptoms—how these symptoms affect the ADL
- Previous history of treatments
- History of falls
- Eye movement and vision assessment

Eye-Head Coordination Test

Ask the patient to track a moving object without moving the head. Observe for how the eyes move and observe whether the head is also moving or not.

Check the acuity of vision by Snellen's chart.

Vertigo Assessment

Dix-Hallpike maneuver:[10] Also known as Nylen-Barany test and is a diagnostic maneuver for BPPV.

Make the patient lie in supine position.

Now ask the patient to sit with the head turned to 45° to one side and extended about 20° backward.

Observe the eyes of the patient. If there is burst of nystagmus, then the test is positive.

Supine Roll Test

The supine roll test is performed by positioning the patient supine and head in neutral position.

The therapist then quickly rotates the head to 90° to one side. Observe the patient's eyes for nystagmus. Then turn the head to neutral position and observe whether the nystagmus is subsided and repeat the same by turning the head to opposite side.

Balance and Gait Assessment

- Various static and dynamic balance tests are used to evaluate (refer to Chapter 7).
- General neurological assessment to rule our underlying neurological causes.
- General musculoskeletal assessments to rule out the musculoskeletal causes and pathology leading to problem.

MANAGEMENT—VESTIBULAR REHABILITATION THERAPY

Based on the assessment, the therapist design the vestibular rehabilitation therapy.

The VRT exercise focuses on decreasing dizziness and visual problems, improving balance and walking functions.

Vestibular Habituation Exercises[11,13]

These exercises are used to deal with dizziness or vertigo in individuals with movement related dizziness. The individuals are asked to repeat the movements which provoke dizziness and vertigo, e.g. turning the head in a certain plane or direction.

These exercises are believed to adopt the brain for repeated exposure to the stimulus which reduces the vertigo. The various exercises are:

a. Make the patient sit on a couch and ask to lie supine. Repeat 5 times.
b. Make the patient lie in supine position and turn to left. Repeat 5 times.
c. Make the patient stand, then ask the patient to right and left alternatively. Repeat for 5 times.
d. Make the patient sit on a chair. Now ask the patient to bend and touch nose to left knee and later to right knee. Repeat for 5 min.
e. Make the patient in a sitting position. Turn head to right and left. Repeat for 5 min.
f. Make the patient to sit. Then ask the patient to stand and then sit. Stand and tilt the head up and down. Repeat for 5 times.

Note: If dizziness occurs during any step, stay in that position until dizziness subsides, then resume exercises.

Balance Retraining Exercises

These exercises are designed to make a person steadier during walking and standing through improvements in coordination of muscle responses and organization of sensory information.

Principles of Exercise Design for Balance Retraining Exercises

- The first step is to find a good time to carry out the exercise. Each exercise may be carried out for 10 min. The exercise should be repeated every day.
- Choose a safe place to practice
- Choose the exercise using a timed exercise scoring sheet.

Timed Exercise Scoring Sheet

1. First carry out shaking exercise which is described below.
2. After doing for 10 min, wait for 10 sec and then note down how dizzy the patient is feeling.

Put an "S" by each score to show the exercise is done in sitting.

Use the scores on the exercises sheet to decide which exercises to do for the next week.

If there are no symptoms, then put a "0" by an exercise. If the patient describes it as mild or moderate symptoms, then grade it as 1 or 2 and the exercise should be repeated daily.

It the symptoms are severe, then it is graded as 3, then you need to practice it more slowly at first, everyday.

Basic Exercises

Shake: Ask the patient to turn the head from right to left and back again 10 times in 10 seconds. Ask the patient twist your head round as far as it will comfortably allows. Wait for 10 sec, after you have done 10 turns, then repeat for 10 more.

Nod: Ask the patient to nod the head up and down for 10 times in 10 sec. Wait for 10 seconds after the patient has completed 10 complete turns, then make the patient repeat 10 more.

Shake with eyes closed: Carry out the shaking exercises with eyes closed. Wait 10 seconds after you have done 10 complete turns, then repeat for 10 more.

Nod, eyes closed: Ask the patient to carryout nod exercises with eyes closed. Wait for 10 sec after 10 complete turns, then repeat another 10.

Shake/stare: Ask the patient to carryout shaking exercises while staring at a finger placed in front of him. Ask the patient not to let out eyes move away from finger. Ask the patient to continue to focus on the finger along with shaking the head. Wait for 10 sec after completing 10 turns and repeat another 10.

Nod/stare: Ask the patient to do nod exercises by staring at one finger pointed in front of the patient. Do not let the patient move from your finger. Wait for 10 seconds after 10 complete turns then repeat 10 more.

Cawthorne-Cooksey Exercises

1. **In bed or sitting:**
   a. Eye movements—at first slow, then quick
      i. Up and down
      ii. From side-to-side
      iii. Focusing on finger moving from 3 feet to 1 foot away from face
   b. Head movements at first slow, then quick, later with eyes closed
      i. Bending forward and backward
      ii. Turning from side-to-side.
2. **Sitting:**
   a. Eye movements and head movements as above.
   b. Shoulder shrugging and circling.
   c. Bending forward and picking up objects from the ground.
3. **Standing:**
   a. Eye, head and shoulder movements as before.
   b. Changing from sitting to standing position with eyes open and shut.
   c. Throwing a small ball from hand-to-hand (above eye level).
   d. Throwing a ball from hand-to-hand under knee.
   e. Changing from sitting to standing and turning around in between.
4. **Moving about (in class):**
   a. Circle around center person who will throw a large ball and to whom it will be returned.
   b. Walk across room with eyes open and then closed.
   c. Walk up and down slope with eyes open and then closed.
   d. Walk up and down steps with eyes open and then closed.
   e. Any game involving stooping and stretching and aiming such as bowling and basketball.

Balancing Exercises

Various balancing exercises like standing heel-toe, walking on heels and then with toes, head moving back and front, up and down, walking on spongy surfaces or sand walking with eyes closed.

Gaze Stabilization Exercises

Ask the patient to look forward at a fixed point usually a card held in front of the patient. Now ask the patient to turn the head slowly to the right and left, by keeping the gaze focussed on the card. Repeat for 5–10 times.

Epley and Semont Maneuvers[12] (Fig. 18.1)

The Semont maneuvers are named after its inventor, Dr Alain Semont and are intended to move debris out of the sensitive part of the ear to a less sensitive location. The over all duration for this maneuver is 15 min.

The patient is rapidly moved from lying on one side to lying on the other (Levrat et al. 2003).

The Epley maneuver is also called the particle repositioning or canalith repositioning procedure. It was invented by Dr John Epley.

The patient is asked to move head in a sequence into four positions. The patient is asked to maintain in each position for roughly 30 sec.

Instructions for patients after Epley or Semont maneuvers: Instruct the patients as

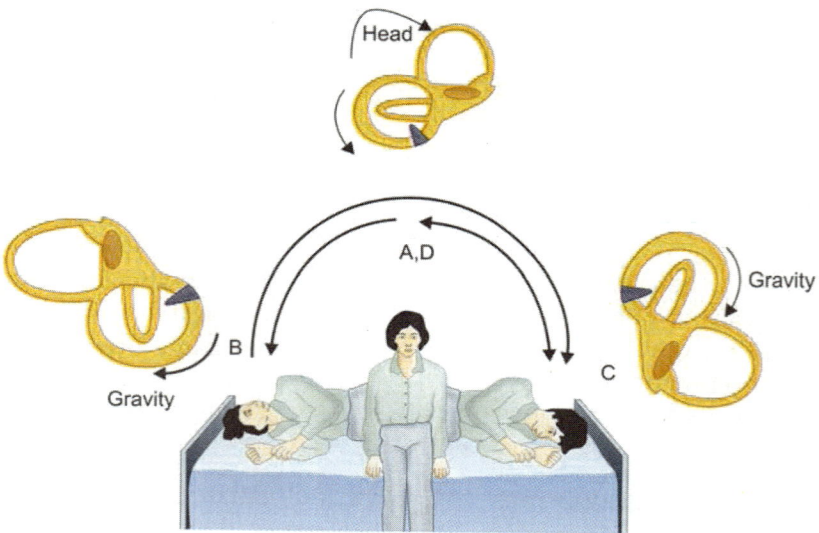

Fig. 18.1: The Epley and Semont maneuvers

follows, which are aimed at reducing the chance that debris might fall back into the sensitive back part of the ear.

a. Wait for 10 minutes after the maneuver is performed before going home in order to avoid quick spins or brief bursts of vertigo. Never allow the patient to self-drive home after the session.
b. Sleep in semi-recumbent position for the next night.
c. For at least one week, avoid provoking head positions.
d. Use two pillows when sleeping and avoid sleeping on bad side.
e. At one week after treatment, ask the patient to put himself/herself in the position that usually makes them dizzy.

Brandt-Daroff Exercises

These are home exercises for treatment of BPPV.

These exercises are performed in three sets per day for two weeks, in each set, one performs the maneuver for five times.
- Make the patient to sit upright in bed.
- Move the patient into side lying position and maintain for 30 sec or until the dizziness subsides.
- Make the patient to go back to sitting position. Stay for 30 sec.
- Repeat the same to the opposite side lying.

REFERENCES

1. Post RE, Dickerson LM. "Dizziness: a diagnostic approach". American Family Physician 2010;82(4): 361–369.
2. Hogue JD. "Office Evaluation of Dizziness". Primary Care 2015;42(2):249–258.
3. McDonnell MN, Hillier SL. Vestibular rehabilitation for unilateral peripheral vestibular dysfunction. Cochrane Database of Systematic Reviews 2015;Issue 1. Art. No.: CD005397.
4. Herdman SJ. Vestibular rehabilitation. Curr Opin Neurol 2013;26:96–101.

5. Strupp M, Thurtell MJ, Shaikh AG, Brandt T, Zee DS, Leigh RJ. "Pharmacotherapy of vestibular and ocular motor disorders including nystagmus". Journal of neurology 2011;258(7):1207–1222.
6. Neuhauser HK, Lempert T. "Vertigo: epidemiologic aspects". Semin Neurol 2009;29(5):473–481.
7. Dieterich, Marianne (2007). "Central vestibular disorders". Journal of Neurology 254(5):559–68.
8. Bladin PF. "History of "epileptic vertigo": Its medical, social and forensic problems". Epilepsia 1998;39(4):442–47.
9. Shepard NT, Telian SA. Programatic vestibular rehabilitation. Otolaryngol Head Neck Surg 1995; 112(1):173–182.
10. Dix MR, Hallpike CS. "The pathology symptomatology and diagnosis of certain common disorders of the vestibular system". Proc R Soc Med 1952;45(6):341–354.
11. Hiller SL, Hollohan V. Vestibular rehabilitation for unilateral peripheral vestibular dysfunction. Cochrane Database of Systematic Reviews 2007;Issue 4. CD005397. Pub. 2.
12. Hilton M, Pinder D. The Epey (canalith repositioning) manoeuvre for benign paroxysmal positional vertigo. Cochrane Database of Systematic Reviews 2004;Issue 2. CD003162. Review, 2004.
13. Herdman SJ, (Ed). Vestibular Rehabilitation. 3rd ed. Philadelphia: F.A. Davis Co. 2007.

19

Physiotherapy in Neurosurgery

LEARNING OBJECTIVES

At the end of this chapter, you will be able to:
- Demonstrate knowledge on various surgical techniques of neurosurgery
- Identify the indications of various neurosurgeries
- Plan and carry out assessment of the patients referred to neurosurgery
- Plan and carry out preoperative and postoperative physiotherapy on the basis of evidence-based physiotherapy

BURR HOLE SURGERY (Fig. 19.1)

A burr hole for subdural hematoma is performed to remove a hemorrhage (blood clot) from around the surface of the brain. The location of the blood clot is beneath the firm covering of the brain known as the dura mater, and is, therefore, called subdural hematoma.

- Generally, when a blood clot is moderately old (at least two to three weeks), it may be drained through a small hole in the skull, and a large craniotomy flap (opening in the skull) might be avoided.

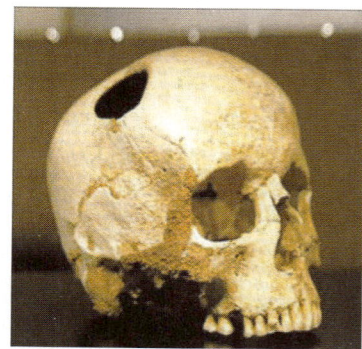

Fig. 19.1: Burr hole surgery

- The patient will be taken to the operating room and put to sleep under general anesthesia. The head will be partially shaved, to expose the area of operation. The head may simply rest on towels, or it may be placed in three fixation points (Mayfield head pins). The area where surgery is to be performed is then "prepped and draped" using an antibiotic solution.
- Next, the surgeon will make an incision, and reflect the scalp over the area of the hematoma. Then, an air powered drill is used to make a hole in the skull. The dura mater (tough covering of the brain) is then opened. The hematoma (blood

clot) is now seen, and the surgeon will irrigate some of it out, and may pass a drain around the brain to provide postoperative drainage. The surgeon will then close the scalp.

Risks

- **Surgical exposure:** The patient is placed in a supine position (on their back). There is risk of non-healing of the scalp postoperatively. Although very uncommon, there can be injury to or tearing of the scalp from the pins on the Mayfield clamp.[4,5]
- **Brain injury:** The surgery involves exposure of the surface of the brain. There is the possibility that there may be injury to the brain. If so, this could result in weakness, seizures, stroke, paralysis, coma or death. There may be residual fluid or blood, requiring additional surgery in the future. If the fluid around the brain is loculated in pockets separated by membranes, then the surgery will be unlikely to remove all the fluid, and may in fact only remove a small portion. This would necessitate additional surgery, possibly a larger craniotomy to remove the membranes and blood.
- **General risks:** These include such general difficulties, such as bleeding, infection, stroke, paralysis, coma and death. Incisions on the low back generally heal well, but if could be tender, or may heal in an unpleasant manner. There is also the possibility that the surgery may not relieve the symptoms for which the procedure was performed. The problem for which the surgery was performed may recur, requiring additional surgery in the future. In addition, although every attempt is made to protect all areas of the body from pressure on nerves, skin and bones, injuries to these areas can occur, particularly with prolonged cases.
- **Risks of anesthesia:** Blood clots in the legs, heart attacks, reaction to the anesthesia, reaction to blood transfusion, if it given.

CRANIOTOMY[1-3]

A craniotomy is a surgical operation in which a bone flap is removed from the skull, to access the brain.

- Craniotomies are often a critical operation performed on patients suffering from brain lesions or traumatic brain injury (TBI).
- Can also allow doctors to surgically implant deep brain stimulators for the treatment of Parkinson's disease, epilepsy and cerebellar tremor.
- The procedure is also widely used in neuroscience for extracellular recording, brain imaging, and for neurological manipulations such as electrical stimulation and chemical titration.
- Human craniotomy is usually performed under general anesthesia but can be also done with the patient awake using a local anesthetic; the procedure generally does not involve significant discomfort for the patient.
- In general, a craniotomy will be preceded by an MRI scan which provides a picture of the brain that the surgeon uses to plan the precise location for bone removal and the appropriate angle of access to the relevant brain areas.
- The amount of skull that needs to be removed depends to a large extent on the type of surgery being performed.
- In a craniotomy, the skin over a part of the skull is cut and pulled back.
- Small holes are drilled into the skull and a special saw is used to cut the bone between the holes.
- The bone is removed, and a tumor or other defect is visualized and repaired.
- The bone is replaced and the skin is closed.

Preoperative Care

- The blood thinning medications must be discontinued at least seven days before the surgery to reverse any blood thinning effects.
- Additionally, the surgeon will order routine or special laboratory tests as needed.
- The patient should not eat or drink after midnight the day of surgery.
- The patient's scalp is shaved in the operating room just before the surgery begins.

Postoperative Care

Craniotomy is a major surgical procedure performed under general anesthesia. Immediately after surgery, the patient's pupil reactions are tested, mental status is assessed after anesthesia, and movement of the limbs (arms/legs) is evaluated.

Shortly after surgery, breathing exercises are started to clear the lungs.

- Typically, after surgery, patients are given medications to control pain, swelling, and seizures. Codeine may be prescribed to relieve headache. Special leg stockings are used to prevent blood clot formation after surgery.

Patients can usually get out of bed in about a day after surgery and usually are hospitalized for 5 to 14 days after surgery.

- The bandages on the skull are to be removed and replaced regularly. The sutures closing the scalp are removed by the surgeon, but the soft wires used to reattach the portion of the skull that was removed are permanent and require no further attention. Patients should keep the scalp dry until the sutures are removed.

Full recovery may take up to two months, since it is common for patients to feel fatigued for up to eight weeks after surgery.

NERVE REPAIR SURGERY

- Usually in a nerve injury, the nerve regeneration occurs roughly at the rate of 1 mm per day or 25 mm per month.
- If there is no sign of motor or sensory recovery in the expected time of recovery, surgery is indicated.
- The appearance of a palpable neuroma at the site of nerve injury is another indication.
- The surgery is done under the general anesthesia and tourniquet control, the nerve is explored by a long incision and the site of injury exposed.

Types of Nerve Repair

Neurolysis (Fig. 19.2)

- This is a process of nerve release.
- If the nerve is in continuity and there is no neuroma at the site on injury, the mere release of the injured segment from surrounding scar tissue is sufficient.

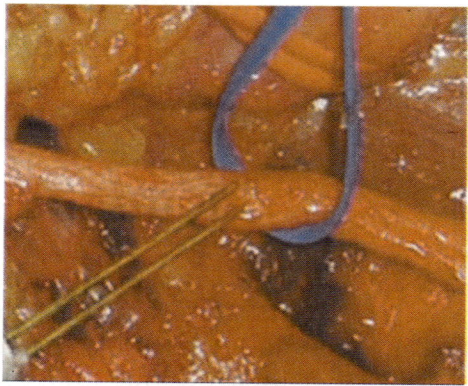

Fig. 19.2: Neurolysis

Neurorrhaphy (Nerve Repair) (Fig. 19.3)

- If a neuroma is found in the course of the nerve, it is excised and the cut ends are sutured. Recently microsurgical techniques are used in suturing nerve ends.

Nerve Suturing (Fig. 19.4)

The two cut ends of the nerve are sutured by a microscopic surgical process.

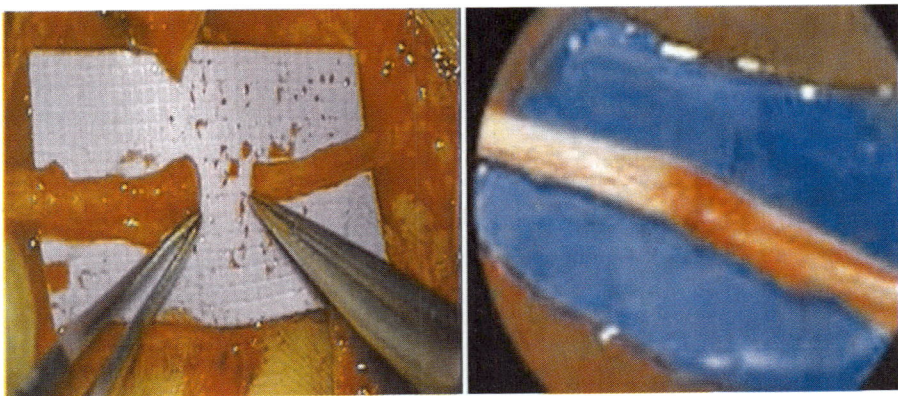

Fig. 19.3: Neurorrhaphy (nerve repair)

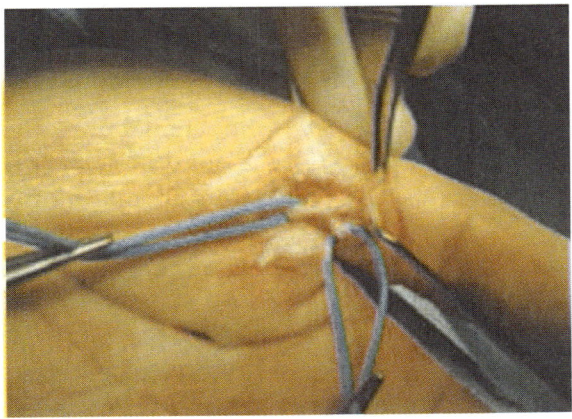

Fig. 19.4: Nerve suturing

Tendon Transfer Surgeries in Peripheral Nerve Injuries

Nerve injury	Tendon transfer	Purpose
Radial nerve	Pronator teres is transferred to substitute the loss of ECRL and ECRB	To restore extension of the wrist
	Flexor carpi radialis tendon is split and the radial half of the tendon is transferred to the abductor pollicis longus tendon	For gaining thumb stability
	Posterior deltoid is transferred	To gain elbow extension
	Flexor carpi ulnaris transfer	For gaining extension of fingers
	Robert Jones transfer: Pronator teres for ECRL and ECRB FCR for EPL/EPB FCU for ED	To stabilize wrist and promote movements of thumb and fingers
Median nerve	Flexor digitorum superficialis transfer	For restoration of thumb opposition
	Brachioradialis insertion tendon transfer to the insertion point of FPL.	For restoration of flexion of DIP of thumb

(Contd.)

(Contd.)

Nerve injury	Tendon transfer	Purpose
Ulnar nerve	Slitting of tendon of flexor digitorum profundus of the ring and little finger and transferring them to the middle and index fingers	To restore the flexion of index and middle fingers
	Biceps transfer	For restoration of pronation
	Paul Brand's transfer: The extensor carpi radialis is detached from its insertion and lengthened by a free tailed graft of plantaris muscle	To correct claw hand deformity
	This is then split into four moved through the lumbrical canal and attached to the extensor aponeurosis.	
Common peroneal nerve palsy	Tibialis posterior transfer by circumtibial route or interosseous route	To compensate the lost muscles due to nerve injury

PHYSIOTHERAPY MANAGEMENT IN NEUROSURGERY

Preoperative Physiotherapy

Aims

- Provide psychological support to the patient
- Maintain clear chest
- Improve ventilatory capacity
- To maintain normal joint range of motion at all joints

The physiotherapist shall reassure the patient about the surgery and provide psychological support by explaining briefly about the surgery and the outcome of it.

Counsel the patient such that he gains confidence and advice him and his family members which will enable them to reduce the overall effect that the condition may have on the lifestyle.

A clear chest is vital for any patient posted for surgery. A thorough chest examination and assessment is done by the therapist and clearance of chest is done by postural drainage and deep breathing exercises.

Note: Head low position is contraindicated in patients posted for cranial surgery as head low position shall increase the intracranial pressure.

Active movements are indicated for all the joints to maintain normal joint range of motion.

Ankle toe movements are done to prevent deep vein thrombosis.

If the patient is in comatose stage, periodic, rhythmical passive movements are given every 2nd hourly.

Postoperative Physiotherapy

Aims

- To keep the chest clear and expand fully
- Maintain muscle power of the extremities
- To promote/assist in the ADL
- To prevent secondary complications like DVT or pressure sores
- To mobilize the patient and promote early ambulation
- In nerve repair, re-education of the nerve/muscle
- To provide home advices at the time of discharge

Deep breathing exercises and coughing huffing techniques are taught to the patient preoperatively and carried put postoperatively.

Ventilatory muscle training exercises, and inspiratory spirometry exercises are given to maintain good normal chest expansion.

Postural drainage to remove the secretions formed because of bedrest and anesthesia.

- The muscle power of the extremities are assessed and exercises are prescribed and carried out accordingly.
- In nerve repair surgeries, faradic re-education is done to preserve/train/feedback for the nerve muscle for a new action or the action that has been forgotten.
- Depending upon the surgery, immobilization period depends.
- Early mobilization is done by building the confidence in the patient.
- Preparation of the patient for getting out of bed includes strengthening exercises for the arm, if walking aids are prescribed.
- In spinal surgeries, a corset is recommended and fitted by the therapist.

Methods Used to Get on–off the Bed

Technique 1: The patient is turned into the prone position and gets out of bed feet first, with the hips supported on the bed.

- He keeps a straight back and comes upright using hip and back extensor muscles switch arm support.
- The bed height is important in this technique.

Technique 2:
- The patient is turned onto the pain-free side, he slides his legs over the side of the bed and comes to sitting position by pushing up with both arms while the legs go down with gravity.
- The arms are then supporting the back.
- The patient moves forward, perching on the bed with his feet firmly on the floor.
- This method helps to overcome initial dizziness while the patient is still supported.

Early Ambulation

- On the first time out of bed, re-education of balance with weight transference using arm support should be practiced.
- If the patient not dizzy, a short walk helps regain confidence.
- The patient then progresses until he is independent with help of tone stick out of doors, if necessary.

ADL Training

- Patients should be advised ADL by the occupational and physical therapist.
- Each patient is given progressive exercise program.
- Isometric and relaxation exercises are given.
- Correct lifting and back care is taught.

REFERENCES

1. Reponen E, Korja M, Niemi T, Silvasti-Lundell M, Hernesniemi J, Tuomien H. Preoperative identification of neurosurgery patients with a high risk of in-hospital complications: a prospective cohort of 418 consecutive elective craniotomy patients. J Neurosurg 2015;123(3):594–604.
2. Rengachary, Setti S, Ellenbogen, Richard G. Principles of Neurosurgery 2nd ed. New York: Elsevier Mosby, 2005.
3. Edward A, Yegge J, Recktenwald A, Jadwsiak L, Kieffer P, Hohrein M, et al. Risk Factors for Craniotomy or Spinal Fusion Surgical Site Infection. Pediatr Infect Dis J 2015.

4. Baerts WD, de Lange JJ, Booij LH, Broere G. Complications of the Mayfield skull clamp. Anesthesiology 1984;16:460–61.
5. Lee M, Rezai AR, Chou J. Depressed skull fractures in children secondary to skull clamp fixation devices. Pediatr. Neurosurg 1994;21:174–77.
6. Rohde V, Graf G, Hassler W. Complications of burr-hole craniostomy and closed-system drainage for chronic subdural hematomas: a retrospective analysis of 376 patients, Neurosurg Rev 2002; 25(1–2):89–94.
7. Kumar JA Maiya AG, Pereira D. Role of physiotherapists in intensive care units of India: A multicenter survey. Indian J Crit Med 2007;11:198–203.
8. Ersson U, Carlson H, Mellstrom A, Ponten U, Hedstrand U, Jakobsson S. Observations on intracranial dynamics during respiratory physiotherapy in unconscious neurosurgical patients. Acta Anaesthesiol Scand 1990;34:99–103.

Model Papers

MODEL PAPER A

PART A: MULTIPLE CHOICE QUESTIONS

1. What is the root value for biceps jerk?
    a. C5, C6
    b. S1, S2
    c. L3, L4
    d. None of the above
2. Which of the following is an acute ascending type of polyneuropathy?
    a. Diabetic polyneuropathy
    b. GBS
    c. Neurosyphilis
    d. All of the above
3. Which of the following can be a complication in GBS?
    a. Lower respiratory tract infections
    b. Deep vein thrombosis
    c. Retention of urine
    d. All of the above
4. Which drugs are used in GBS?
    a. Vasopressor and immunoglobulins
    b. Antibiotics
    c. CNS depressants
    d. Anticholinergic drugs
5. Which of the following is NOT a technique to prevent pressure sores?
    a. Frequent change of position
    b. Use of water or air bed
    c. Making creases on the bed
    d. Application of powder and making the area clean and dry
6. What is the condition in which the syrinx is formed in brainstem?
    a. Syringomyelia
    b. Syringobulbia
    c. Syringobrainia
    d. All of the above
7. Which of the following is a clinical manifestation in autonomic nervous system is disturbed in the syringomyelia?
    a. Erectile dysfunction and bladder dysfunction
    b. Loss of muscle power in lower limbs
    c. Loss of sensations in the upper limbs
    d. None of the above
8. Which of the following is a clinical manifestation of syringobulbia?
    a. Ataxia
    b. Loss of sensation over the C5 and C6 dermatomes
    c. Inflammation
    d. Vocal cord paralysis
9. What is the lower motor neuron leision of facial nerve known as?
    a. Bell's palsy
    b. Erb's palsy

c. Klumpke's palsy
d. Facial palsy
10. Which electrodiagnostic test is done with graphical representation of strength and duration in peripheral nerve injuries?
   a. FG test
   b. SD curves
   c. NCV
   d. EMG
11. What is the condition in which there is inflammation involving across a single or several spinal cord segments?
   a. Transverse myelitis
   b. GBS
   c. Syringomyelia
   d. All of the above
12. What is the complete transection of nerve is known as?
   a. Neurotemesis
   b. Neuropraxia
   c. Axonotemesis
   d. Cyanosis
13. What happens, if there is anterior horn cell damage in the transverse myelitis?
   a. Muscle paralysis
   b. Sensory loss
   c. Loss of reflexes
   d. All of the above
14. How to maintain physiological memory of movement in brain in the transverse myelitis?
   a. Passive movements
   b. Chest physiotherapy
   c. SWD application
   d. IFT application
15. Who is/are the members of multidisciplinary approach in the transverse myelitis?
   a. Neurologist
   b. Physiotherapist
   c. Nurse
   d. All of the above
16. Which surgery is performed to remove subdural hematoma?
   a. Burr hole surgery
   b. Laminectomy
   c. Nerve fusion
   d. Craniotomy
17. Which surgery involves removing a bone flap from the skull?
   a. Craniotomy
   b. Laminectomy
   c. Spinal fusion
   d. All of the above
18. Which of the following is a risk of anesthesia?
   a. Loss of sensation
   b. Loss of reflexes
   c. Blood clots
   d. All of the above
19. What is the capacity of nerve regeneration?
   a. 1 mm per day
   b. 2 mm per day
   c. 1 cm per day
   d. No regeneration capacity
20. Which is the common nerve used in nerve grafting?
   a. Sciatic nerve
   b. Femoral nerve
   c. Cranial nerves
   d. Sural nerve
21. What is the paralysis seen, if a patient has complete transection at cervical segments?
   a. Hemiplegia
   b. Quadriplegia
   c. Paraplegia
   d. Monoplegia
22. What are the symptoms of Horner's syndrome?
   a. Ptosis
   b. Anhydrosis
   c. Miosis
   d. All of the above
23. What is the damage to only right or left half of spinal cord known as?
   a. Coning
   b. Anterior cord syndrome
   c. Brown–Sequard syndrome
   d. Both a and b
24. What is the type of edema seen in patients with lack of muscle activity?
   a. Non-pitting
   b. Pitting
   c. Nonadhered
   d. Both a and c

25. Which of the following is an infection to CNS which is sexually transmitted?
    a. Tuberculosis
    b. Myasthenia gravis
    c. Neurosyphilis
    d. None of the above
26. What is the causative organism for neurosyphilis?
    a. *Treponema pallidum*
    b. *Mycobacterium*
    c. Bacillus
    d. Virus
27. Which nerve injury can cause wrist drop?
    a. Ulnar nerve
    b. Sciatic nerve
    c. Radial nerve
    d. All of the above
28. Which deformity is seen with medial nerve injury?
    a. Wrist drop
    b. Foot drop
    c. Ape thumb deformity
    d. All of the above
29. What is the significance of Hitchhiker sign?
    a. Radial nerve injury
    b. Ulnar nerve injury
    c. Median nerve injury
    d. Hip dislocation
30. What splint is used to correct wrist drop?
    a. Cock-up
    b. Milwaukee
    c. L splint
    d. All of the above

PART B: SHORT ESSAY

1. a. Define myasthenia gravis.
   b. Describe the classification of myasthenia gravis.
2. a. Define transverse myelitis.
   b. Illustrate the clinical features of transverse myelitis involving the C8–T4 segments.
3. a. Classify peripheral nerve injuries.
   b. Describe briefly the physiotherapy management of radial nerve injury.
4. a. Define syringomyelia.
   b. Describe the physiotherapy management of syringomyelia involving C4–C8 spinal segments.
5. a. List out the causes for spinal cord injuries.
   b. List out the various myotomes of upper limb and lower limb.

PART C: LONG ESSAY

1. A 30-year-old female patient came with right wrist drop after an accident. She is nondiabetic, but hypertensive. On performing FG test, the results showed galvonic +ve.
   a. List out the various causes of radial nerve injuries.
   b. Anticipate the clinical features of radial nerve injuries.
   c. Desribe in detail physiotherapy management for radial nerve injuries.
2. A 24-year-old male patient non-diabetic and non-hypertensive, non-smoker and non-alcoholic has been referred to physiotherapy department for treatment and advice for left side Bell's palsy.
   a. Define Bell's palsy.
   b. What are the muscles affected in Bell's palsy?
   c. Describe the principles of physiotherapy management.
   d. Give home advices for the patient with Bell's palsy.
3. A 50-year-old female patient was referred to physiotherapy department after a diagnosis of neurosyphilis.
   a. What are the causes for neurosyphilis?
   b. What are the common clinical features of tabes dorsalis?
   c. Describe the physiotherapy management of neurosyphilis.
4. A 50-year-old male patient was referred to physiotherapy department after a craniotomy surgery.
   a. What are the indications of craniotomy?

b. What are the common complications of craniotomy?
c. Describe the physiotherapy management of craniotomy.
5. A patient was diagnosed with syringomyelia extending the complete spinal cord at C4 until C7.
   a. Describe the detailed clinical features seen in this patient under the following headings:
      I. Reflexes
      II. Sensations
      III. Motor tone and power
      IV. Bladder and bowel
      V. Gait
   b. Describe in detail physiotherapy management for the same.

MODEL PAPER II

PART A: MULTIPLE CHOICE QUESTIONS

1. What is the root value for quadriceps jerk?
   a. C5, C6
   b. S1, S2
   c. L3, L4
   d. None of the above
2. Which of the following scale is used to measure level of consciousness?
   a. Berg Scale
   b. Barthel Index
   c. Folestein Scale
   d. Glasgow Coma Scale
3. Which of the following can be a complication in stroke?
   a. Unilateral neglect
   b. Heart failure
   c. Retention of urine
   d. All of the above
4. Which of the following is an etiological factor for stroke?
   a. Smoking
   b. High fiber diet
   c. Antihypertension drugs
   d. Anticholinergic drugs
5. Which of the following is NOT a technique to prevent pressure sores?
   a. Frequent change of position
   b. Use of water or air bed
   c. Making creases on the bed
   d. Application of powder and making the area clean and dry
6. What is the scale used to assess the level of ADL?
   a. Glasgow Coma Scale
   b. Barthel Index
   c. Glass Coma Scale
   d. All of the above
7. The inability to judge the distance in cerebellar ataxia case, known as
   a. Dysmetria
   b. Dyssynergia
   c. Dysarthria
   d. None of the above
8. Which of the following is a motor aphasia?
   a. Wernicke's aphasia
   b. Broca's aphasia
   c. Global aphasia
   d. Vocal cord paralysis
9. What is the lower motor neuron leision leads to?
   a. Spasticity
   b. Rigidity
   c. Flaccidity
   d. Clasp knife mechanism
10. Expand CT scan.
    a. Computerised tomography
    b. Computed tomography
    c. Clear test scan
    d. None of the above
11. Which exercises best fit for patients with sensory ataxia?
    a. Frenkel exercises
    b. Aerobic exercises
    c. Gymnastics
    d. Progressive resisted exercises
12. What is the type of gait seen in parkinsonism patients?
    a. Festinating gait
    b. Wide-based gait
    c. Circumductory gait
    d. None of the above
13. Which of the following is NOT a tumor of brain?
    a. Meningioma
    b. Glioma

    c. Ewing's sarcoma
    d. Medulloblastoma
14. How to maintain physiological memory of movement in brain in flaccid paralysis stage of stroke?
    a. Passive movements
    b. Chest physiotherapy
    c. SWD application
    d. IFT application
15. Who is/are the members of multidisciplinary approach in the rehabilitation of stroke?
    a. Neurologist
    b. Physiotherapist
    c. Nurse
    d. All of the above
16. What is most common cause of head injuries?
    a. Gunshots
    b. Blow or hit on the head
    c. Road traffic accidents
    d. Fall from a height
17. The state of confusion in head injuries is known as
    a. Stupor
    b. Coma
    c. Spinal shock
    d. All of the above
18. Episode of abnormal hyperactive brain electrical activity is known as
    a. Parkinsonism
    b. Epilepsy
    c. Stroke
    d. All of the above
19. The inflammation of meninges of brain is known as
    a. Meningitis
    b. Carditis
    c. Encephalitis
    d. Graves' disease
20. Which is the common investigation done in infections of nervous system?
    a. Complete blood picture
    b. CSF analysis
    c. CT scan
    d. EMG studies
21. What is the paralysis seen in a patient with stroke most commonly?
    a. Hemiplegia
    b. Quadriplegia
    c. Paraplegia
    d. Monoplegia
22. The stiffness of neck in encephalitis is known as
    a. Nuchal rigidity
    b. Rigidity
    c. Miosis
    d. All of the above
23. In which condition, patient shall exhibit "Masked face"?
    a. Motor neuron disease
    b. Parkinsonism
    c. Brown–Sequard syndrome
    d. Both A and B
24. What is the type of edema seen in patients with lack of muscle activity?
    a. Non-pitting
    b. Pitting
    c. Nonadhered
    d. Both a and c
25. What is the condition characterized by the progressive degeneration of anterior horn cells?
    a. Tuberculosis
    b. Myasthenia gravis
    c. Neurosyphilis
    d. Motor neuron disease
26. What scale is used for the measurement of spasticity?
    a. Modified Ashworth scale of spasticity
    b. Barthel index
    c. Berg scale
    d. None of the above
27. The abnormal movement pattern seen in stroke patients is known as
    a. Incoordination
    b. Stroke
    c. Synergy
    d. All of the above
28. What is the speech disorder seen in pakinsonism?
    a. Broca's aphasia
    b. Wernicke aphasia
    c. Dysarthria
    d. All of the above
29. What is the disorder in which there is patch areas of demyelination occur in CNS?

a. Multiple sclerosis
   b. Motor neuron disease
   c. Meningitis
   d. Encephalitis
30. What procedure is done to do CSF analysis?
   a. Lumbar puncture
   b. Craniotomy
   c. Nerve grafting
   d. All of the above

PART B: SHORT ESSAY

Answer any three (3) of the following:
1. a. Define stroke.
   b. Describe the flexion and extension synergy seen in stroke patients.
2. a. Define cerebellar ataxia.
   b. Illustrate the clinical features of cerebellar ataxia.
3. a. Define meningitis.
   b. Describe briefly the physiotherapy management of meningitis.
4. a. Define parkinsonism.
   b. Describe how you manage a Parkinson's patient with physiotherapy.
5. a. List out the causes for head injuries.
   b. Describe the physiotherapy management of a patient in coma.

PART C: LONG ESSAY

Answer any two (2) of the following:
1. A 30-year-old female patient came with right stroke. She is non-diabetic, but hypertensive.
   a. List out the various causes of stroke.
   b. Anticipate the clinical features for middle cerebellar sysndrome.
   c. Describe in detail physiotherapy management for this patient in flaccid state.
2. A 24-year-old male patient non-diabetic, non-hypertensive, non-smoker and non-alcoholic has been referred to physiotherapy department for treatment and advice for motor neuron disease.
   a. Define motor neuron disease.
   b. What are the various clinical types of motor neuron disease.
   c. Describe the principles of physiotherapy management.
3. A 10-year-old female patient was referred to physiotherapy department after a diagnosis of meningitis.
   a. What are the causes for meningitis?
   b. What are the common clinical features of meningitis?
   c. Describe the physiotherapy management of meningitis (2 + 5 + 13 = 20 marks)?
4. A 20-year-old male patient came with post-head injury. He is non-diabetic, but hypertensive.
   a. List out the various causes of head injuries.
   b. What are the clinical features of contusion types of brain injuries?
   c. Desribe in detail physiotherapy management for this patient in flaccid state?
5. A 42-year-old female patient, non-diabetic and non-hypertensive, non-smoker and non-alcoholic has been referred to physiotherapy department for treatment and advice for multiple sclerosis.
   a. Define multiple sclerosis.
   b. What are the various clinical types of multiple sclerosis.
   c. Describe the principles of physiotherapy management.
6. A 10-year-old female patient was referred to physiotherapy department after a diagnosis of encephalitis.
   a. What are the causes for encephalitis?
   b. What are the common clinical features of encephalitis?
   c. Describe the physiotherapy management of encephalitis.

Index

Abdominal reflex 28
Abductor lurch 121
Acetylcholine 202
Action tremors 138
Agnosia 59
Akathisia 140
Alcoholic neuropathy 240
Alternating electric field
    therapy 108
Amnesia 80
Amyotrophic lateral
    sclerosis 165, 168
Ankle jerk/TA jerk 31
Apgar scoring 261
Aphasia 58
Apraxia 59
Arnold-Chiari malformation 195
Asphyxia 118
Astrocytoma 102
Asymptomatic neuro-
    syphilis 122
Ataxia 126
Athetoid 258
ATNR 265
Autoimmune
    disorder 201
    encephalitis 112
Axonal degeneration 238
Axonopathy 237

Babinski reflex 185
Baclofen 170
Balint's syndrome 59

Barthel index 65
Benign paroxysmal positional
    vertigo (BPPV) 24, 283
Berg balance scale 85
Biceps jerk 30
BMI 120
Bobath approach 35
Brachioradialis jerk 30
Brachytherapy 106
Bradykinesia 139
Brandt-Daroff exercises 286
Brown-Sequard syndrome 154
Brudzinski's sign 110
Brunhilde 117
Brunnstrom approach 37
Bulbar 118
    palsy 167
Burr hole 288

Calisthenics 149
Camptocormia 139
Carr–Shepherd 67
Cerebral palsy 257
Cervical spondylosis 168
Charcot
    disease 168
    joint 197
    triad 178
Chemotherapy 107
Chromatolysis 166
Circle of Willis 51
Circumduction gait 61

Clasp knife mechanism 56
Cogwheel 139
Coma 80
Concussion 78
Contusion 78
Corneal reflex 27
Coughing 114
Countrecoup injury 78
Coup injury 78
Craniopharyngioma 102
Craniotomy 289
C-reactive protein 110
Cremasteric reflex 28
Crocodile tears 233
Cryokinetics 161
Cryotherapy 96

Dandy-Walker malfor-
    mations 255
Deep brain stimulation (DBS)
    surgery 142
Deep vein thrombosis 158
Demyelination 176
Dermatitis 141
Dermatomes 186
Diabetic neuropathy 247
Diffuse axonal injury 78
Dix-Hallpike
    maneuver 283
    test 24
Dizziness 283
Doll's eye reflex 265
Drooling 273

Duchenne muscular dystrophy (DMD) 274
Dynamic gait index 181
Dysarthria 81
Dysdiadokinesia 128
Dysesthesia 240
Dysmetria 128
Dysphagia 167
Dyspnoea 88
Dyssynergia 128
Dystonia 140
Dystrophin 274

EMG (electromyography) 185
Encephalitis 111
　lethargic 112
　lethargica 138
Epileptic vertigo 283
Epley and Semont maneuvers 285
Equilibrium tests 32
Extensor lurch 121
Extensor—abductor lurch 121

Fasciculation 167
Festination 139
Floppy kids 260
Foley's catheter 163
Friedreich's ataxia 127
Functional electrical stimulation (FES) 160
Functional re-education exercises 95
Fungal meningitis 110

Gag reflex 28, 118
Gait training 71
GAITRite portable walkway system 187
Gallant's reflex 265
Gardner's hydrodynamic theory 196
General paralysis of insane 122
Gerstmann's syndrome 59
Glasgow coma scale 79
Gliosis 176
Gower's sign 275
Guillain-Barré syndrome 238
Gumma 123

Hand to knee gait 121
Head injury 77
Hemangioblastoma 102
Hemorrhage 78
Hoehn and Yahr scale 143

Homolateral limb synkineses 58
Hospital anxiety and depression scale 65
Hydrocephalus 195, 253
Hydrocolloid dressing 174
Hydromyelia 195
Hyperpathia 240
Hypomimia 140

ICU 94
Idiopathic 183
Illingworth scale 261
Immunomodulatory therapy 185
Infective encephalitis 112
Internal tremor 138
International classification of functioning 72
Intramedullary pulse pressure theory 196
IPG device 142
Ischemic penumbra 53

Johnstone 65
Joint kinesthetic sense 27
Joint position sense 26, 185
Jolt accentuation maneuver 110

Katz index 65
Kernig's sign 110
Knee jerk/quadriceps jerk 31

Lanchi 117
Landry syndrome 239
Lasegu's test 226
Lead neuropathy 241
Lead pipe 139
Leon 117
Lewy body disease 137
Lhermitt's test 227
Limbic encephalitis 112
LMN type 241
Locked in syndrome 55
LOKOMAT 163
Lou Gehrig's disease 168
Lumbar puncture 105

Madras motor neuron disease 166
Margaret Rood 38
Mayfield head pins 288
Medulloblastoma 102
Meniere's disease 282
Meningioma 102

Meningitis 109
Meningocele 252
Meningococcal meningitis 110
Meninovascular neurosyphilis 122
Micrographia 140
Micturition 13
Milestones 263
Modified Ashworth scale of spasticity 31
Mollaret's meningitis 110
Moro's reflex 265
Motor neuron disease 165
Motor relearning 99
Motor unit potential with polyphasia 240
MRI 129
MRP theory 43
Multiple sclerosis 176
Muscle girth 32
Muscular dystrophies 274
Myasthenia gravis 201
Myelinopathy 238
Myelocele 253
Myelogram 106
Myelomeningocele 252

Nebulizations 114
Neck rigidity 110
Neonatal reflexes 265
Nervous tissue 1
Neural tube defect 251
Neurogenic bladder 13
Neurolysis 290
Neuromyelitis optica (Devic's disease) 184
Neuron 1
Neurophysiological stimulation (NPF) of respiration 88
Neuroplasticity 45, 100
Neurorrhaphy 290
Neurosyphilis 122
Nissl's granules 2
Non-equilibrium tests 32
Nudge test 33
Nylen-Barany test 283
Nystagmus 126

Ocular flutter 128
Oldfield's theory 196
Oligodendroglioma 102
Ophthalomoplegia 184
ORLAU parawalker 163

Orthotic supports 96
Oscillopsia 282
Otologic dizziness 282

Pallidotomy 142
Palpebral reflex 27
Parasitic meningitis 110
Parasympathetic 12
Parkinsonism 137
Penicillin G 123
Peto approach 38
Photophobia 110
Plantar reflex 28
Play therapy 122
PLEX (plasma exchange) 185
PNF 39
   techniques 96
Poliomyelitis 117
Polyneuritis 237
Polyneuropathy 237
Positioning 68
Positron emission tomography 105
Postural drainage 94
Postural or sustention tremors 138
Postural sway test 33
Pressure sores 158
Prevalence 152
Primary lateral sclerosis 168
Progressive bulbar palsy 167
Progressive muscular dystrophy 166
Pseudodiabetic form of ataxia 241
Pseudohypertrophy 276
Ptosis 202
Pupillary reflex 27

Raccoon's eye 83
Radiation therapy 106
Raimiste's phenomenon or sign 58
Rappaport disability scale 85
Reflex 12
Rehabilitation 114
Rest tremor 138
Rigidity 77
Riluzole 169
Rinne test 24
Robot assisted walking therapy 163
Romberg test 32
Rood approach 271
Rooting reflex 265

Saccadic dysmetria 128
Scalenus anticus syndrome 60
Schwab and England activities of daily living scale 144
Schwannoma 102
Seizures 104
Sensory evoked potentials 180
Sensory re-education exercises 116
Shaking palsy 137
Sharpened Romberg test 33
Shunt system 256
SLR 226
Slump test 226
Snellen's chart 283
Souques' phenomenon or sign 58
Spina bifida 195, 251
   cystica 252
   occulta 252
Spinal cord 165
Spirometric 114
STNR 266
Stoop's test 228
Stupor 80
Subdural hematoma 288
Supine roll test 283
Sympathetic 12
Synergy 56

Syringobulbia 197
Syringomyelia 168, 195
Syrinx 195

Tabes dorsalis 122
Tectum 5
Temple Fay and Doman approach 271
Tendon transfer surgeries 291
Tetrapelgia 193
Thalamotomy 142
Tinetti gait and balance assessment 181
Transfer techniques 191
Transient ischemic attack (TIA) 50
Transverse myelitis 183
Tremor 138
Trendelenburg sign 61
Triceps jerk 30
Tuberculous meningitis 109
Tumor 104

Unified Parkinson's disease rating scale 144
Ureteroileostomy 253

VDRL test 122
Vegetative state 80
Ventriculostomy 256
Vertigo 282
Vestibular rehabilitation therapy 282
Viral meningitis 110
Visual evoked potentials 180
Vojta 44
Vojta's reactions 262
Voluntary control 156

Water-hammer 196
Weber test 23
William's theory 196
Wilson's disease 127